MICROBIOLOGY IN PRACTICE

Individualized Instruction for the Allied Health Sciences   2nd EDITION

Mary Alice Bruce

Lois Beishir
Antelope Valley College

Canfield Press cṕ San Francisco
A Department of Harper & Row, Publishers, Inc.
New York   Hagerstown   London

Cover photograph: *Serratia marcescens* on potato agar; 48 hour growth.
Cover photograph and part opening photographs and photomicrographs:
R. Gordon Clifgard.

Design: Karen Emerson
Biological art: Emily Chronic
Typesetting: Typeset Services

ACKNOWLEDGMENTS

I wish to give recognition to Donna Newton Dinger for writing and revising a major portion of this publication. I also wish to thank Kathryn Jones for her role as coordinator of the entire project and to express my appreciation to Gladys Baird for her constant and willing help.

MICROBIOLOGY IN PRACTICE: Individualized Instruction for the Allied Health Sciences, Second Edition

Copyright © 1974, 1977 by Lois Beishir

78 79 10 9 8 7 6 5 4 3 2

Library of Congress Cataloging in Publication Data

Beishir, Lois
   Microbiology in practice.

   Bibliography: p.
   1. Microbiology—Laboratory manuals. I. Title.
QR63.B44 1977 616.01 77-9255
ISBN 0-06-380463-8

Contents

To my parents,
Mr. and Mrs. S. J. Beishir, Sr.

Introduction to the Student

You are about to embark on a course in which you will use a new and exciting method of learning microbiology. Difficult-to-grasp techniques and concepts will be almost impossible *not* to learn if you make an earnest effort. The entire presentation is designed with your success in mind and to facilitate your learning and, especially, your understanding of microbiology. This new approach to learning depends upon your being honest in evaluating yourself, since your instructor will only infrequently measure your mastery of the material. Therefore, your honest self-evaluation of your mastery of each module, by means of the post test, is vital to your success. Do not be satisfied with partial understanding. Nobody knows better than you do whether you are "getting it."

The format of each module is as follows:

1. *Prerequisite skills* are taught as the course proceeds. Skills and techniques learned in early modules must be understood before later materials can be grasped.

2. *Materials* needed for the activities are listed at the beginning of each module.

3. An *Overall objective* for each module states clearly the general principle or procedure that you are expected to learn.

4. *Specific objectives* set down explicit terms to be defined, particular concepts to be grasped, and descriptions to be made.

5. *Discussion* provides explanation of theories and procedures, describes specific aspects of each module's topic, explains each objective, and prepares you for performing the activities.

6. *Activities* are techniques or experiments that support the discussion and the objectives. These are the core of the module.

7. *Figures* are used throughout the text to give visual explanations of complex techniques, schemes, and expected results. *Tables* show you how to collect data and record them.

8. The self-evaluating *Post test* allows you to determine whether you have successfully completed the module and thereby fulfilled the objectives. The post test also helps determine which parts of the module you have not learned and, therefore, should review or repeat.

9. *Related experiences* provide optional activities that will add to your understanding and improve your skills.

These components support and reinforce the objectives and make this presentation a self-taught course. This method gives you the choice of working at a slower or faster pace, without feeling pressured. Speed of performance is not a measure of solid learning.

Each module essentially contains a single technique or concept that is explicated by the discussion, activities, and the post test. In those modules dealing with difficult techniques, the activities may seem at first to be overdone, but they will be time saving to you in the long run. For example, the aseptic tube transfer of bacteria is a technique that is difficult to learn, but it is the single most important technique to learn in microbiology since it protects you, your neighbor, and your pure cultures from contamination. If you do the practice activities thoroughly, you can learn such difficult-to-grasp techniques in one lab period.

As you perform the activities in the modules, you will be asked to place descriptions, drawings, and some tables in a file folder. Your folder is not to be a collection of data sheets. Draw a conclusion for every activity even if one is not asked for. Your instructor can then determine if you have learned the procedures that the activities were designed to teach you. Your file folder must be complete at the end of the semester for you to be successful in the course. Your instructor may also make an unannounced spot check of the folders periodically.

Periodically in the course, a MODS will be scheduled. A MODS (Module Objective Discussion Session) is an informal discussion period, conducted by a student leader, where you explain the specific objectives to the other students and discuss them together. If you prepare for the MODS by reviewing your completed modules and file folder, these discussions are a very useful learning aid.

HELPFUL HINTS

1. The biggest mistake you can make with materials that you submit to your file is merely to copy the figures in each module. Many of these figures are included to help you determine whether you are seeing the correct organisms and/or structures through your microscope. It is to your advantage to make your drawings as you see them so that you will be able to recognize and identify them when you see them through the microscope again. In summary, don't dry lab. It is very easy for your instructor to determine whether you are giving some thought and effort to your drawings and descriptions.

2. It is advisable to supplement the information in each module with reference reading. At the end of this book, you will find a list of references. This supplementary reading will help you attain a better overall view of microbiology.

3. If you do not score 100 percent on your post test, review or repeat those parts of the module that will enable you to achieve a perfect score. As another aid to learning, go back to the post test a week after you have evaluated yourself, and see if you can still pass it with 100 percent. If you can do this, you have indeed learned the information.

4. In those modules without a related experience (designed to expand your learning), try to invent one yourself. Discuss this related experience with your instructor before you proceed.

5. Materials and cultures for each lab session will be made available to you in a designated area. Less frequently used materials and equipment will be kept in specified storage areas.

LABORATORY RULES

Certain rules should be followed while you are working in the laboratory. Some are listed here. Your instructor may add other appropriate rules or suggest more safety procedures to remember.

1. Never eat or drink in the laboratory, and avoid putting objects in your mouth. Remember that you are working with living microorganisms, most of which are harmless, but others, if ingested, can cause you physical discomfort.

2. Disinfect your working area, and wash your hands thoroughly at the beginning and end of each laboratory period. Wash your hands before you leave the lab for any reason, even a coffee break.

3. If you spill living microorganisms, cover the spilled material with paper towels, and pour your laboratory disinfectant over the towels and the contaminated area. Wait fifteen minutes before you clean it up.

4. If you are injured (most often burns or cuts), notify your instructor immediately.

A drawer, tray, cabinet, or other storage area will be assigned to you for the materials you will be using almost daily. You will also be assigned a microscope. It is your responsibility to keep it clean and to report any malfunctions to your instructor.

Suggested materials to supply for yourself:

1. protective garment such as a lab coat or apron
2. glass marking pencil
3. 1 pound coffee can for storage and incubation of culture tubes
4. hot pads or pot holders
5. colored pencils (red, blue, and green)
6. typing paper for drawings

You are expected to read the assigned modules before attending each laboratory period. This will allow you to use your time efficiently, which is important since you will be performing different stages of two or three different modules during the same lab period. That is, you may be preparing the media and reagents for one module, inoculating for another module, and collecting the results for still another module.

These self-instructional modules are presented in such a manner that you should be able to proceed with the activities (if you have preread the module) without a lengthy discussion and explanation from your instructor. Upon your arrival at the lab, you should be ready to go to work.

PART ONE

Use of Equipment

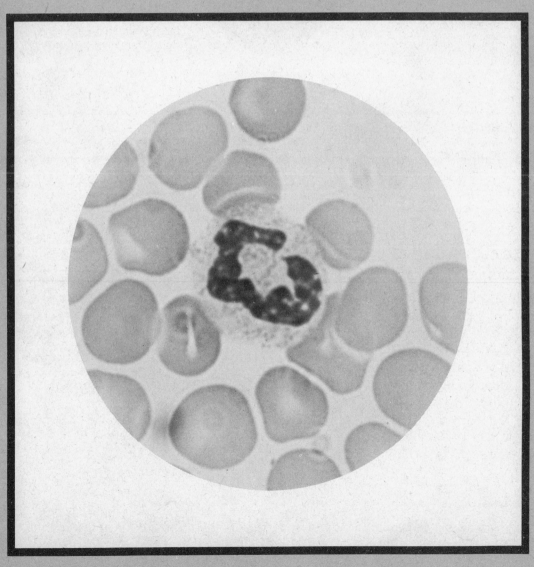

Human blood smear showing leucocytes among
the erythrocytes
(1000 x).

MODULE 1

Triple Beam Balance

PREREQUISITE SKILL

Working knowledge of the metric system of length, mass, and volume as commonly used in the science laboratory.

MATERIALS

triple beam balance
glazed weighing paper
150 ml beaker

spatula or weighing spoon
clean sand

OVERALL OBJECTIVE

Demonstrate your ability to weigh a granular substance accurately using a triple beam balance.

Specific Objectives

1. Demonstrate your understanding of the relationship between the three beams of the triple beam balance.
2. Describe how to correct for the weight of the paper or container used to hold the substance being weighed.
3. Demonstrate your ability to weigh a large amount and a small amount of a granular substance.
4. Define the term *tare*.

DISCUSSION

Weighing powders and granular substances accurately is essential to the preparation of microbiological media. When you reconstitute dehydrated media, you add a specific number of grams of the powdered medium to distilled water and heat this to dissolve the powder completely. If you have not weighed the powder and measured the water accurately, you will not have the proper proportion of nutrients in the finished medium. In the case of solid media, if you do not weigh accurately, you run the risk of having your medium turn out semisolid or too dry.

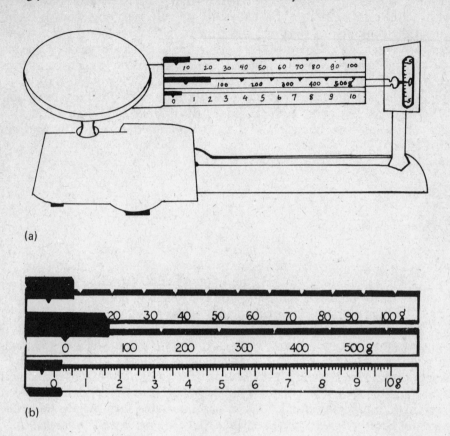

FIGURE 1-1
(a) Triple beam balance. (b) Detail of beam
scales and weights.

Although not as accurate as an analytical balance, the triple beam balance can weigh quantities larger than 1 gm with sufficient accuracy for your purposes if used carefully. Familiarize yourself with the balance shown in Figure 1-1. Notice the three beams, each with a separate scale. These are shown in more detail at the bottom of the figure. Each beam has a weight that can be moved along the scale. The first scale, the one closest to you, is graduated in grams from 0 to 10. Each major division equals 1 gm, and the smaller lines mark off 1/10 gm. The second scale has no subdivisions but is graduated from 0 to 500 gm in 100 gm increments. There are notches in this beam to mark the position of these 100 gm increments. The third beam also has notches to mark the increments but on a scale from 0 to 100 gm. Be sure the sliding weight rests *in* the desired notch. When you move one of the weights to a given number, it counterbalances the platform with that amount of weight. For example, if you move the weight on the third beam to the first notch (10 gm), you could add material to the platform until the pointer swings to zero, which would mean that you have placed 10 gm of the material you are weighing on the platform. This concept is difficult to grasp just from a description, so you should go over to the balances and look at the beams and their weights. Move the weights back and forth, and weigh your pencil, keys, or some other object until you feel that you understand the scales and their relationship to each other.

When you weigh powders or granular materials, you must never put them directly on the weighing platform of the balance. You would find that very soon every substance you weigh would be contaminated with the last powder weighed on the same balance. Even traces of certain chemicals could affect the medium enough to interfere with the growth of microbes. So for amounts up to 10 to 15 gm, you should use glazed weighing paper to hold the material you are weighing, and for amounts larger than this you should use a small beaker.

Of course, you will have to correct for the weight of the paper or beaker, or you will not have the right amount of the substance you are weighing. This deduction of the weight of the container or weighing paper from the total weight to determine the weight of the contents is called *taring* or finding the *tare*. Therefore, in order to tare, you must first weigh the container and *add* this weight to the weight of the powdered medium. For example, if your glazed paper weighs 0.6 gm and you want to weigh 8.2 gm of dehydrated nutrient agar, you would add the two figures for a total of 8.8 gm. Then move the weights on the beams of the balance to the setting that corresponds to this total. Now you can add the powdered medium carefully until the pointer swings to zero.

ACTIVITIES

Activity 1: Using Weighing Paper

For practice using the triple beam balance and correcting for the weight of the glazed paper, you should now go to one of the balances and weigh 5.5 gm of clean sand. Near the balance or in an appropriately labeled place, you will find weighing spoons which you should always use to place the substance you are weighing onto the paper on the platform. It will be helpful for you to make a note of the weight of the glazed paper since you will find that the size and weight are fairly consistent from one piece to the next. For accuracy, however, you should recheck the tare for each sheet of weighing paper. It is a good idea to take a piece of scratch paper with you to the balance and write the weight of the glazed paper on it, as well as the amount of material you wish to weigh. If you add the two on the scratch paper, you are less likely to make an error than if you try to do it mentally. Remember to add the material to the platform slowly so that you do not add a surplus and overbalance the weights. Patience is a virtue!

Activity 2: Using a Weighing Container

Remove the glazed paper and sand from the platform, return the weights to zero, and place a 150 ml beaker on the platform. Weigh the beaker and write its weight on scratch paper. Now add 43 gm to this and move the weights to this total weight. With your weighing spoon, carefully add clean sand to the beaker until the pointer swings to zero. How many grams of sand are in the beaker? If you said 43 gm, you have grasped the concept of correcting for the weight of the container and have learned how to tare. Return the sand to the container from which you obtained it.

Write down your weighing calculations from Activities 1 and 2, and place them in your file.

Take the post test next.

POST TEST

The post test is a self-evaluation. It is not used for a grade. It is designed only to let you decide if you have successfully completed this module.

True or False

_____ 1. It is best to make it a habit to calculate your tare mentally.

_____ 2. The weight of the glazed weighing paper varies considerably from one sheet to the next.

_____ 3. It is best to weigh amounts less than 15 gm on glazed paper and larger amounts in a small beaker.

—— 4. Each major division of the first scale equals 1 gm and is subdivided into tenths of a gram.

—— 5. Because the glazed paper is so light, you can disregard its weight as negligible.

—— 6. To weigh 315 gm you must set the weight on the second beam at 300 gm, the third beam weight at 10 gm, and the first beam weight at 5 gm.

—— 7. The scales on all three beams have subdivisions between the major divisions.

—— 8. You should not attempt to weigh amounts smaller than 1 gm on a triple beam balance.

—— 9. The triple beam balance should only be used to weigh powdered or granular materials.

—— 10. Traces of foreign chemicals would not be likely to affect a large amount of prepared media.

MODULE 2

Preparing and Dispensing Media

PREREQUISITE SKILL

Ability to weigh materials accurately using a triple beam balance.
Suggestion: Do this module and Module 3, "Sterilization of Media and Equipment," in the same lab period.

MATERIALS

500 ml graduated cylinder
hot plate stirrer
   or Bunsen burner and ring stand
stirring magnet
   or glass stirring rod
stir bar retriever
triple beam balance
600 ml beaker
dehydrated nutrient agar
250 ml beaker

dehydrated nutrient broth
distilled water
8 oz screw-cap bottle
   or similar size flask
Salvarsan burette
test tubes and capalls (30)
   10 tubes for slants
    6 tubes for deeps
   14 tubes for broths
wire baskets (2)

OVERALL OBJECTIVE

Properly reconstitute dehydrated nutrient media and dispense them in standard quantities and containers for various uses.

Specific Objectives

1. Demonstrate your ability to calculate from the grams/liter directions on the medium bottle the amount of powdered medium necessary to make less than a liter of medium.

2. Demonstrate your ability to measure liquids accurately with a graduated cylinder.

3. Use a magnetic stirrer-hot plate combination, or the available equipment, to prepare media that require heat to dissolve completely.

4. Prepare a liquid medium and a solid medium.

5. Demonstrate your ability to use a Salvarsan burette to dispense measured amounts of medium into the appropriate containers.

6. Describe the correct and complete labeling of the various containers of medium according to standard format.

7. Name the type of balance most commonly used to weigh microbiological materials.

8. Name the containers used to weigh less than 15 gm and more than 15 gm.

9. List the preferred amounts of medium and the containers used to sterilize the medium for each of the following: slant, broth, stab, deep, culture plate, 8 oz bottle.

DISCUSSION

Essential to the preparation of microbiological media is the accurate weighing of powders and granular substances. When dehydrated media are reconstituted, a specific number of grams of the powdered medium is added to distilled water and heated to dissolve the powder completely. If you have not weighed the powder and measured the water accurately, you will not have the proper proportion of nutrients in the finished medium. If you do not weigh solid media accurately, you run the risk of having the medium turn out semisolid or too dry.

A triple beam balance is generally used to weigh media since large amounts are involved. When you weigh powders or granular materials, you must never put them directly on the weighing platform of the balance. For amounts up to 15 gm, you should use glazed weighing paper to hold the powdered media you are weighing, and for amounts larger than this you should use a small beaker.

The directions on the labels of most commercially prepared media bottles give you the amount of powdered medium to be rehydrated in 1000 ml (1 liter) of water. The amount of powdered medium per liter of water varies for each medium. Therefore, be sure to *read the label* for each type of medium. You rarely need to make 1000 ml of medium, and so you need to calculate for smaller amounts.

Obviously if you need 500 ml of medium (½ liter), you simply divide both water and powdered medium by 2. If you need 250 ml (¼ liter), you divide both liquid and powder by 4. The easiest way to determine amounts necessary for various other quantities of medium is to find the amount needed for 100 ml by simply moving your decimal point one place to the left, which is actually dividing by 10 (1000 ÷ 10 = 100.0). Then multiply the number of grams necessary to make 100 ml by the number of hundreds (400 = 4 x 100) of milliliters you wish to make. For example, if the label on the bottle directs you to add 15 gm of medium to 1000 ml of distilled water, how many grams of powdered medium should you weigh to make 200 ml?

$$gm/1000 \ ml \div 10 = gm/100 \ ml$$

Moving the decimal point one place to the left would give

$$15 \ gm/1000 \ ml = 1.5 \ gm/100 \ ml$$

the amount needed for 100 ml. Now multiply by the number of 100 ml you wish to make: 2 x 100 or

$$1.5 \ gm/100 \ ml \ x \ 2 = 3.0 \ gm/200 \ ml$$

So you see if you need 1.5 gm/100 ml but you are making 200 ml, you will need 3.0 gm/200 ml.

Another method for figuring how much powdered medium you need is by using a direct proportion. In this method you must set up a ratio between the grams/liter directions on the label and the amount of medium you wish to make. Using the previous example, your proportion is

$$\frac{15 \text{ gm}}{1000 \text{ ml}} = \frac{x \text{ gm}}{200 \text{ ml}}$$

To solve for x:

$$1000x = (15)(200)$$

$$x = \frac{3000}{1000}$$

$$x = 3 \text{ gm}$$

You may use either method of calculation. Use the one that you understand better. Most important is that you are confident that your calculations are accurate.

ACTIVITIES

Activity 1: Calculating Grams of Media Necessary

A thorough understanding of this simple arithmetic operation is necessary if you are to make media properly. So for more practice, try these next few calculations. After working out the following problems, check your calculations with the answers given at the end of this activity.

A. The label on the bottle directs you to add 42 gm of medium to 1000 ml of distilled water. How many grams should you weigh to prepare 300 ml of medium? Remember to move the decimal one place to the left and multiply by the number of 100 ml you are making.

B. The medium label calls for 25 gm/liter, but you only want 400 ml. How many grams of medium must you weigh?

C. This is a slightly trickier problem: You need 12 grams of medium/liter of distilled water. You want to rehydrate 350 ml of medium. How many grams must you weigh?

D. Now try the following calculations:
   (1) You wish to make 600 ml of medium. The label calls for 30 gm/liter. How many grams must you weigh?
   (2) You need 400 ml of nutrient agar, which requires 23 gm/liter. How many grams do you need?
   (3) To make 150 ml of a medium that requires 60 gm/liter, how many grams of dry medium must you weigh?

The following are the answers for the calculations in this activity. Feel free to use proportion if you wish. The answers are underlined.

A. 42 gm/1000 ml ($\div$ 10) = 4.2 gm/100 ml
   4.2 gm/100 ml (x 3) = 12.6 gm/300 ml

B. 25 gm/1000 ml ($\div$ 10) = 2.5 gm/100 ml
   2.5 gm/100 ml (x 4) = 10 gm/400 ml

C. 12 gm/liter = 1.2 gm/100 ml
   1.2 gm/100 ml (x 3.5) = 4.2 gm/350 ml

D. (1) 30 gm/liter = 3.0 gm/100 ml
       3.0 gm/100 ml (x 6) = 18 gm/600 ml

(2)　23 gm/liter = 2.3 gm/100 ml
　　　2.3 gm/100 ml (x 4) = 9.2 gm/400 ml
(3)　60 gm/liter = 6.0 gm/100 ml
　　　6.0 gm/100 ml (x 1.5) = 9.0 gm/150 ml

Put the calculations that you have worked out in your file if you feel you have grasped this principle and can work easily with it. If you do not feel comfortable with this concept, you should do several more calculations of this type. You may make up your own problems for practice or ask your instructor for help. The important thing is to practice until you feel you have mastered this type of calculation.

Activity 2: Making Nutrient Agar

The graduated cylinder is the basic measuring device for liquids. If you need to measure 10 ml or less, it is best to use a pipet of appropriate size. The various cylinders are graduated differently, depending on their sizes and total volumes. For example, a 10 ml cylinder is graduated in 1 ml quantities with subdivisions of 1/10 ml, but a 500 ml cylinder has major graduations of 50 ml with 10 ml subdivisions. Go to your storage area for glassware, and examine the various cylinders and their graduations now. Whenever you use a graduated cylinder, you should first inspect the graduations to determine exactly what you can measure accurately with it. For instance, if you want to measure 142 ml, you need to use a 250 ml cylinder that has 2 ml subdivisions, rather than a 500 ml cylinder that has 10 ml subdivisions.

Agar media must be heated to boiling for several minutes before they will dissolve completely. This can be accomplished by setting up a ring stand or tripod with a wire gauze square over a Bunsen burner, as shown in Figure 2-1. You must stir the media almost constantly with a glass stirring rod to prevent burning or boiling over.

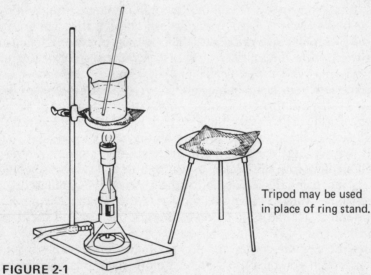

Tripod may be used
in place of ring stand.

FIGURE 2-1
A ring stand or tripod is set up over a Bunsen
burner to boil agar media.

If you are fortunate enough to have one, the hot plate stirrer is an extremely useful aid in making media. Any medium that requires heating to dissolve can be made on the hot plate stirrer, which can also be used simply to stir without heat. Study Figure 2-2 to familiarize yourself with the magnetic stirrer.

You must place a teflon-coated magnet of appropriate size in the beaker with the measured water and powdered ingredients. When you turn the stirrer on (left switch), a magnetic field is created, which causes the magnet in the beaker to spin and, in so doing, stirs the medium. If you are preparing an agar medium that dissolves slowly, or if the label directs you to use heat in broth preparations, you should turn on the right switch, which controls the hot plate. You must *always* bring *agar* media to a *full boil*

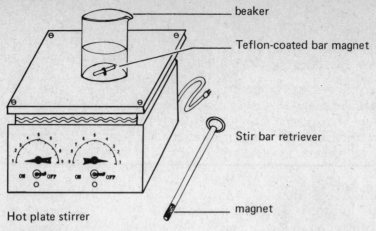

beaker

Teflon-coated bar magnet

Stir bar retriever

magnet

Hot plate stirrer

FIGURE 2-2
The hot plate stirrer is used to make media.

before they will go into solution. When the medium has come to a boil and is ready to dispense, it will no longer be turbid.

Now, at last, we can get to the actual preparation of your nutrient agar (NA). Obtain a 500 ml cylinder, a hot plate stirrer, a teflon-coated magnet (or a ring stand and burner), and a 600 ml beaker. With the 500 ml graduated cylinder, measure 300 ml of distilled water. Remember that you must always have the *bottom* of the meniscus (the curved line you see at the top of a liquid) on the graduation for the quantity you wish to measure. In this case you should have the bottom of the meniscus just touching the 300 ml graduation line. Weigh 6.9 gm of dehydrated nutrient agar. Now pour about 2/3 of the distilled water into a 600 ml beaker and, if you are using a hot plate stirrer, put a teflon-coated magnet in also. Always select a beaker that will hold about twice as much liquid as you will be working with. This allows room for stirring and helps avoid boiling the agar over the top of the beaker onto the surface of the hot plate stirrer or ring stand and your table top. This makes a mess and can cause burns.

Place the beaker over your heat source *very gently* since it can break with the slightest impact. Now slowly turn the left control knob (stirrer knob) until the water is moving vigorously. You should also turn the hot plate on (right knob) and set the control at 9 or 9½. If you are using a ring stand and burner, you should light the burner and adjust the flame. Now add the 6.9 gm of dehydrated nutrient agar. After you have poured the powdered medium into the beaker of water, hold the glazed weighing paper over the beaker, and, using the remainder of the water in the graduated cylinder, rinse any remaining powder from the paper into the beaker. The medium must be stirred continuously as it heats to a boil to prevent burning and sticking to the bottom of the beaker.

Warning: Once the medium comes to a boil remove it immediately from the hot plate stirrer. Even though you turn the heat switch off, there is enough heat remaining in the hot plate to cause the temperature in the beaker to continue to rise.

Activity 3: Dispensing Nutrient Agar

Examine the Salvarsan burette setup shown in Figure 2-3. You will see that it is essentially a graduated cylinder with an inverted scale and a pinch clamp to control the flow of a liquid.

By watching the graduations on the burette or by using an accurately measured guide tube each time, you will soon be able to dispense roughly equivalent amounts of agar into each of several tubes or other containers. As a general rule of thumb for the most commonly used culture tube (16 mm x 150 mm), you will need to dispense 10 ml of agar medium into each tube for slants, 12 ml into each tube for deeps, 7 ml into each tube for stab inoculations, and from 100 to 200 ml into each 8 oz bottle for pour plates. If you have not obtained a Salvarsan burette setup yet, get one now and take it

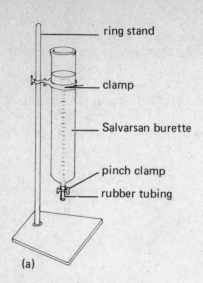

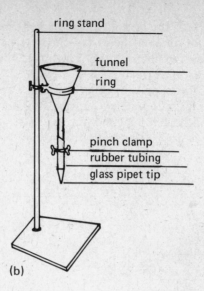

FIGURE 2-3
(a) The Salvarsan burette setup. (b) Alternate dispensing funnel setup.

back to your table. Examine the graduations and operate the pinch clamp to become familiar with the burette.

Check your nutrient agar to see if it has come to a full boil yet. Remember that it is ready to dispense when it has boiled until it looks transparent. Then remove it from the ring stand or the hot plate stirrer, and take the stirring magnet out with a stir bar retriever, as shown in Figure 2-2. Turn off the controls of the hot plate stirrer, and rinse the magnet and stir bar retriever at once. Now pour the hot nutrient agar into the burette carefully. *Be sure that the pinch clamp is closed before you pour the agar.* It is a good idea to bring pot holders or a kitchen towel from home since you will be handling hot containers from time to time during the course.

Arrange your test tubes in your tube rack, and dispense 10 ml of nutrient agar into each of 10 tubes for slants. After you fill the first tube or two, you can watch the liquid in the tube you are filling and close the pinch cock when the level is about the same as in the first tube. That is, you can use an accurately measured tube of medium as a guide and fill the others accordingly. To do this you will be holding a guide tube and an empty tube beside each other in one hand. The guide tube allows you to disregard the graduations on the Salvarsan burette or to use an ungraduated dispensing funnel.

Now dispense agar in six tubes for deeps; remember that you need 12 ml for a deep. All the agar must be dispensed before it can solidify in the burette.

When you have dispensed medium into the 16 tubes as directed, allow the remaining agar to run into an 8 oz bottle. This bottled agar, after autoclaving, can then be poured into several sterile petri dishes to make culture plates. A culture plate contains approximately 20 to 25 ml of sterile agar medium. Media are *never* sterilized in the autoclave while in the petri dish since the agar would bubble out of the plate (petri dish). Now remove the burette from the ring stand, and rinse it immediately with *hot* water. Return the burette to the ring stand, and leave the pinch clamp open so that it can drain. You should form the habit of rinsing with hot water all containers that have held agar *as soon as they are emptied.* This is especially important for dispensing burettes and pipets because the narrow openings can become clogged as the agar solidifies.

Now that you have dispensed all your agar medium, place the screw cap *loosely* (backed off a turn or two) on the 8 oz bottle. When media are sterilized in a steam sterilizer, they are superheated under pressure, that is, the boiling point is raised as the steam pressure increases in the sealed sterilizer. Under these conditions a tightly stoppered container could explode when the medium expands because of the extreme heat. You must always leave about 2 inches of air space between the agar level and the neck of the bottle for this same reason. Even with the cap loose, the expanding agar can boil over the top of the bottle if you do not leave enough air space. If you are using capall-covered or cotton-stoppered flasks, the problem is removed because the air escapes freely.

Put capalls or other tube closures on or in all your tubes, and put a small piece of tape with your name, date, "NA slant," and media preparation on *each* of the 10 tubes with 10 ml of medium, as shown in Figure 2-4; label your deeps "NA deeps" and your bottle simply "NA." All labels should have this information clearly indicated.

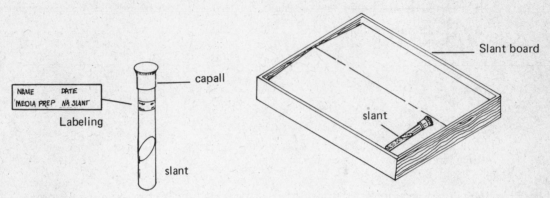

FIGURE 2-4
Tubes of media must be labeled accurately and later placed on a slant board to solidify.

Put all your nutrient agar (NA) slant tubes together in a wire basket, and your NA deep tubes in another basket; label the baskets with a piece of tape with the same information listed, plus the type of agar, module title, and activity number. Put a similar label on your 8 oz bottle of medium. All this agar is now ready to be sterilized.

After your tubes are sterilized and while the medium is still liquid, the tubes for slants should be placed on the slant board (see Figure 2-4) and allowed to solidify undisturbed. All sterile media should be refrigerated until you are ready to use them.

Activity 4: Making Nutrient Broth

Obtain a bottle of dehydrated nutrient broth (NB), and go to the triple beam balance, taking a piece of scratch paper with you. After reading the label on the broth bottle carefully, calculate how much powdered medium you will need to prepare 150 ml of NB. If you figured 1.2 gm, you are absolutely right! Next, place a piece of glazed weighing paper on the weighing platform. With a clean spatula or weighing spoon, weigh out the proper amount of powdered medium. Measure 150 ml of distilled water, and mix your broth in a 250 ml beaker using the same method as in Activity 2. It is rarely necessary to boil broth. In fact, the medium will often dissolve completely with no heat at all.

When your broth has dissolved completely, gather together a wire basket, 14 or 15 tubes, and a dispensing burette or funnel. Broths are normally dispensed in 5 to 10 ml amounts in tubes. Dispense your broth in 10 ml amounts into the tubes. It is a good idea to make an extra tube in case of breakage or other error. You may discard any extra broth. Rinse the burette and allow it to drain. Put capalls or other closures on the tubes, and place them in the wire basket which you have labeled with your name, date, type of medium, and the module and activity. These tubes are now ready for sterilization.

After the broth tubes are sterilized, they should be stored in the refrigerator with the nutrient agar until you are ready to use them.

Table 2-1 is a summary of the amounts of medium to be used in various culture preparations. Some common medium containers are depicted in Figure 2-5.

TABLE 2-1 Amounts of medium to be used

Medium	Amounts	Autoclaving container
Broth	5-10 ml	Culture tube
Agar slants	10 ml	Culture tube
Agar deeps	12 ml	Culture tube
Agar stabs	7 ml	Culture tube
Agar plates	200 ml (20-25 ml/petri dish)	8 oz bottle (2/3 full)

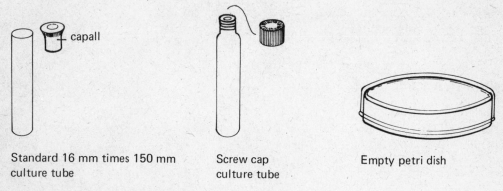

Standard 16 mm times 150 mm culture tube

Screw cap culture tube

Empty petri dish

FIGURE 2-5
Some types of culture containers.

POST TEST

The post test is a self-evaluation. It is not used for a grade. It is designed only to let you decide if you have successfully completed this module.

Part I: True or False

_____ 1. To reconstitute 1 liter of nutrient agar you must weigh 23 gm. If you only need 250 ml of NA, you need to weigh 5.75 gm of dehydrated medium.

_____ 2. The graduated cylinder is the basic measuring device for more than 10 ml of liquid.

_____ 3. Heating agar media to 50°C is sufficient to dissolve it.

_____ 4. You should put powdered medium into the beaker in which you will prepare it before you add the water.

_____ 5. You should prepare media in a beaker with a capacity just slightly greater than the volume of medium you need.

_____ 6. You should dispense roughly equal amounts of medium into each tube for slants.

_____ 7. It is important to screw the cap tightly in place on a bottle of medium that is about to be sterilized.

_____ 8. You should be careful to leave about 2 inches of air space in each bottle of medium.

_____ 9. Media should be sterilized in boiling water for 30 minutes.

_____ 10. Sterile media should be refrigerated until used.

_____ 11. An analytical balance is most often used to weigh microbiological media.

_____ 12. The container used to weigh more than 15 gm is glazed weighing paper.

Part II

1. List the amounts of media used for the following:

    a. slant _____

    b. deep _____

    c. broth _____

    d. stab _____

    e. 8 oz bottle _____

    f. culture plate _____

2. Name the container used to sterilize media for the following:

    a. broth _____

    b. slant _____

    c. deep _____

    d. culture plate _____

MODULE 3

Sterilization of Media and Equipment

PREREQUISITE SKILL

Successful completion of Module 2, "Preparing and Dispensing Media."

MATERIALS

dry-air oven
petri dishes (4-6)
sterilizing can for petri dishes
1 ml pipets (2)
10 ml pipets (2)
pipet can (2)
cotton
portable steam sterilizer and/or autoclave
distilled water

500 ml beaker
media from Module 2, "Preparing and Dispensing Media" (tubes of broth, tubes, and bottle of agar)*
demonstration setup of membrane filter assembly
bacteriological filters, several types

*Media will be prepared for you if your instructor chooses not to perform Module 2.

OVERALL OBJECTIVE

Understand and become adept in the performance of the major methods of sterilization used for microbiological media and equipment.

Specific Objectives

1. Explain the principal use of the dry-air oven and why it is preferred for this use.
2. Discuss the theory of the autoclave.
3. List three types of items routinely sterilized by autoclaving.
4. Discuss the theory of sterilization by bacteriological filtration.
5. Define the terms *filtrate* and *membrane filter*.
6. Describe the necessary correction for the temperature and pressure used for autoclaving at different altitudes.

7. List the length of time and temperatures used for dry sterilization and autoclave sterilization at sea level.

8. Give the pressure and temperature of autoclaving used at your lab because of altitude changes if there are any.

9. List two pieces of microbiological equipment that are sterilized in the dry-air oven and describe how they are placed in the oven.

DISCUSSION

Microbes are abundant in your environment. They are on your body, in the air, and on every surface and object that you encounter. This fact will be vividly demonstrated as you proceed through this course.

The systematic study of selected, isolated microorganisms would be impossible without some method of destroying unwanted organisms. Without effective sterilization procedures, aseptic surgery, other vital medical techniques, food preparation, and preservation methods would not be possible.

A variety of effective sterilization methods is available to us. No single method is ideal for every type of object to be sterilized; that is, methods using heat are unsatisfactory for thermometers and tend to dull cutting instruments, while chemical methods are often corrosive to metal objects. All known methods of sterilization are based on either removal or destruction of bacterial cells. In this course, you will confine your study of sterilization to those methods *commonly* used in the microbiological laboratory.

At some schools the media are dispensed and sterilized for you throughout the entire course. If this is the situation in your case, it would still be prudent to read Modules 2 and 3 since work in the field may require that you know these procedures.

Chemical sterilization in the microbiology laboratory is usually confined to the disinfection of your working area. Containers of disinfectant are supplied in the laboratory, usually beside the sinks. You should make it a habit to clean the table tops in your working area with this disinfectant *every lab period* both *before* you begin work and *after* you finish. This habit helps to control contamination by greatly reducing indigenous microorganisms and by destroying microbes you might inadvertently scatter as you work.

Dry sterilization is used for glassware, principally petri dishes and pipets, both of which are sterilized in cans. Dry sterilization is preferred for glassware because there is no condensation of moisture. The dry-air oven that is used for dry sterilization is simply a household oven on a larger scale. High temperatures must be maintained for long periods of time to achieve sterilization with dry heat. The routine sterilization of glassware in the dry-air oven requires a temperature of 180°C (350°F) for two hours.

Wet sterilization in the microbiology laboratory is commonly confined to sterilization by steam in the *autoclave* and/or portable steam sterilizer. Culture media, solutions, and cotton (in fact, most objects sterilized in the microbiology lab except glassware or solutions decomposed by heat) are sterilized by steam in the autoclave. Hospitals, for instance, sterilize surgical packs, obstetrical packs, towels, and patient drapes in the autoclave.

The autoclave is essentially a large pressure cooker. When the autoclave is closed and the heat is turned on, it becomes a sealed container. As the water heats, pressure builds up in the sealed autoclave. As the pressure increases above normal atmospheric pressure, the boiling point of the water is raised. For example, at 15 pounds per square inch (psi) above normal atmospheric pressure at sea level, water boils at 121°C (250°F) instead of 100°C (212°F) as it does at normal atmospheric pressure at sea level. Normally, all forms of life are killed by maintaining a temperature of *121°C for 15 minutes,* which is the routine sterilizing cycle for the autoclave. Because 15 psi at sea level raises the boiling point of water to 121°C, this cycle is often expressed as *15 psi for 15 minutes.* You must remember, however, that this only applies *at sea level.* If your laboratory is not at sea level, it will be necessary to determine, by careful observation of your pressure gauge and thermometer, whether you must maintain more or less pressure to attain the necessary temperature. For example, at 2700 feet above sea level, it is necessary to maintain 20 psi to reach a temperature of 121°C. Even if the directions on a medium bottle say "sterilize at 15 psi for 15 minutes," what they really mean is to *sterilize at 121°C for 15 minutes.*

Some media, such as certain fermentation broths, nutrient gelatin, and litmus milk, deteriorate at 121°C. They must be autoclaved at lower temperatures and, therefore, at lower pressures. Table 3-1 shows pressures and temperatures that should help you to determine the pressure that you must maintain at your altitude in order to achieve *the necessary temperature (121°C) for sterilization.*

TABLE 3-1 Autoclave Steam Pressure and Corresponding Temperatures*

Steam pressure in psi at sea level	Temperature °C	°F	Steam pressure in psi at sea level	Temperature °C	°F
0	100.0	212.0			
1	101.9	215.4	16	122.0	251.6
2	103.6	218.5	17	123.0	253.4
3	105.3	221.5	18	124.1	255.4
4	106.9	224.4	19	125.0	257.0
5	108.4	227.1	20	126.0	258.8
6	109.8	229.6	21	126.9	260.4
7†	111.3	232.3	22	127.8	262.0
8	112.6	234.7	23	128.7	263.7
9	113.9	237.0	24	129.6	265.3
10†	115.2	239.4	25	130.4	266.7
11	116.4	241.5	26	131.3	268.3
12	117.6	243.7	27	132.1	269.8
13	118.8	245.8	28	132.9	271.2
14	119.9	247.8	29	133.7	272.7
15†	121.0	249.8	30	134.5	274.1

*Figures are for *steam pressure only* at *sea level.* The presence of any air in the autoclave invalidates temperature readings for any given pressure from this table.
†Common sterilizing settings.

For example, if your directions tell you to autoclave at 7 psi for 15 minutes, you should check the table to determine the temperature corresponding to 7 psi at sea level (111.3°C). Then you must determine by careful inspection of your own autoclave pressure gauge and thermometer the pressure at your altitude that corresponds to 111.3°C.

Solutions that decompose at autoclave temperatures are sterilized by *bacteriological filtration.* This method of sterilization is based on the purely mechanical process of removing all "life forms" from the solution by passing it through a filter with openings so small that bacterial cells and larger microorganisms are retained on the filter. The filtrate is collected in a sterile container and aseptically pipetted to sterile tubes or other containers for use. This filtrate is quite sterile without treatment by heat or chemical action.

Several types of bacteriological filters will be available for your inspection in the lab although you will not actually use them in this module. If you choose or your instructor so indicates, you may use a membrane filter in a future module dealing with the action of bacterial enzymes on urea. Urea decomposes at autoclave temperatures and so must be sterilized by bacteriological filtration.

ACTIVITIES

Activity 1: Dry Sterilization

Gather four to six petri dishes from the storage area in the laboratory and place them in one of the cans in which they are to be sterilized. Most of these cans hold 12 petri dishes, so share one with another student if you are both working on this module. *Do not* label the can. The label will burn in the prolonged intense heat necessary for dry sterilization. Place your petri dish can in the dry-air oven, and set the thermostat at 180°C (350°F) for two hours.

Now gather together two 1 ml pipets, two 10 ml pipets, and a small piece of cotton. Pull off a very small bit of cotton, and roll it between your fingers until it can be inserted into the bore of the mouth end of one of your pipets with the aid of a stylet, as shown in Figure 3-1. You should not insert it very far, just far enough so that no cotton fibers protrude from the mouth end. The cotton plug should be approximately ½ inch long and should be inserted so that it is about 2 mm from the mouth end of the pipet. Have your instructor approve your pipets. The most common error your instructor will find is the protrusion of cotton fibers. Repeat this for each of your pipets.

Put your 1 ml pipets into one pipet can and your 10 ml pipets into another pipet can. Place the pipets in the cans tip end first so that you can remove the sterile pipets for use by handling the mouth end only. When all the pipets in the class have been placed in the cans, the cans should be closed and placed in the dry-air oven at 180°C (350°F) for two hours. This entire procedure is often avoided by using sterile, disposable pipets and petri dishes.

Precaution: Make it a habit to line up the seams in the cans and their lids when you close them. They are much easier to open if you have taken this simple precaution. This applies to both petri dish cans and pipet cans.

The petri plates you are sterilizing, if kept in the sterile cans, can be used for subsequent modules. So place them in your sterile storage area.

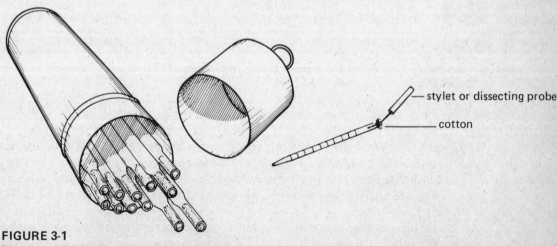

FIGURE 3-1
Pipets are sterilized in a pipet can. A stylet is used to insert a bit of cotton in the mouth end of each pipet.

Activity 2: Wet Sterilization Using the Portable Steam Sterilizer

The portable steam sterilizer can probably be most useful to you when you need to work independently, as in the identification of unknown organisms. Then you may need small amounts of several media for specific tests, and you can work independently. It is to your advantage to become thoroughly familiar with this piece of equipment if available.

Study Figure 3-2 to familiarize yourself with the working parts of the portable steam sterilizer. Examine the sterilizer itself, and find the parts that are labeled in the figure. When you feel that you are familiar with the sterilizer, take some of your media from Module 2, "Preparing and Dispensing Media," or the media provided, to the sterilizer.

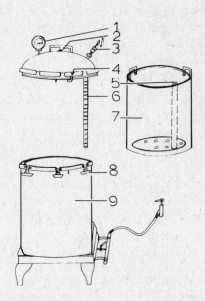

FIGURE 3-2
Portable Sterilizer
1. pressure gauge
2. safety plug
3. safety valve with toggle
4. index arrow
5. guide channel
6. flexible exhaust tube
7. aluminum inner container
8. toggle bolt with wing nut
9. sterilizer body.

Follow this check list to sterilize media in the portable steam sterilizer.

1. Pour 500 ml of clean distilled water into the sterilizer body.
2. Place media bottles and/or baskets of tubes in the inside aluminum container, allowing room for free movement of steam between the bottles or tubes during sterilization. Place the container in the sterilizer body.
3. Put the cover on the sterilizer body, matching index arrows and inserting the flexible tube into the guide channel on the inside wall of the aluminum container. Swing the toggle bolts up into the notches on the cover, and tighten the wing nuts evenly, bringing them down to a uniform pressure with the cover accurately centered. This is best done by tightening the two nuts directly opposite each other at one time, then the next two, and so forth. *Caution:* Tighten only by hand. Keep the tapered sealing surfaces on both the cover and the rim of the sterilizer body clean, and protect them from scratching and marring.
4. Open the safety valve toggle. When open, the toggle should be straight up. Light the hot plate and turn the flame high. Steam generated in the bottom of the sterilizer body will travel up around the inside container and then down between the bottles and tubes to the bottom of the container, forcing air out from the bottom up through the flexible exhaust tube and safety valve.
5. When steam escapes vigorously (making a hissing sound), close the safety valve by pushing the toggle down to a horizontal position. Pressure will then rise in the sterilizer and will be indicated on the pressure gauge.
6. When the pressure gauge shows 15 psi (or the pressure necessary at your altitude to reach 121°C or 250°F), reduce the heat as necessary to maintain this pressure without the safety plug blowing.

7. Sterilize your media by maintaining the necessary pounds per square inch for 15 to 20 minutes. *Watch the pressure when sterilizing.* The safety plug directly behind the handle will automatically release should the pressure reach 35 psi, in which case a new one must be installed. *Caution:* For media such as litmus milk, fermentation broths, and nutrient gelatin, which require lower temperatures, check Table 3-1 in this module to obtain the desired temperature at the altitude of your laboratory. Maintain the necessary pressure for 15 to 20 minutes in all cases unless otherwise instructed.

8. At the end of the sterilizing period, turn off the heat, and allow the pressure to reutrn to zero. *Do not* open the safety valve until the pressure has returned to zero, as a quick release of pressure may cause the media to boil over and closures to blow off.

9. When the gauge shows zero, open the toggle valve and allow the small amount of remaining steam to escape. Next loosen the wing nuts, swing the bolts from the slots, twist the cover, and remove it. *Do not* loosen the wing nuts until the gauge shows zero. If the sterilizer is allowed to cool after the gauge shows zero without opening the safety valve, a vacuum may form within the chamber, and the cover cannot be removed until the vacuum is broken.

Start each load with clean, fresh water. When you are finished using the sterilizer, drain any water remaining in the chamber, and dry all parts thoroughly. Do not immerse the gauge in water while you are cleaning the lid. Do not add very cold water when the sterilizer is hot.

Activity 3: Wet Sterilization with the Autoclave

Examine the autoclave. All significant parts necessary to operate the autoclave should be clearly labeled. Some parts, such as the thermometer and pressure gauge, are self-evident and therefore will not be labeled. Locate all appropriate parts that correspond to those on the portable steam sterilizer. Take the remainder of the media from Module 2, or the media provided, to the autoclave.

Follow this procedure to sterilize media in the autoclave.

1. Place items to be sterilized into the autoclave chamber, allowing room for free passage of steam between them.

2. Close the autoclave door and fasten it securely to assure a tight seal. Some autoclaves have a wheel to tighten the seal; others have doors that simply roll closed and then must be lifted and pulled as far to the front as possible.

3. If your autoclave has a sight glass, check the water level in the sight glass. The water level should be 1 inch from the top of the glass. In recent models the addition of water is automatically controlled.

4. Close the exhaust valve.

5. Turn on the timer (Electro Stopswitch) to 45 minutes.

6. When the pressure gauge reaches 5 to 7 pounds, open the exhaust valve allowing the air to escape from the chamber until the pressure drops 2 pounds. This is essential to attain the temperature necessary for sterilization.

7. Close the exhaust valve.

8. When the pressure gauge reads 15 psi (or the pressure you require for your altitude) and the temperature is 121°C (250°F), set the timer (Electro Stopswitch) at 15 minutes.

Caution: Do not open the exhaust valve after maximum pressure has been reached or until the pressure has come down to zero after autoclaving is finished.
If lower pressures are desired, check Table 3-1 in this module and set the pressure selector dial to give the correct temperature. Many expensive autoclaves

are equipped with controls that automatically remove the air before the steam pressure is allowed to rise to 15 psi. However, in older, less expensive, or smaller autoclaves used in private offices this must be done manually by opening the exhaust valve and allowing the steam to force out the air.

Activity 4: Examination of Several Types of Bacteriological Filters

A membrane filter assembly is set up in the laboratory. Examine this equipment, which is one of the more recent improvements in bacteriological filters and one of the easiest to use. Inside the filter funnel is a membrane made with tiny pores of an exactly specified and controlled size. These pores are usually 0.45 μm (micrometers), which is equivalent to 0.00045 mm. The average bacterial cell is larger than this and so cannot pass through the filter membrane.

The filter funnel assembly and filter flask are wrapped and autoclaved before use. When they are set up, the solution to be sterilized is poured into the filter, and a vacuum pump is attached to the side arm of the flask. As the flask is evacuated, the solution is drawn rapidly through the filter membrane. The bacterial cells that are too large to pass through are retained in the funnel. (See Figure 3-3.)

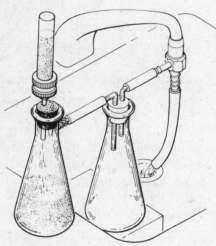

FIGURE 3-3
Apparatus for bacteriological filtration.

All other bacteriological filters operate on the same principle regardless of the substance used for the filter. The Seitz filter, which employs an asbestos pad, is in common use. Other filters use filter candles made of sintered glass or porcelain, instead of a membrane or pad. Some of these filters are available for your inspection near the membrane filter assembly. Most are no longer used because of the difficulty in cleaning them.

Take the post test now.

POST TEST

The post test is a self-evaluation. It is not used for a grade. It is designed only to let you decide if you have successfully completed this module.

Part I

Select a tube from the center of one basket sterilized in the portable steam sterilizer and one from a basket from the autoclave. Incubate these tubes in the 30°C incubator for 24 hours. If there is no apparent growth in the tubes after incubation, your sterilization has been successful.

Part II: True or False

_____ 1. Petri dishes must be autoclaved before they are used to culture bacteria.

_____ 2. The filtrate is what remains on the filter after a solution has been passed through it.

_____ 3. In sterilization by autoclaving, the proper temperature is more critical than any specific pressure.

_____ 4. It is more correct to say "sterilize at 121°C for 15 minutes" than "sterilize at 15 psi for 15 minutes."

_____ 5. Empty glassware is best sterilized in the dry-air oven at 180°C (350°F) for two hours.

_____ 6. It is unnecessary to sterilize a bacteriological filter apparatus before use because all bacteria will be trapped on the filter.

_____ 7. You should always clean your working area with disinfectant before and after you work with living bacteria.

_____ 8. Every medium used in the microbiology lab should be autoclaved at 15 psi (121°C) for 15 minutes before it is used.

_____ 9. You should take care to leave enough space between objects in the autoclave to allow free passage of steam between them.

_____ 10. Air must be carefully exhausted from the autoclave chamber in order to attain the temperatures and pressures necessary for sterilization.

_____ 11. At higher altitudes the temperature and pressure to sterilize by autoclaving are decreased.

_____ 12. At higher altitudes the temperature reading of the autoclave remains the same, and the pressure varies.

_____ 13. Petri dishes are sterilized in the dry-air oven by placing the bottom and the lid upside down on the oven rack.

_____ 14. Pipets are sterilized in the dry-air oven by standing each pipet up with the mouth end down on the oven rack.

_____ 15. The use of the dry-air oven for sterilization of pipets and petri dishes can be avoided by using these items in disposable form.

MODULE 4

Compound Microscope for the Study of Microbes

PREREQUISITE SKILL

None.

MATERIALS

compound light microscope with oil-immersion objective

light source

microscope slides (2)

concentrated salt solution

prepared slides of stained blood smear

stained smear of yeast, *Bacillus* sp. (any species of the genus *Bacillus*), and *Escherichia coli*

prepared slide of mixed bacteria types

OVERALL OBJECTIVE

Demonstrate your ability to utilize a compound light microscope using all powers of magnification.

Specific Objectives

1. Make drawings as requested in the activities using the various objectives.
2. Define the terms *magnification at eyepoint, object*, and *parfocal*.
3. Name the optical parts of the microscope and their functions.
4. Name the movable parts of the microscope not in the optical system and also name their functions.
5. Name the three parts of the microscope most critical in light control.
6. Label all the described parts of your microscope.
7. Relate the magnification of the various objectives to the approximate size of the field.
8. Describe the manipulation of the mirror, diaphragm, and condenser for improving the light and defining the object.

9. Describe the proper care of the microscope.
10. Describe how a microscope should be carried.
11. Describe how the different powers of your objectives are marked for identification on your microscope. This objective is done by inspection.

DISCUSSION

Learning to use your microscope correctly is one of the accomplishments most valuable to you in your study of microbes. In microbiology the size of the organisms studied requires that you become an expert microscopist. You can accomplish this only with much practice or experience. This module is designed to show you how to use a microscope, and not why a microscope magnifies to the extent it does or how the image is formed.

As a beginning microbiologist, the amount of skill you develop in using the microscope can make this course either interesting or not interesting to you. Therefore, to reemphasize, it is not as important to understand why or how a microscope functions as it is for you to see what you are supposed to see through your microscope. If you wish to learn the mechanism by which your microscope magnifies an object, you need only read any microbiology text or most laboratory manuals. These sources explain in detail resolving power, numerical aperture, real image, virtual image, and refractive index. If you plan to major in microbiology or do advanced work in it, you should learn these functions of the microscope. It is the purpose of this module, however, to allow you to enjoy microbiology by teaching you to use your microscope to its optimum.

Two types of microscopes are illustrated in this module. See Figures 4-1 through 4-3, and study the figure that resembles your microscope. You should be able to name all the labeled parts and explain their functions. (See the appropriate tables accompanying the microscope illustrations.) It is important at this point to study these figures and tables.

Each of the three objectives on your microscope has a number etched on it: 10x, 44x, or 97x. This number indicates the number of times an object is magnified or enlarged by the objective. The 10x objective is the low-power objective, and it magnifies an object 10 times its actual size. The 44x objective (high power) magnifies an object 44 times its actual size, and the 97x objective (oil immersion) magnifies 97 times the actual size. Some microscopes are equipped with objectives that vary from the usual 10x, 44x, and 97x combination. The most common variation is a 10x, 45x, and 100x combination. Some microscopes are equipped with a 3.5x scanning lens in addition to the other three objectives.

The objectives are often marked with incised bands around the lower part of the objective to allow you to distinguish one from another quickly and easily without finding the magnification number each time. This is especially important for the high-power and oil-immersion objectives, which are nearly the same length. The high-power objective (44x) is meant to be used dry, and the outer lens could be damaged by immersing it in oil accidentally. If the objectives are coded with bands, the 10x usually has one band, the 44x has two bands, and the oil-immersion objective (97x) has three bands. The bands are sometimes colored for still easier recognition, in which case the oil-immersion objective is usually banded in red.

Your ocular or eyepiece also magnifies, usually 10 times. Thus, the magnification at eyepoint (that is, what you see in the microscopic field while looking through the ocular) is magnified 10 times more than the magnification marked on the objective you are using.

Ocular		Objective		Magnification at eyepoint
10x	X	10x	=	100x
10x	X	44x (45x)	=	440x (450x)
10x	X	97x (100x)	=	970x (1000x)

The objectives on your microscope, and on most microscopes, are *parfocal*. That is, they are mounted so that when an object is in sharp focus with one objective, it will be in approximate focus with the other objectives when they are rotated into working position. Thus, if you have an object in sharp focus on low power, you can rotate the high-power objective (44x) into working position and achieve sharp focus with only a slight turning of the fine adjustment knob.

Despite the fact that the objectives are parfocal, the novice microscopist frequently "loses" the object when switching to a higher power objective. This happens because each increase in power of magnification *decreases* the microscopic field by about one-half. The illuminated field *appears* to be the same size at all powers, but the area actually being magnified shrinks each time you switch to an objective that magnifies to a higher power. So if the object is not in the center of the field when you switch to a higher power, you can easily lose it from the resulting diminished field. Your task can be made much simpler if you remember always to center the object before you switch to a higher power.

In general, the more the magnification is increased, the more light must enter the optical system. This module contains a practice activity designed to help you master light control, which is critical to good microscopy. It is to your advantage and well worth your time to repeat the practice activity until you master light control with your microscope.

Binocular Microscope

Table 4-1 lists the parts and their functions of an AO Spencer series 50 binocular microscope. (The nosepiece moves to focus. In other brands, the stage moves for focusing.) Figures 4-1 and 4-2 show the parts of this microscope.

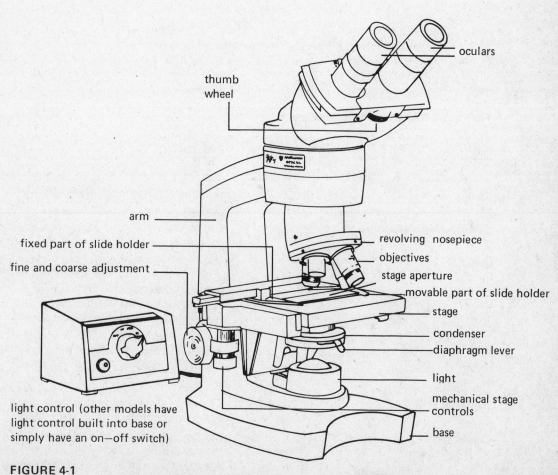

FIGURE 4-1
AO Spencer series 50 binocular microscope (left front view).

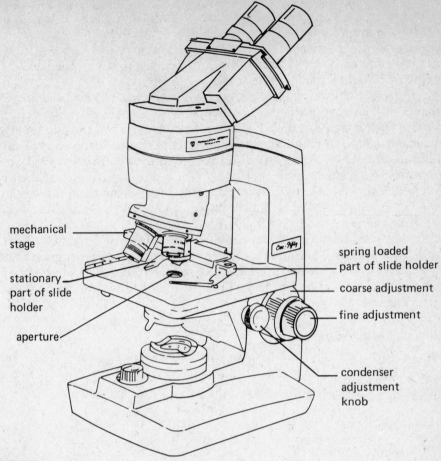

mechanical stage

stationary part of slide holder

aperture

spring loaded part of slide holder

coarse adjustment

fine adjustment

condenser adjustment knob

FIGURE 4-2
AO Spencer series 50 binocular microscope
(right front view).

TABLE 4-1 The Binocular Microscope

Part	Function
Oculars (eyepieces)	A series of lenses that usually magnify 10 times.
Thumb wheel	Interpupillary adjustment; adjust for *your* eyespan
Revolving nosepiece	Can be rotated to change from one objective to another and is raised and lowered in focusing.
Objectives	Usually three magnifications (if no scanning lens is present): 10x, low power; 44x or 45x, high power; and 97x or 100x, oil immersion. Powers of the objectives are distinguished by different colored bands on the objectives.
Slide holder	Spring loaded portion allows placing the microscope slide in the mechanical stage where it is tightly held.
Nosepiece	Raised and lowered in focusing.
Diaphragm lever	Opens and closes the diaphragm to control the amount of light striking the object.
Condenser	Condenses light waves into a pencil shaped cone, thereby preventing the escape of light waves. Also controls light intensity when raised and lowered.

Condenser adjustment knob	Raises and lowers condenser.
Mechanical stage	Allows the slide to be moved.
Mechanical stage controls	Moves the slide on two horizontal planes, that is, back and forth and side to side.
Base	Supports entire microscope.
Light intensity control	Turns light source on and off and controls light intensity.
Arm	Supports upper half of the microscope.
Coarse adjustment	Moves the nosepiece up and down rapidly for purposes of approximate focusing.
Fine adjustment	Moves the nosepiece up and down very slowly for purposes of definitive focusing.

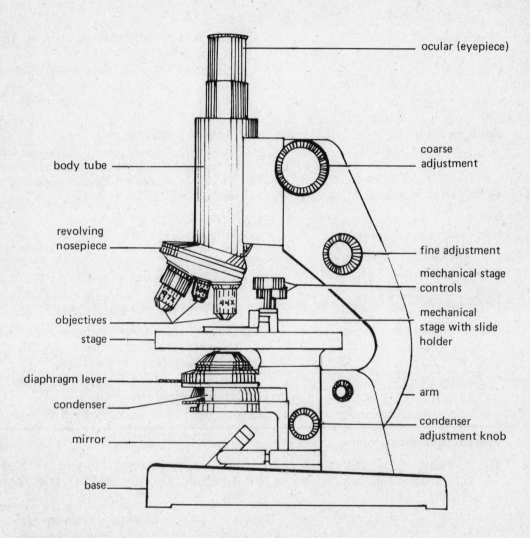

FIGURE 4-3
Monocular microscope (Graf-Apsco or Bausch & Lomb type).

Monocular Microscope

Table 4-2 lists the parts and their functions of a monocular Graf-Apsco or Bausch & Lomb type of microscope. (The body tube moves to focus.) Figure 4-3 shows the parts.

TABLE 4-2 The Monocular Microscope

Part	Function
Ocular (eyepiece)	A series of lenses that usually magnify 10 times.
Body tube	Raised and lowered in focusing.
Revolving nosepiece	Can be rotated to change from one objective to another.
Objectives	Usually three magnifications (if no scanning lens is present): 10x, low power; 44x or 45x, high power; and 97x or 100x, oil immersion. The powers are distinguished by different colored bands on the objectives or different numbers of black bands.
Stage	Supports the mechanical stage and microscope slide.
Diaphragm lever	Opens and closes the diaphragm to control the amount of light striking the object.
Condenser	Condenses light waves into a pencil-shaped cone, thereby preventing the escape of light waves. Also controls light intensity when raised or lowered.
Condenser adjustment knob	Raises and lowers condenser.
Adjustable mirror	Reflects light waves from the light source into the condenser. Always use the flat (plane) side of the mirror when the microscope has a condenser.
Base	Supports entire microscope.
Mechanical stage with slide holder	Spring-loaded portion of slide holder allows placing the microscope slide in the mechanical stage where it is tightly held. The mechanical stage allows the slide to be moved.
Mechanical stage controls	Move the slide on two horizontal planes, that is, back and forth and side to side.
Arm	Supports upper half of the microscope.
Coarse adjustment	Moves body tube up and down rapidly for purposes of approximate focusing.
Fine adjustment	Moves body tube up and down very slowly for purposes of definitive focusing.

Care of Your Microscope

The following instructions and precautions will be most important to you in regard to the rapidity with which you become a skilled microscopist. It would be advisable to memorize them.

1. Always use both hands when you carry your microscope. Grasp the microscope arm firmly with one hand, and lift it carefully. Place your other hand under the base of the microscope for support as you carry it. Keep your microscope vertical since the oculars could fall out if the microscope is tilted.

2. Each time you use your microscope, clean the optical system (ocular lens, objectives, condenser lens, and mirror if present) before and after use. This is especially necessary if you must share your microscope with another student in another lab section. Use *only* optical lens tissue to clean the optical system. To remove oil or dust from other portions of your microscope use a soft cloth or facial type tissue. *Always keep your microscope immaculate!*

3. *Never* remove any parts of the microscope without consulting your instructor.

4. When you have finished using your microscope for the day and have cleaned it properly, if your microscope does not have an autofocus stop, put your low-power objective into working position because it is shorter than the 44x and 97x objectives and, therefore, less likely to be damaged by striking the mechanical stage accidentally. Replace the dust cover before returning your microscope to the storage cabinet.

5. If your microscope does not have an autofocus stop, *never* focus downward while looking through the eyepiece. To prevent breaking slides and possibly damaging the objective you should always turn the objective down to its lowest point, while watching from the side, before you look through the eyepiece and begin to focus. Do not touch the lenses of the eyepiece. Skin oils can mar the polished glass surface of the lens.

Practice Activity for Light Control

1. Turn on the light source. If your microscope does not have a built-in light source, you must now position the lamp as shown in Figure 4-4 and adjust the mirror for optimum light. Remember that the mirror is movable. It is designed so that you can adjust it to reflect the maximum amount of light on the object. If you have a built-in light source, then omit Step 2.

2. Now look through the ocular with the low-power objective in place. Be sure the plane (flat) side of the mirror is up. The mirror pivots in all directions, as shown in Figure 4-5. As you look through the ocular, move your mirror in all the indicated directions until you visually determine that the maximum amount of light is illuminating the microscopic field. Study the check list that follows this list for proper adjustment of your Koehler illuminator.

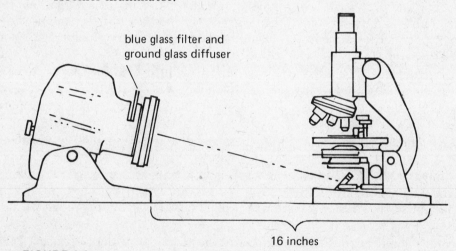

blue glass filter and ground glass diffuser

16 inches

FIGURE 4-4
The proper position of a Koehler illuminator.

FIGURE 4-5
The mirror is on a pivot and moves in all directions. Move your mirror in all the arrow-indicated directions until you visually determine that the maximum amount of light is passing through the aperture in the stage.

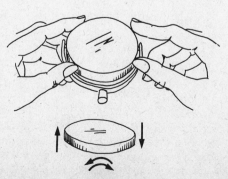

3. Open and close the iris diaphragm by moving the diaphragm lever as you look through the ocular. Observe the effect this has on the brightness of the microscopic field. Do this several times. Adjust the iris diaphragm to optimum brightness. Optimum brightness does not necessarily mean maximum brightness.

4. Using the condenser adjustment knob, raise and lower the condenser, still looking through the ocular, and note what effect the position of the condenser has on the intensity of the light. When the ground glass (roughness) appears, raise the condenser until the roughness just disappears. Do this several times, adjusting the condenser for optimum illumination. Do not hesitate to readjust the mirror, open and close the diaphragm, or raise and lower the condenser. Once the condenser has been adjusted to its optimum, do not change the setting. The amount of light must now be controlled with the iris diaphragm. Repeat this practice activity several times since light control is most critical to good microscopy.

Check List: Proper adjustment of the Koehler illuminator

1. Place the illuminator about 16 inches from the microscope and directly in front of it. See Figure 4-4 for the correct position of the illuminator.

2. Turn on the illuminator, and remove the blue glass filter and the ground glass diffuser. (See Figure 4-4.)

3. Align the lamp by adjusting the knob on the right side of the illuminator *(as you face it)* so that the light strikes the center of the mirror.

4. Tilt the mirror to direct the light rays into the condenser. Be sure the plane (flat) side of the mirror is up.

5. Close the iris diaphragm.

6. Stand or lean over behind the illuminator, and, looking into the mirror, use the small knob on the right side of the lamp *(from behind it)* to focus an image of the lamp filament on the lower surface of the leaves of the iris diaphragm. You should see the coils of the filament distinctly.

7. Replace the blue glass filter and the ground glass diffuser in the space provided in the illuminator. It does not matter which you place in front.

8. Looking through the microscope, tilt the mirror to center the light source in the field of view.

9. Focus your 10x objective on a slide upon which you have marked an x with a wax pencil.

10. Turning the condenser adjustment knob slowly, bring the image of the etching of the ground glass diffuser into focus. This is difficult to see. Look for a change of texture of the *glass* beside the wax pencil mark. It will appear rough when the etching is in focus.

11. Carefully roll the condenser upward until the etching is just out of focus. This is usually the best condenser position for stained smears. However, the condenser is raised and lowered for other preparations.

12. Open the iris diaphragm. Now, while looking through the microscope, slowly close the iris diaphragm until the glare just disappears.

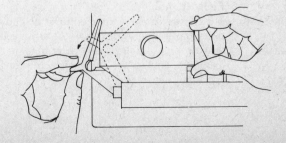

FIGURE 4-6
Placing the slide in the slide holder. There are variations of this slide clip, but all slide clips work on the same principle; they are spring loaded and put tension on the slide to keep it in place as it moves with the mechanical stage.

Your illuminator is now in focus and will give you better definition as you use your microscope. It is not necessary to adjust your Koehler illuminator each time you use it. However, you should make a careful adjustment the first time you use your illuminator and anytime you have unusual difficulty obtaining proper illumination and definition.

ACTIVITIES

Activity 1: Low-Power Observation of Salt Crystals

Put a small drop of concentrated salt solution on a microscope slide, spread it out, and allow it to dry. Place the slide in the slide holder as shown in Figure 4-6. Position the crystals under the objective by using the mechanical stage controls as you watch from the side. Using the coarse adjustment, move the 10x objective as close to the slide as possible, still watching from the side, until you reach the stop. Looking through the ocular now, slowly turn the coarse adjustment, moving the objective away from the slide until the salt crystals come into approximate focus. To achieve definitive focus you must next use the fine adjustment.

Adjust the optical system as you did in the practice activity until you obtain the best definition of the salt crystals. Adjust and readjust the mirror, iris diaphragm, and condenser. It will probably be necessary to reduce the light intensity in this case, so experiment with the iris diaphragm and condenser until you find the optimum position of each that allows you to see the most definition of the salt crystals. Do not be satisfied with seeing just any image; get the best possible.

Make a composite drawing of the salt crystals. Select a crystal from several different fields so that you must use the mechanical stage. Submit this drawing to your file.

Activity 2: High-Power Observation of a Stained Blood Smear

Place a prepared slide of a stained blood smear in the slide holder. Be sure the smear side is up. Position the stained smear under the objective by using the mechanical stage as you watch from the side.

With your 10x objective still in place, adjust for maximum light. Repeat Activity 1 until the blood cells are in sharp focus. Rotate your revolving nosepiece now to bring your high-power objective (44x) into working position. Adjust for the sharpest possible focus with the fine adjustment. Sketch representative blood cells at 440x, and submit the sketch to your file. Do not dwell upon the different types of cells here. Wait until you do the next activity to study cell types.

Activity 3: Demonstration of Parfocality Using Oil-Immersion Objective

With the stained blood smear from Activity 2 still in focus and the 44x objective still in place, adjust for maximum light. Rotate your revolving nosepiece until the *space* between the 44x objective and the 97x objective is directly over the center of the smear; no objective is in the working position. Carefully place one or two large drops of immersion oil on the smear, being careful not to get any on the microscope stage. Rotate the revolving nosepiece now until the 97x objective clicks into working position. The tip of the objective should be immersed in the oil but not quite touching the slide.

Look through the ocular, and focus critically with the fine adjustment. Draw several blood cells as they appear at 970x, and submit them to your file. See Figure 4-7 for blood cell types. Include erythrocytes and leucocytes in your drawing. Repeat this activity using a stained smear of yeast and a stained smear of bacteria (*Bacillus* sp.) as shown in Figure 4-7.

After you have become proficient in employing parfocality to use the oil-immersion objective, and after you are satisfied that you are getting the maximum amount of definition by using the optimum amount of light, proceed to the next activity.

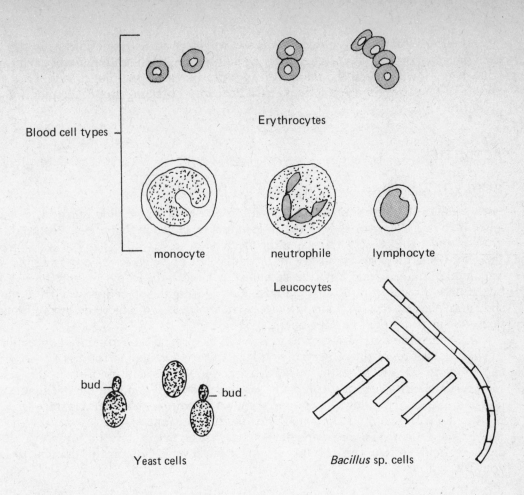

Blood cell types

Erythrocytes

monocyte   neutrophile   lymphocyte

Leucocytes

bud

bud

Yeast cells   *Bacillus* sp. cells

FIGURE 4-7
Blood cell types, yeast cells, and bacteria cells as
viewed with the oil-immersion objective (not to
scale).

Activity 4: Direct Use of the Oil-Immersion Objective

As you become an expert microscopist, and as you begin to study bacterial structures,
you will find it expedient to use only the oil-immersion objective. This can be done
without damaging the microscope slide containing the object or the oil-immersion
lens by lowering the objective until it just barely touches the slide. Do this while
looking closely from the side. If you happen to have an AO Spencer series 50 microscope, your oil-immersion objective has an autofocus stop, and, after applying oil to
the slide, you can lower the coarse adjustment rapidly to its positive stop. There is no
danger of touching the slide with the oil-immersion objective, and the object will
be in focus with a slight adjustment of the fine adjustment knob.

    Caution: If you are using a microscope similar to the one depicted in Figure 4-3,
then you must lower the oil-immersion objective with great care to just barely touch
the slide. *There is no automatic stop; therefore, the objective lens can be damaged.*

    Place a simple stain of yeast cells in your mechanical stage. Put a large drop
or two of oil on the slide. Using the coarse adjustment, immerse the oil-immersion
objective into the oil while looking from the side. Now gently lower the objective until
the slide-objective contact has been made or until the stop is reached, depending upon
your microscope. Do not force the coarse adjustment knob down at this point of
contact. Now looking through the ocular, *slowly* turn the coarse adjustment back, that
is, move the oil-immersion objective *slowly* away from the object until it comes into
focus. When you can see the object vaguely, use the fine adjustment for definitive
focus.

Precaution: Your fine adjustment should be in the middle of its track before you begin this activity. This means that it should not be turned back to where there is no more focus adjustment.

Increase your light for optimum definition. Make a drawing of several yeast cells, and submit it to your file.

Repeat the above procedure using stained smears of *Bacillus* sp. and *Escherichia coli.* Reread the instructions in this activity if necessary. If you take a lot of precautions and time at this point, you will become an expert microscopist much sooner. Time, practice, and mastery of the use of your microscope can make microbiology much more interesting to you.

Make drawings of numerous cells from both the *Bacillus* sp. slide and the *Escherichia coli* slide, and submit them to your file.

Activity 5: Continued Practice Using the Oil-Immersion Objective

Use a prepared smear of mixed bacterial types or any other available smears, and examine them with your oil-immersion objective. This is a repetition of Activity 4.

Permanent commercially prepared slides have cover slips over the smear. Place the immersion oil directly on the cover slip, and immerse the oil-immersion objective carefully. *Caution:* The cover slip is easily broken, so be sure to look from the side as you immerse the objective. This will prevent breaking the cover slip and getting the mounting medium from the prepared slide onto the objective lens.

Draw several representatives of each bacterial cell type, and submit them to your file. Figures 4-7 and 6-1 may help you identify cell types.

When you feel that you are very well acquainted with your microscope and its parts, take the self-evaluating post test. If you are not satisfied with your results, review this module before going on. Your success in this course depends on your being honest with yourself.

POST TEST

The post test is a self-evaluation. It is not used for a grade. It is designed only to let you decide if you have successfully completed this module.

Part I: True or False

_____ 1. When using the oil-immersion objective, the image you perceive at eye-point is magnified 97x to 100x larger than the actual size of the object.

_____ 2. If two objectives are parfocal, one will be in approximate focus when the other is in sharp focus.

_____ 3. The two functions of the optical parts of your microscope are to project light waves on the object and to form a magnified image of the object.

_____ 4. The iris diaphragm, condenser, and ocular are the three parts of the microscope that are most critical in light control.

_____ 5. The size of the microscopic field remains constant regardless of the magnification of the objective you are using.

_____ 6. It is safe to use facial type tissue to clean the optical parts of your microscope.

_____ 7. The plane side of the mirror should always be used if the microscope has a condenser but lacks a built-in light source.

_____ 8. Before you store your microscope, you should clean it thoroughly, rotate the low power objective into working position, and replace the dust cover.

_____ 9. Magnification at eyepoint equals the magnification of the objective multiplied by the magnification of the ocular.

_____ 10. Once you have achieved an image, you should move the optical parts that control the light striking the object as little as possible.

Part II

List four major points to be remembered when carrying a microscope.

1. _____

2. _____

3. _____

4. _____

Part III

Match the parts of the microscope with their functions.

_____ 1. Ocular

_____ 2. Thumb wheel

_____ 3. Revolving nosepiece

_____ 4. Objectives

_____ 5. Slide holder

_____ 6. Stage aperture

_____ 7. Nosepiece

_____ 8. Diaphragm lever

_____ 9. Condenser

_____ 10. Condenser adjustment knob

_____ 11. Mechanical stage

_____ 12. Mechanical stage controls

_____ 13. Base

_____ 14. Arm

_____ 15. Coarse adjustment

_____ 16. Fine adjustment

_____ 17. Body tube

_____ 18. Mirror

a. Opens and closes diaphragm to control amount of light striking object.

b. Reflects light waves from light source into condenser.

c. A series of lenses that usually magnify 10 times.

d. Condenses light waves into pencil shaped cone, thereby preventing escape of light waves; also controls amount of light striking object.

e. Raised and lowered in focusing some microscopes.

f. Usually has three magnifications.

g. Raises and lowers condenser.

h. Supports upper portion of microscope.

i. Rotates to change from one objective to another.

j. Adjusts interpupillary distance for your eyespan.

k. Moves nosepiece or body tube up and down rapidly for purposes of approximate focusing.

l. Allows the slide to be moved.

m. Hole in stage to allow light waves to strike the object.

n. Supports entire microscope.

o. Spring loaded portion allows placing the slide in the mechanical stage where it is tightly held.

p. Moves nosepiece or body tube up and down very slowly for definitive focusing.

q. Moves the slide on two horizontal planes.

Part IV

List the three parts of both types of microscope critical in light control and consequent good definition.

1. _____

2. _____

3. _____

Part V

Describe the markings on all your objectives:

1. Low power _____

2. High power _____

3. Oil immersion _____

PART TWO

Basic Microbiological Techniques

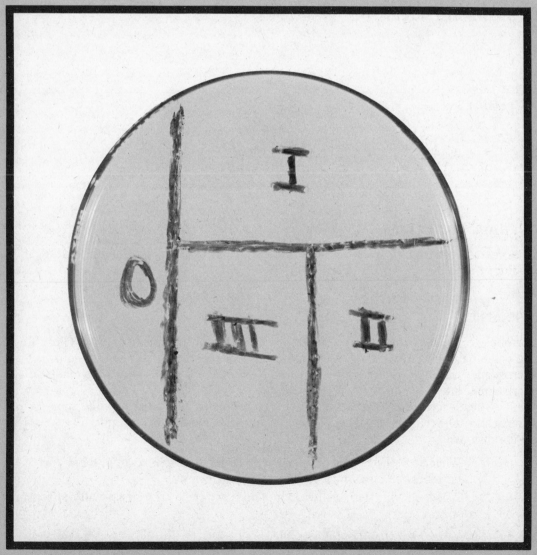

Sectors for streaking for isolation.

MODULE 5

Cleaning Microscope Slides

PREREQUISITE SKILL

None.

MATERIALS

microscope slides (6)	bar of Bon Ami cleanser
95% alcohol in squeeze bottles (isopropyl)	disposable wipers (e.g., Kim Wipes) or paper towels

---

OVERALL OBJECTIVE

Demonstrate your ability to clean microscope slides using the procedure in this module.

DISCUSSION

Microbes are so small that they can be observed only by using a microscope. There are two principal ways of preparing microbial specimens for observation with a light microscope: wet mounts and stained smears. Both of these preparations are made on microscope slides.

Since bacterial cells are so small, special problems arise in their microscopic observation. The two principal problems that must be avoided occur under the following conditions:

1. When scratched or dirty microscope slides are used, scratches and dust particles can be confused with microorganisms.
2. When working with a slide that is not *very clean*, the smear will not remain spread out.

If you use the following slide-cleaning procedure faithfully, you can save time and achieve better definition as you look at your microscopic preparations. It is advisable to use this procedure to clean slides for every smear or wet mount you make in this laboratory course.

ACTIVITY

Activity 1: Slide-Cleaning Procedure

1. Near the sink in your laboratory, you should find a bar of Bon Ami cleanser.
2. Adjust the faucets to warm water and wet your first two fingers, that is, your forefinger and middle finger.
3. Rub your wet fingers on the bar of cleanser until an obvious amount of cleanser is adhering to them.
4. Clean the slide on both sides by rubbing it with your Bon Ami covered fingers.
5. Rinse the Bon Ami cleanser off the slide with warm water.
6. You may find it necessary to repeat Steps 4 and 5 if you are cleaning a previously stained slide.
7. Apply several drops of 95% isopropyl alcohol to both sides of the slide. You should find the alcohol in a polyethylene squeeze bottle beside the sink.
8. Let the excess alcohol drain off the slide, and wipe it thoroughly, before the alcohol dries, with disposable wipers or paper towels.

You now have an immaculately clean slide which you can use successfully to make a good wet mount or bacterial smear. Take care to handle the clean slide by the edges to avoid depositing your skin oils on the clean flat surface that you will be using for the bacterial preparation.

Using the slide-cleaning procedure as outlined above, clean six slides now. The cleanliness of these slides will be evaluated in the post test, which you should now be ready to take.

POST TEST

The post test is a self-evaluation. It is not used for a grade. It is designed only to let you decide if you have successfully completed this module.

Place a drop of tap water on each slide. Spread the drop out to the approximate size of a quarter with your inoculating loop. If the water does not coalesce and it remains spread out, then you have clean slides and have passed the post test.

MODULE 6

Preparing a Wet Mount

PREREQUISITE SKILL

Completion of Module 4, "Compound Microscope for the Study of Microbes," and Module 5, "Cleaning Microscope Slides."

MATERIALS

hay infusion*

broth cultures (pour into 20 ml beakers so medicine droppers can be used)

  Bacillus subtilis

  Staphylococcus epidermidis

  Rhodospirillum rubrum

  unknown organisms

chemically clean microscope slides (6)

glass cover slips (6)

medicine droppers (5)

petroleum jelly

toothpicks

*To be prepared a week in advance of this module. Place these ingredients in a 1000/ml or 1500 ml beaker: 600 to 700 ml water and dry hay or dry grass (several handfuls). Bubble air through this hay infusion via rubber tubing attached to a slightly turned-on air jet. Allow air to bubble through this hay infusion constantly (2 to 5 days). One hay infusion is sufficient for the entire class.

OVERALL OBJECTIVE

Develop the skill of making a wet mount preparation for microscopic examination of living microbes.

Specific Objectives

1. Draw and describe three different microbes in a hay infusion wet mount.
2. Draw and describe spherical, cylindrical, and spiral shaped bacteria from a wet mount preparation.
3. Given an unknown broth culture; demonstrate your ability to prepare a wet mount by drawing and describing the organisms in it.
4. Name two types of preparation used to observe living microbes.
5. List, in order, the steps of the procedure used to make a wet mount as described in this module.

6. List the three reasons that wet mounts are used in the study of microbes.
7. Give the name of the "father of microbiology."
8. List the ingredients of a hay infusion.
9. List five groupings (arrangements) of spherical bacteria.
10. List three arrangements of bacilli.
11. Describe coccobacillus.
12. Draw and describe a few (two to three) organisms in a wet mount of pond water.*

DISCUSSION

It is important that you review Activities 1 and 2 of Module 4 before beginning the activities in this module. There are two principal ways of preparing a microbial specimen for observation with your light microscope:

1. A wet mount preparation
2. A stained smear.

The second preparation, a stained smear, is more commonly used for microscopic observation of microorganisms. However, a stained smear of microbes allows you to observe only dead organisms.

This module tells you how to make a wet mount preparation so that you can look at living microbes. In order for microbes to stay alive while you make your microscopic examination, they must be kept in a liquid environment since most bacteria die rapidly if no moisture is available to them.

Two types of preparation are used to look at living bacteria:

1. A wet mount
2. A hanging drop,

With both types you use a single drop of liquid containing living microbes. In this lab course you will be using only the wet mount preparation. If you are curious about a hanging drop preparation, refer to a textbook or laboratory manual in microbiology.

Wet mount slide preparations are used in microbiology because they allow you to see the following:

1. The size and shape of individual organisms
2. The characteristic arrangement or groupings of bacterial cells
3. Whether an organism is motile or nonmotile.

The stained smear preparation you will be using most often in microbiology can distort the size and arrangement of bacteria and make observation of motility impossible. Therefore, becoming adept at making a wet mount and being able to use your microscope correctly to observe the bacteria in the wet mount are important in your study of microbes because of the three observations listed above.

As you observe and make drawings of the microorganisms in the following activities, refer to Figure 6-1 for the descriptive names of the various shapes and groupings of bacteria. Figure 6-2 depicts a few of the most common animal and plant cells that you are likely to see in the hay infusion or pond water wet mount. If you find organisms in the hay infusion that are not shown in Figure 6-2, *do not concern yourself with their identification.* Just make the drawings of them now; an entire module will be devoted to their identification later. The skill you are to master from this module is the ability to prepare a wet mount. It is not the intent of this module to enable you to name all types of protozoa, algae, bacteria, and multicellular organisms.

Microscopic examination of a hay infusion provides an excellent demonstration of the ubiquity of microorganisms. Hay infusions are prepared by simply adding hay to an open container of water and aerating it.

After two or three days of rehydration, the water becomes turbid because it is teeming with various kinds of microscopic organisms. The one-celled microscopic organisms you are most likely to find are bacteria, protozoa, and algae.

*Specific Objective 12 is to be fulfilled if you perform the related experience.

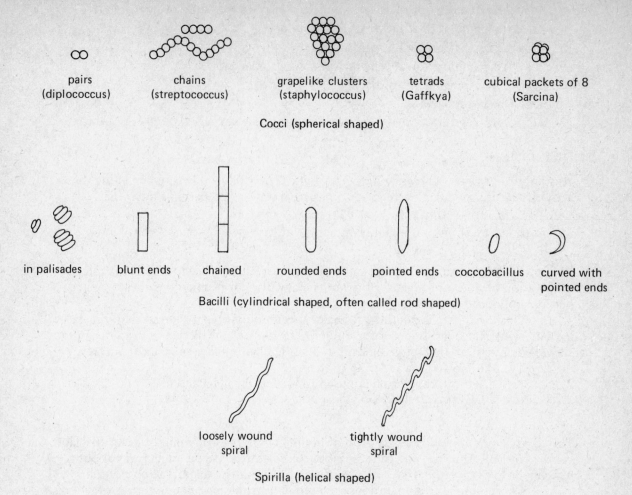

FIGURE 6-1
Bacterial shapes and groupings.

Anthony van Leeuwenhoek, the father of microbiology, first used hay infusions in the late 1600s. He used them to study shapes, groupings, and movements of microbes. You can obtain similar results by using dry grass instead of hay. It is peculiar that even though dry grass is used instead of hay, it is still called a hay infusion.

You should find a hay (grass) infusion already prepared for you. Leave it where it is in the lab since all your classmates will be using drops from it.

Precaution: Before you begin making a wet mount, read the following carefully to avoid the usual errors. First, it is advisable to use the high-power objective (44x) for your drawings even though the magnification is less than with the oil-immersion objective. The reason for this is that the oil makes the cover slip cling to the objective, causing the cover slip to move up and down as you turn the fine adjustment knob back and forth. Hence, your wet mount preparation will not stay in focus. Second, wet mounts dry out fairly rapidly. Therefore, you must make your microscopic observations rapidly. Drying out usually happens in about 10 minutes. Third, the unstained bacteria of your wet mount have almost the same refractive index as glass and the liquid environment they are suspended in. Therefore, the amount of light must be reduced in order to distinguish the microorganisms from their environment. The amount of light is reduced by closing the iris diaphragm. If the amount of light passing through your wet mount is still too much, you may wish to lower your microscope condenser also. Finally, it is easy to get petroleum jelly on the objectives, which obscures the field of vision. If this happens, a thorough cleaning of the objectives must ensue. Keeping these problems in mind, begin Activity 1.

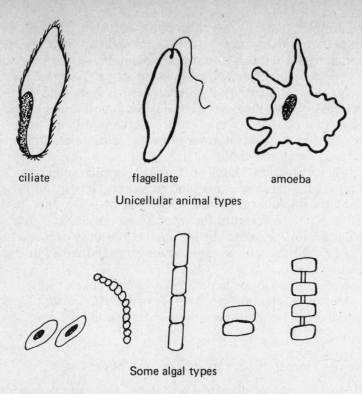

ciliate   flagellate   amoeba

Unicellular animal types

Some algal types

FIGURE 6-2
Some of the most common plant and animal
cells to be seen in the hay infusion wet mount.

ACTIVITIES

Activity 1: Preparation of a Wet Mount Using Hay Infusion

Using a toothpick as a dispenser, make a petroleum jelly square the size of the outside
edge of a cover slip on a clean microscope slide. Make the petroleum jelly square
approximately 1/16 of an inch high and as narrow as possible. Refer to Figure 6-3,
Part A.

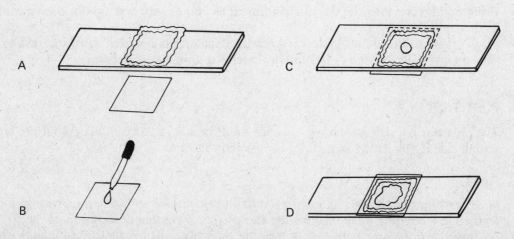

A

B

C

D

FIGURE 6-3
Preparing a wet mount.

Using a medicine dropper, place a single drop of hay infusion in the center of your cover slip as shown in Part B of Figure 6-3. Invert your slide so that the petroleum jelly square is down, and gently touch it to the cover slip. Be sure to keep the drop in the center of the petroleum jelly square as shown in Part C of Figure 6-3. Next pick up the slide, petroleum jelly, and cover glass preparation, and, holding the slide at both ends, turn it over, that is, turn the cover glass side up. It will look like Part D of Figure 6-3. If the cover slip is not level or is not touching the petroleum jelly around the entire square, gently tap the loose cover slip edges with a pencil. The hay infusion drop will spread out, which is all right.

You have now made a wet mount. Look at it with your microscope. Place the wet mount preparation on your microscope stage, and reduce the amount of light as instructed above. Locate the microbes with your low-power objective first. When you have them in focus under low power, turn the revolving nosepiece to high power. Each time you switch from the low-power to the high-power objective you will have to readjust the amount of light striking the wet mount preparation by readjusting the condenser and/or the iris diaphragm.

When you have the organisms in the hay infusion wet mount as visible as possible, satisfy Specific Objective 1 by making drawings and written descriptions of three different organisms. Make these drawings now. Figures 6-1 and 6-2 may be helpful here.

Activity 2: Drawing and Description of *Bacillus subtilis* Wet Mount

Using a broth culture of *Bacillus subtilis,* prepare a wet mount using the same procedure as in Activity 1. Make a drawing, and give a written description of the cells in this wet mount also.

Activity 3: Drawing and Description of *Staphylococcus epidermidis* Wet Mount

Prepare another wet mount using a broth culture of *Staphylococcus epidermidis.* Examine it under your microscope as before. Make a drawing, and give a written description of these bacteria.

Activity 4: Drawing and Description of *Rhodospirillum rubrum* Wet Mount

Using a broth culture of *Rhodospirillum rubrum,* make another wet mount and drawing, and give a written description of these bacteria.

Put in your file the six drawings required in this module, draw conclusions about wet mounts, and then take the post test.

Related Experience

Bring pond water into the lab and examine it as you did the wet mount preparations in this module.

You will find pond water, especially stagnant pond water, most interesting since many different one-celled and multicellular animals are present.

POST TEST

The post test is a self-evaluation. It is not used for a grade. It is designed only to let you decide if you have successfully completed this module.

Part I

Make a wet mount of an unknown bacterial broth culture applying the procedure you learned in Activity 1. Make a drawing, and prepare a written description of the organism or organisms as they appear under your microscope. Be sure to include the exact labeling on the unknown broth culture with your drawing and description. Show this to your lab instructor for approval before you proceed to the next module.

Part II

List the three major advantages of using a wet mount preparation instead of a stained smear to observe bacteria.

1. _____

2. _____

3. _____

Part III

List four problems that you may encounter when using wet mount preparations for microscopic observation of microbes.

1. _____

2. _____

3. _____

4. _____

Part IV

List the three different bacterial shapes.

1. _____

2. _____

3. _____

Part V: True or False

_____ 1. Wet mounts are used more often than stained smears in the study of microbes.

_____ 2. The two most common arrangements of bacilli are single cells and chain formation.

_____ 3. It is difficult to determine the shape of a coccobacillus.

_____ 4. Louis Pasteur is called the "father of microbiology."

_____ 5. Bacilli can be straight or curved rods.

MODULE 7

Ubiquity of Microorganisms

PREREQUISITE SKILL

Completion of Module 5, "Cleaning Microscope Slides," and Module 6, "Preparing a Wet Mount."

MATERIALS

---

sterile tube of nutrient broth (10 ml)*
wet mount materials
sterile nutrient agar plates (5)
trypticase soy agar plate (TSA)
tube containing two sterile swabs and
  2 ml sterile saline

For related experiences:
  sterile nutrient agar plates (2 to 3)
a surgical scrub pack†

*To be prepared by the student if the instructor so indicates, or a tube of broth from Module 2, "Preparing and Dispensing Media," can be used.
†Usually obtained by student.

OVERALL OBJECTIVE

Show that microbes are everywhere in your environment.

Specific Objectives

1. Demonstrate the presence of microorganisms in different areas of your environment, such as in the soil, in the air, on the table top, on your hands, and in your mouth.
2. Describe, with examples, the ubiquity of microorganisms in your environment.
3. Define the terms *ubiquitous* and *omnipresence*.
4. Describe the location of labels on petri dishes and the position of culture plates when placed in the incubator.
5. Give the temperature of incubation for organisms that make up the normal flora inside your body.

DISCUSSION

Although this module is designed to show that microbes are ubiquitous, it also introduces the necessity of using aseptic technique when you are working with bacteria in the laboratory. In Module 8 you will learn the skill of aseptically transferring bacteria from one tube to another. It is in this next module that you will actually manipulate living bacteria correctly, without fear of self-contamination or contamination of a pure culture.

After you complete this module, you will realize that the omnipresence of microorganisms is intimately associated with the use of aseptic technique while transferring bacteria. This will become even more apparent to you after you grow many different kinds of microorganisms from the different sources of your environment as listed under Specific Objective 1. You will believe beyond a doubt that microbes are indeed ubiquitous or omnipresent, which means that *they are everywhere!* The are at the bottom of the ocean, on ice-capped mountain tops, in hot sulfur springs, in milk, in drinking water, that is, in every possible place on this planet.

You should become truly bacteria-conscious when you realize that you are surrounded by bacteria, fungi, protozoans, and other microorganisms. The activities in this module will prove to you that tiny, invisible, ever-reproducing, living microbial cells are all around you, on you, and inside you. This should prepare you for the next module as you will appreciate the real need for aseptic technique when you work in the microbiology lab.

Now gather together the materials you will need to perform the following activities.

Precaution: You will be working with sterile petri dishes containing two different agar media. Notice that they look alike. *Do not* remove your plates from the supply area until you have labeled them on the bottom so that you can distinguish between the two media. Petri dishes are always labeled on the bottom instead of the lid so that there will be no errors caused by interchanging the lids.

ACTIVITIES

Activity 1: Growing Microorganisms from Environmental Sources

A.   Growing Microorganisms from Soil

Label your tube with your name, date, and activity number. Step outside your lab; obtain one pinch of moist soil, and put it in the tube of nutrient broth. Put your broth tube in a coffee can or test tube basket in the 30°C incubator for 48 hours, or incubate it at room temperature for a longer period of time. Make a wet mount and submit a written description of the results to your file.

B.   Growing Microorganisms from the Air

Label a sterile nutrient agar plate carefully as you did your nutrient broth tube. Be sure to label the *bottom* of the petri dish. Microbial cells are so small and so light that they are constantly being wafted around you by air currents, or they can be hitching a ride on dust particles. To show this, open your petri dish containing sterile nutrient agar. Let the petri dish remain open for 20 minutes. When you expose the microbial nutrient in the petri dish, the airborne bacteria and fungi will settle on it and develop into colonies after incubation.

Put your nutrient agar plate where many students are moving about or outside the lab where air currents are maximal. You will, however, find that your lab work table is also a good place to find bacteria. You decide where you would like to open your plate.

After your plate has been exposed to the air for 20 minutes, close it and put it in the 30°C incubator upside down for 48 hours. *Always* incubate culture plates in an inverted position unless otherwise instructed.

In your next lab session, draw several different colony types. Not all colonies will be bacteria; light, fluffy, cotton-like colonies are fungi. Submit your drawings to your file.

C.   Growing Microorganisms on Your Table Top

After you have grown the bacteria on your table top, you will understand why you should wipe down your working area with disinfectant at the beginning and at the end of each laboratory period. From now until the end of the course, make a habit of disinfecting your work table before you begin your lab work and again after you have finished.

Label the bottom of a nutrient agar plate carefully with your name, date, activity number, and type of medium. To show the importance of disinfecting your working area, remove one saline soaked swab from the tube, and rub the cotton tip on your table top before it is disinfected. Try to include such places as around gas jets and corners where dust particles are present. Use the contaminated swab to inoculate a nutrient agar plate. It does not matter how you inoculate the nutrient; simply rub the cotton swab gently over the surface of the agar. Use as much of the surface of the nutrient as you can for the inoculation. Return the cover to your plate. Once again you must wait 48 hours for the colonies to form. So invert the plate, and place it in the 30°C incubator for 48 hours.

In your next lab session, draw the growth on the plate. Submit the drawing to your file. You may wish to repeat this exercise after you have disinfected your table top. This second plate should show a reduced amount of growth if disinfection has been effective.

D.   Growing Bacteria *on* You

Label a nutrient agar plate as in the preceding activities. You can grow bacteria from any surface of your body at any time. To demonstrate this, place your fingers lightly on the surface of the sterile nutrient agar in a petri dish, and drag your fingers gently back and forth across the plate two or three times. Close the plate, and incubate it, upside down, at 30°C for 48 hours. Wash your hands with soap and rinse well, but do not dry them; repeat this activity with a second plate.

In your next lab session, make drawings of the growth of these skin bacteria, and submit them to your file. This will require two drawings, one before and one after washing your hands.

E.   Growing Bacteria *in* You

Label a trypticase soy agar plate (TSA) carefully and completely. TSA is preferred to NA as a nutrient medium for this activity because it is especially formulated to support the growth of streptococci and other organisms commonly present in your body which have more complex nutritive requirements (fastidious organisms).

All orifices of your body contain many different types of microbes. To demonstrate the bacteria in one of your body openings, touch your tongue to the sterile surface of a TSA plate. The more surface area of your tongue that touches the agar the more growth you can expect. Close the plate and incubate it at 37°C for 48 hours. Note the higher temperature of incubation. This higher incubation temperature has a logical reason, that is, microbes that flourish in you usually grow best at body temperature, which is 37°C or its equivalent 98.6°F.

In your next lab session, make drawings of the colonies arising from your mouth microorganisms, and submit them to your file.

Activity 2: Your Conclusions on the Ubiquity of Microorganisms

After growing microbes from several different sources of your environment, summarize your findings in a short written form. Turn your results and conclusions in to your file.

Take the post test. Be sure that you are completely satisfied with your results before proceeding to the next module.

Related Experiences

1. Repeat Activity 1D, that is, Growing Bacteria *on* You, by doing a surgical scrub on your hands. Did the number of bacteria decrease?
2. Press your slightly open lips against the surface of a nutrient agar plate. Incubate for 48 hours at 30°C. Are bacteria transmitted while kissing?

POST TEST

The post test is a self-evaluation. It is not used for a grade. It is designed only to let you decide if you have successfully completed this module.

True or False

_____ 1. While transferring bacteria, aseptic technique is intimately associated with the ubiquity of microorganisms.

_____ 2. The omnipresence or the ubiquity of microorganisms means that they are everywhere.

_____ 3. Microorganisms are not truly ubiquitous because they cannot be isolated from the bottom of the ocean or from ice-capped mountain tops.

_____ 4. Disinfection of your working area helps to counterbalance the ubiquity of microorganisms.

_____ 5. In microbiology a bacterial inoculation means the introduction of bacteria into a culture medium.

_____ 6. Washing your hands with soap removes all bacteria.

_____ 7. Disinfection of your working area is a necessary precaution to reduce possible contamination of pure cultures.

_____ 8. No bacteria can be found in drinking water.

_____ 9. Bacteria can be grown from all your body openings.

_____ 10. Soil bacteria are incubated at a temperature below body temperature.

_____ 11. A petri dish is labeled on either the lid or the bottom of the dish.

_____ 12. Inoculated culture plates are placed in the incubator upside down unless otherwise instructed.

_____ 13. The incubation temperature for bacteria that are normal inhabitants inside the body is 30°C.

_____ 14. 30°C is equal to 98.6°F.

MODULE 8

Aseptic Transfer of Microbes

PREREQUISITE SKILL

None.

MATERIALS

test tube rack

empty test tubes with capalls or other closures (2)

slant culture of
  Serratia marcescens

sterile nutrient agar slant (3)*

sterile tube of nutrient broth*

inoculating equipment:
  inoculating loop
  Bunsen burner or Fisher burner
  burner striker

*To be prepared by the student if the instructor so indicates; or the media from Module 2, "Preparing and Dispensing Media," can be used.

OVERALL OBJECTIVE

Master the technique of aseptic transfer of bacteria from one tube to another.

Specific Objectives

1. Define the terms *aseptic technique* and *pure cultures*.
2. List reasons why aseptic technique is important when you transfer bacteria.
3. Practice to become adept at the method used in transferring bacteria from one tube to another by using empty test tubes or "dry runs."
4. Be able to demonstrate a "dry run" for your instructor.
5. Aseptically transfer living bacteria growing on a nutrient agar slant to three uninoculated slants using the manual manipulations learned in Specific Objective 3, and obtain pure cultures.
6. Be able to identify any errors or omissions in this technique while observing another person doing aseptic transfers.

7. Demonstrate your ability to transfer bacteria aseptically from a slant culture to broth.

DISCUSSION

Since microbes are omnipresent, which means they are present everywhere, it is necessary to use extreme care not to introduce unwanted organisms into a pure culture. Unwanted bacteria may be introduced by direct contact with contaminated surfaces or your hands, that is, by actually touching the media or inner surfaces of the tube with any object that has not been sterilized. Since bacteria are also airborne, they can enter your tubes via air currents. There are certain tested techniques that you can use to keep outside microorganisms from contaminating your transfer culture. These tested techniques are called aseptic techniques. These same techniques will also protect you from self-contamination and those around you from contamination in the laboratory. Laboratory accidents do not happen very frequently if you make it a habit always to practice good aseptic technique. If you learn to handle bacteria correctly, you will then have only the desired organisms growing in your transfer culture *and* since you transferred a pure culture, you will get only this particular bacteria to grow. You will *not* have contaminated your pure culture. A pure culture is one in which only one single species is to be found.

This module shows, in a step-by-step description, how to manipulate pure culture transfers without adding outside, contaminating bacteria or infecting yourself and contaminating the lab. In transferring bacteria from one tube to another, you must be especially aware of airborne bacteria so that they do not contaminate the material you are working with.

ACTIVITIES

Activity 1: Practice of Aseptic Tube Transfer

When you master a certain tested technique, you will be able to transfer bacteria aseptically. You will be using these basic manipulations in most of your laboratory experiments throughout the laboratory portion of this course. Therefore, it will be to your advantage to master this technique *now*.

This technique requries much observation and practice. Carefully study Figures 8-1 through 8-15, which illustrate the approved technique. Also study the instructions accompanying the figures.

Next imitate the figures by doing several "dry runs," using the two empty test tubes that are in your test tube rack. These figures are presented so that you can do one step at a time and check yourself. After you have imitated the entire series of figures, one step at a time, you will know what is meant by aseptic technique in tube transfer.

The check list following the figures summarizes the manipulations represented by the series of figures. Once you have learned the correct handling of the equipment shown in the figures, you may need only to refer to the check list for a reminder of the step-by-step procedure.

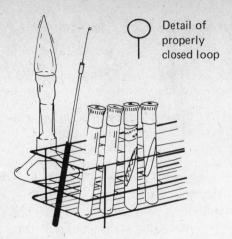

Detail of properly closed loop

FIGURE 8-1
The necessary equipment. Be sure the loop is closed, that is, a circle.

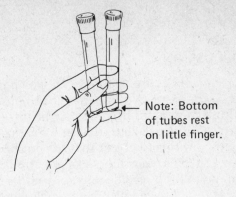

Note: Bottom of tubes rest on little finger.

FIGURE 8-2
Hold both test tubes in your left hand. (Hold tubes in your right hand if you are left handed.) The tubes should *not* be held vertically once the capalls are removed.

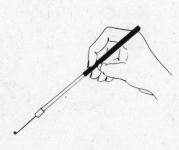

FIGURE 8-3
Hold your inoculating loop in your right hand. (Hold loop in your left hand if you are left handed.) The loop should be held like a pencil.

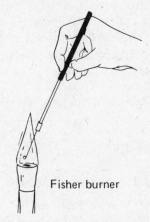

Fisher burner

FIGURE 8-4
Flame the inoculating loop in the Bunsen (or Fisher) burner holding the loop upright so that all the nichrome wire gets red hot. Allow the loop to cool so that you do not cremate the living bacterial cells you are about to transfer.

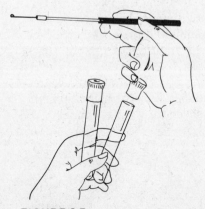

FIGURE 8-5
Remove the capalls or other closures one at a time from both test tubes. Do this by wrapping the little finger of your right hand around the capall of the tube nearest to your right hand. Grasp only the uppermost portion of the capall so that the open end does not touch the heel of your hand.

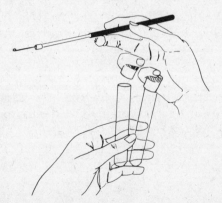

FIGURE 8-6
Now remove the second capall with the finger next to your little finger in the same manner, that is, by wrapping this finger around the second capall. Approach this tube by reaching between the two tubes. *Do not* attempt to reach behind the tubes.

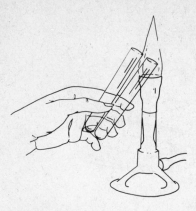

FIGURE 8-7
Flame the necks of the uncovered tubes by passing them back and forth through the flame twice. Be careful to hold the tubes in a nearly horizontal position.

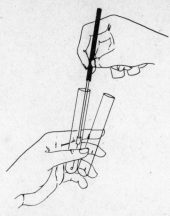

FIGURE 8-8
Insert the inoculating loop into the pure culture (or the practice tube substituting for it), and remove a small amount of bacteria. Note the capall position in the right hand.

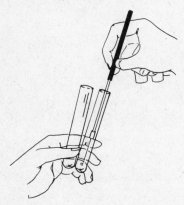

FIGURE 8-9
Transfer this inoculum (the small amount of bacteria) to the surface of the uninoculated slant (or the practice tube substituting for it).

FIGURE 8-10
Reflame the neck of both tubes.

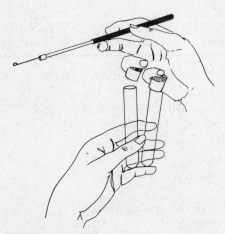

FIGURE 8-11
Recap the tubes, putting the capalls on the same tubes from which they came. It is best to return the cap to the tube you uncapped first, i.e., the capall held in your little finger.

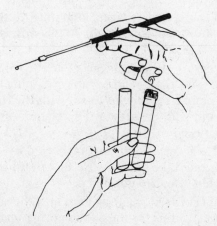

FIGURE 8-12
Then recap the second tube, putting the capall on the remaining uncapped tube.

FIGURE 8-13
Reflame the loop, killing all the bacteria on it.

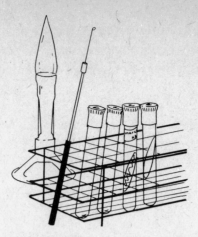

FIGURE 8-14
Return the tubes and the loop to the test tube rack.

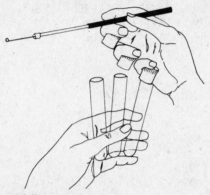

FIGURE 8-15
Three tubes and capalls can be manipulated simultaneously when you master *this* technique. One inoculum is sufficient to inoculate both tubes.

Check List

1. Hold both tubes in your left hand and the inoculating loop in your right hand, as shown in Figures 8-2 and 8-3.*

2. Flame the inoculating loop until all the nichrome wire is hot. Allow the loop to cool for approximately 30 seconds to avoid cremating the bacteria you are about to transfer.

3. Remove the capalls one at a time from both test tubes. Remove the capall closest to your right hand first by wrapping the little finger of your right hand around it. Remove the second capall with the finger next to the little finger on your right hand with the same wrapping motion. Approach the second tube by reaching between the two tubes. *Do not* attempt to reach behind the tubes.

4. Flame the neck of the uncovered tubes by passing the tubes back and forth through the flame twice. Be careful to hold the tubes at less than a 45° angle while they are uncapped.

5. Insert the inoculating loop into the stock culture (or the empty practice tube substituting for it), and remove a *small* amount of bacteria.

6. Transfer this inoculum (that is, the *small* amount of bacteria) to the surface of the uninoculated slant (or the empty practice tube substituting for it).

7. Reflame the neck of both tubes by passing the tubes back and forth through the flame twice.

*Everything is reversed if you are left handed.

8. Recap the tubes, putting the capalls on the same tubes from which they came. It is best to recap the tube nearest your right hand first.

9. Flame the inoculating loop.

10. Return the tubes and the loop to the test tube rack.

Activity 2: Demonstration for a Classmate (optional)

Using the same empty test tubes, have another student observe you as you demonstrate your aseptic technique. Use the step-by-step procedure, referring to the figures and the instructions accompanying them as you proceed. Stop after each step so that your classmate can note on the check list whether your technique for that step is correct. Repeat the instructions in the check list until you can follow the procedure in a flowing succession of motions. This activity will help prepare you for the next.

Activity 3: Dry Run Demonstration for Laboratory Instructor

Practice enough dry runs so that the manipulations of aseptic tube transfer become a natural technique for you. Now demonstrate your aseptic technique for your laboratory instructor and have it approved before proceeding to Activity 4.

Activity 4: Aseptic Transfer of Living Bacteria

You should now be ready to work with living microbes. As you know, *most* bacteria are nonpathogenic. However, those types that *do* cause disease can be very harmful. Therefore, you need to protect yourself and your classmates by handling living bacteria correctly.

In your test tube rack place a slant of red pigment-producing, living bacteria and three uninoculated slants of nutrient agar. Using the technique you learned in your "dry run" practice, transfer a small amount of living bacteria (inoculum) to the nutrient agar slants. Take care not to cut or gouge the agar surface. Allow your loop to glide over the agar surface from the bottom of the tube to the top of the slanted surface with a slight side to side motion.

Each and every time you make an inoculation you must label it carefully and completely. An adequate label should include your name, the date of inoculation, the name of the organism, and/or source of inoculum if appropriate, such as a throat swab, also the type of medium, the temperature of incubation, and the name of the module and/or activity. See Figure 8-16 for an example.

Label your newly inoculated nutrient agar slants, and put them in the 30°C incubator for 48 hours. During your next lab period, examine your transfer slants for colony morphology and pigment production. If your aseptic technique was done correctly, the growth will appear smooth and confluent; also only red pigment producers will be present. Your aseptic technique has been successful if you have just one species of bacteria growing on your transfer slants, which is a pure culture of that species.

Activity 5: Aseptic Transfer from a Slant Culture to Broth

Place a tube of sterile nutrient broth in your test tube rack with the slant culture of red pigment-producing bacteria. Aseptically transfer a small amount of living bacteria to the tube of sterile broth. Immerse your loop, with the inoculum, in the broth, and shake it vigorously two or three times to deposit a few cells in the sterile broth. Label the tube, and place it in a coffee can or wire basket in the 30°C incubator for 48 hours. During your next lab period, examine the broth tube, and notice the turbidity and red pigment. Submit your conclusions on aseptic transfer to your file folder.

Now take the post test.

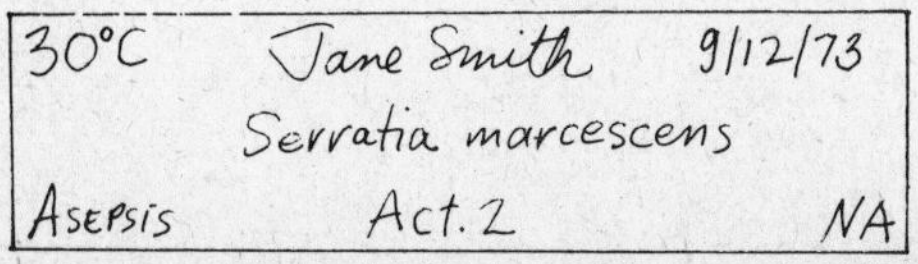

FIGURE 8-16
Labeling.

30°C	Jane Smith	9/12/73
Serratia marcescens		
Asepsis	Act. 2	NA

POST TEST

The post test is a self-evaluation. It is not used for a grade. It is designed only to let you decide if you have successfully completed this module.

Part I: Demonstration of Technique

If you received approval from your instructor on Activity 3, then you have passed this part of the post test, and you have mastered the correct *manipulation* necessary for aseptic tube transfer.

Part II: Application of Correct Manipulations

To pass this part of the post test you must satisfy Specific Objective 5 by successful completion of Activity 4. That is, after incubation your slants will show smooth, confluent growth, and only red pigmented bacteria will be present. If you are not satisfied with your results in this part of the post test, then you probably have not completely mastered the technique of aseptic tube transfer. Therefore, you must review the check list and the figures. You may want to practice another dry run also. Do whatever you must until your cultures appear pure. This part of the post test is probably the most important of all your self-evaluations because it will protect you from contamination throughout this entire course.

Part III: True or False

_____   1. Aseptic technique is important when transferring bacteria, both to prevent contamination of the bacteria you are transferring and to prevent self-infection.

_____   2. Once you have become adept at transferring bacteria aseptically, it is still best to work very slowly, describing each step of the procedure mentally.

_____   3. An inoculum is a small amount of bacterial growth which is used to inoculate sterile media.

_____   4. During aseptic transfer of bacteria, contamination by air currents is as important as contamination by direct contact.

_____   5. You should take care to return the capalls to the tubes from which they came.

Part IV

Define the following:

1.   Pure culture _____

_____

2.   Omnipresence _____

_____

3.   Dry run _____

_____

     If you did not get all the answers correct, you should go back and review this module.

MODULE 9

Aseptic Use of a Serological Pipet

PREREQUISITE SKILL

Mastery of Module 8, "Aseptic Transfer of Microbes."

MATERIALS

Bunsen burner

test tube rack

1 ml, clean pipet

1 ml, sterile, cotton-plugged pipet (disposable or in pipet cans)

10 ml, clean pipet

10 ml, sterile, cotton-plugged pipet (disposable or in pipet cans)

capped, empty test tube

capped test tube containing 5 ml of tap water*

cotton-plugged flask containing tap water*†

empty, sterile petri dish*

sterile tube of nutrient broth*

sterile flask of nutrient broth (50 ml of cotton-plugged broth)*†

nutrient broth culture of *Escherichia coli*

*To be prepared by the student if the instructor so indicates.

†See Figures 16-7, 16-8, and 16-9 in Module 16 for procedure on making cotton plugs.

OVERALL OBJECTIVE

Develop the technique of pipetting so that you can transfer broth cultures aseptically in specified amounts.

Specific Objectives

1. Describe the differences represented by the markings on a 1 ml serological pipet and a 10 ml serological pipet.
2. Describe the functional difference of a pipet that has etched rings on the mouth part and one that has none.

3. Demonstrate the correct method of using and controlling a serological pipet.
4. Use a test tube containing tap water to demonstrate the accurate pipetting of 0.6 ml of the water to an empty test tube.
5. Use a flask of tap water to demonstrate the accurate pipetting of 6.0 ml of water and fractions of 7.0 ml to an empty tube.
6. Transfer aseptically 0.1 ml of a broth culture to a tube of sterile nutrient broth.
7. Transfer aseptically 2.0 ml of a broth culture to a flask of sterile nutrient broth.
8. Transfer aseptically 1 ml of broth culture to an empty petri dish.
9. Explain the purpose of the cotton plug in a pipet.
10. Describe aseptic transfer using a pipet and explain why it is necessary.
11. Describe how to determine the correct pipet to use to transfer different amounts of liquid.
12. Demonstrate how to make and handle a cotton plug closure while pipetting.

DISCUSSION

In microbiology, it is often necessary to transfer measured amounts of liquid broth cultures. This is done by using a sterile serological pipet. A serological pipet differs from a volumetric pipet you may have used in chemistry in that a serological pipet has lined graduations on it enabling you to transfer different amounts.

There are several different sizes and types of pipets. This module explains only two sizes in detail. Each size is accurate within a certain range. For example, a 1 ml pipet is used to transfer amounts of 1 ml or less. It is *not* used to transfer 3 ml of liquid by filling it up three times. If you want to transfer more than 1 ml, you use a larger sized pipet. A 5 ml pipet is used for transferring amounts ranging between 1 ml and 5 ml. For amounts over 5 ml, a 10 ml pipet is the correct size to use.

The correct holding of the pipet and the culture tubes is extremely important in making an aseptic transfer. The manner in which you hold the culture tube and manipulate the capall or cotton plug is similar to the technique you developed to make an aseptic tube transfer of bacteria with your inoculating loop. The major differences are:

1. You will be able to hold easily only one culture tube at a time.
2. You will grasp the pipet differently from the way you held the inoculating loop.

The manner in which you must hold the pipet to control the delivery of broth culture does not allow you to remove both capalls as easily as when using your inoculating loop. It is recommended, therefore, that you handle just one tube and its capall at a time.

Before you begin to use a serological pipet, you should learn the differences in the various sized pipets. Since this is not a quantitative chemistry course, you will be using only graduated serological pipets. Most glassware supply companies (where the pipets are purchased) color-code pipets of different sizes for rapid identification. The color codes used by different companies may vary.

In this module you will be using 1 ml pipets and 10 ml pipets. The 1 ml pipet will have a different-colored patch near the mouth end than the 10 ml pipet has. Obtain a 1 ml pipet and a 10 ml pipet for examination. Notice the different-colored patches, memorizing the color for each.

Distal to the color code from the mouth end on a 1 ml pipet, you will see "1 in 1/100." This is an abbreviation for "1 ml divided into one hundredths." As you look carefully at the graduations (division lines) on this pipet, you can see that there would be 100 small graduations if the narrowed tip of the pipet were marked to the very end of the tip. So each small mark represents 1/100 (0.01) of a milliliter. The longer lines that go almost completely around the pipet are numbered 0, 0.1, and so forth, through 0.9. Between each of these numbered lines is 1/10 (0.1) of a milliliter. That is, 0 to 0.1 is 1/10 of a milliliter, and 0.1 to 0.2 is another 1/10 of a milliliter.

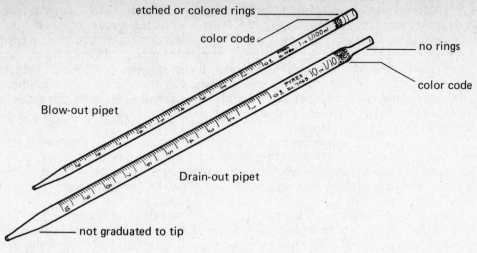

etched or colored rings

color code

no rings

color code

Blow-out pipet

Drain-out pipet

not graduated to tip

FIGURE 9-1
A 1 ml pipet and a 10 ml pipet.

Hence there are ten 1/10 divisions in this pipet, as shown in Figure 9-1. Inspect your 1 ml pipet, and compare it with Figure 9-1 until you understand all the markings and graduations.

Next, examine the 10 ml pipet just as closely. The numbers "10 in 1/10" on this pipet are an abbreviation for "10 ml divided into one tenths." Each small division in this case is 1/10 (0.1) of a milliliter. Now look at the longer numbered lines. The volume between each holds the entire amount contained in the 1 ml pipet. Once again, compare this 10 ml pipet with Figure 9-1 until you understand all the graduations.

Other marks you should look for on all pipets are the rings on the most proximal part of the mouthpiece. If the pipets are pyrex glass, then the rings will be etched. If they are disposable pipets, then the rings are usually colored. In either case, when the rings are present, it means that they are blow-out pipets, and to measure 1 or 10 ml accurately you must blow out the small amount of fluid that remains in the tip after the pipet has been drained. If these rings are not present, you will note that the last inch of the tip is not graduated and therefore cannot be used for measuring. Thus the fluid left in this unmarked portion must be discarded.

If you have doubt about any of the markings or graduations on either pipet, question your lab instructor about them. It is most important that you understand the pipets before you proceed to the activities presented in this module.

To perform the practice activities (Activities 1 and 2), which are designed to make you adept at using pipets, it is not necessary to use a sterile pipet even though you will be observing aseptic technique because you are just practicing.

Pyrex glass pipets are sterilized in pipet cans. Therefore, you must open a pipet can and set aside the lid, as shown in Figure 9-2. When you remove a pipet, use your thumb and forefinger to extract the pipet from the others that are in the can, taking care to touch only the mouth part of the pipet you are removing. This also applies to disposable pipets. Observe the cotton plug in both the glass and disposable pipets.

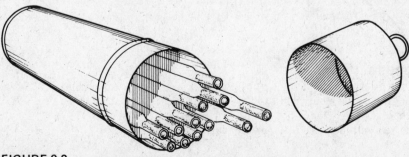

FIGURE 9-2
Pyrex glass pipets in a pipet can.

After removing the pipet from the can, close the can immediately to prevent airborne bacteria from contaminating the remainder of the sterile pipets. If you are using disposable pipets carefully seal over the opened corner of their plastic container after removing a pipet. For singly wrapped disposable pipets, always peel the wrapper on the "mouth" end of the pipets and then pull the wrapper downward. When you have taken the pipet out of the can, wrapper or plastic bag and the container has been closed, you should use the pipet immediately since airborne bacteria can settle on it also. Once you have the pipet in your hand, you must not put it down on any surface since all surfaces are contaminated.

Precaution: It is in removing the pipet and in closing the pipet can, with the pipet still in your hand, that you are most likely to contaminate the pipet. Contamination usually occurs by allowing the tip of the pipet to touch your body or clothing, or the table top. With practice you will be able to close the can while holding the pipet. If contamination of the pipet occurs, discard the pipet, and extract another one from the pipet can. Be conscious of contamination and asepsis at all times.

Pipetting becomes simple with practice. Even though you may already be experienced in pipetting with your mouth, you will still need to perform the practice activities since it is necessary to incorporate aseptic technique into the procedure. Therefore, if you already are able to use and control a pipet, you will need to practice only those manipulations that are so important to asepsis. *Reminder:* While pipetting, line up the bottom of the meniscus with the graduated marks on the pipet.

Now read all the steps in the check list, and perform Activity 1. Repeat Activities 1 and 2 as many times as is necessary to make the transfer correctly. When you no longer need to refer to the check list, *only then* are you ready for Activities 3, 4, and 5, in which you will be pipetting living bacterial cells.

Check List

In order to use a pipet it is important that you learn the following steps of the procedure:

1. Hold the capped test tube in your left hand.*
2. Hold the pipet in your right hand. (See Figure 9-3 for the correct way to grasp the pipet.)
3. Take the cap off the tube with the little finger of your right hand by wrapping your little finger around the capall.
4. Flame the neck of the open tube if asepsis is required. It would be well to practice this step even though you may not be working with sterile materials.

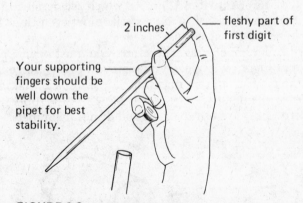

2 inches

fleshy part of first digit

Your supporting fingers should be well down the pipet for best stability.

FIGURE 9-3
Hold the pipet with the fleshy part of your forefinger over the mouth end to allow accurate control of delivery. Notice that the tube cap can be held with the little finger of the same hand.

*Everything is reversed if you are left handed.

FIGURE 9-4
Dry lips are placed around the mouth end of the
pipet while holding the forefinger of the right
hand in readiness.

5. Immerse the tip of the pipet well into the fluid contained in the test tube. *Precaution:* If the tip does not remain immersed throughout the pipetting process, you will get fluid in your mouth. This can be dangerous when you are pipetting bacterial cultures.

6. With the forefinger of your right hand (pipetting finger) out of the way, place your dry lips around the mouth end of the pipet as shown in Figure 9-4. *Caution:* Do not get the mouth end wet.

7. Slowly draw the fluid up the pipet by creating a gentle suction with your mouth and lips.

8. With the fluid drawn up *above* the zero marked line on the pipet, quickly remove your mouth, and immediately put the end of the forefinger of your right hand over the hole in the mouth end of the pipet as shown in Figure 9-3. *Never use your thumb* instead of your forefinger to control the pipet. *Reminder:* If you get the flat surface of the mouth end wet (usually from mucous secretion of your mouth), you will have difficulty controlling the flow of the fluid out of the pipet with your pipetting finger.

9. The forefinger should next be relaxed just enough to let the fluid drain down to the zero-marked line. As you do this, let the fluid coming out of the tip of the pipet return into the tube from which it came. Now you are ready to transfer a measured amount of fluid to the empty test tube. Remember to read the bottom of the meniscus.

10. Flame the neck of the opened test tube containing water before recapping. This is necessary only when making an aseptic transfer, but it should be practiced.

11. With your pipetting finger still tightly pressed over the hole at the mouth end of the pipet, put the cap back on the test tube from which it came. You should still be holding the test tube in your left hand.

12. Return the capped tube to the test tube rack with your left hand. Now that your left hand is free, pick up the tube into which you will transfer the fluid contained in the pipet. *Precaution:* While you are doing this, take care not to tilt the pipet past horizontal since the fluid will run out of the mouth end onto your hand.

13. Remove the capall or plug with the little finger of your right hand.

14. Flame the neck of the open tube if asepsis is required.

15. Insert the tip of the pipet into the opened tube, and let the desired amount of fluid drain out of the pipet by again relaxing the controlling forefinger. (If the tube you are transferring to happens to contain fluid, as it does in Activity 3, do not insert the tip of the pipet into the fluid.

16. Reflame the neck of the open tube.
17. Recap the tube.
18. Place the pipet and any remaining broth culture it contains into the appropriate container (with disinfectant).
19. Return the tube containing the transferred fluid to your test tube rack.

Caution: Never pipet by mouth if a pathogenic organism is suspected. A pipetting bulb should *always* be used to transfer pathogens or suspected pathogens. Mouth pipetting *Escherichia coli,* however, is not dangerous since it is a common intestinal organism.

Now gather together the materials listed at the beginning of this module. Next perform Activity 1. As you practice, refer to the check list as often as is necessary.

ACTIVITIES

Activity 1: Practice Transfer of Water Using a 1 ml Pipet

From the capped test tube containing water, use a 1 ml pipet to transfer 0.6 ml of the water to an empty, capped tube. Follow all the steps in the check list. Repeat the transfer again and again until you become very accurate and very conscious of asepsis as you make the transfer. When you are satisfied with this and are able to control the fluid in the pipet, proceed to Activity 2.

Activity 2: Practice Transfer of Water Using a 10 ml Pipet

Transfer 6 ml of water from a cotton-plugged flask to an empty, capped test tube, using a 10 ml pipet. See Figures 16-6, 16-7, and 16-8 in Module 16. When you can do this readily, transfer the following amounts: 6.1, 6.3, 6.5, and 6.9 ml. Since you will be using only one tube to receive these various amounts, it will be necessary to empty the tube between each transfer. Remove and replace the tube closures between each transfer. Practice all steps of asepsis.

Precaution: The bore of a 10 ml pipet is much larger than that of a 1 ml pipet; therefore, the fluid will run out faster and be more difficult to control. So this activity may require even more practice. Once again remember to strive for accuracy and asepsis.

Activity 3: Aseptic Transfer of a Measured Amount of *E. coli* Broth Culture Using a 1 ml Pipet

Since you are now working with living bacteria, the bore of the mouth end of the pipet should have a cotton plug for this activity. If it is necessary for you to cotton-plug your pipets, read Module 3, "Sterilization of Media and Equipment." The cotton plug filters out only airborne bacteria. It will *not* filter out all the bacteria if it becomes saturated with broth culture.

Reminder: Be sure to keep the tip of the pipet immersed in the broth to avoid getting the bacterial culture in your mouth. If you should get some in your mouth, rinse your mouth out several times with 50% ethyl alcohol. Do not swallow the alcohol rinse! Since *E. coli* is normal intestinal flora, infection is unlikely.

Now disperse the bacteria evenly throughout the culture by gently shaking the tube back and forth. *Do not* shake it up and down, or the culture will come out of the tube. Transfer 0.1 ml of the culture to a tube of sterile nutrient broth using a sterile 1 ml pipet. This necessitates removing the pipet from a pipet can without contaminating it. *Precaution:* Asepsis must again be observed. If even a drop of the bacterial culture is spilled, it must be flooded with disinfectant for a few minutes before being wiped up. If you feel you must repeat this transfer, obtain another sterile nutrient broth tube.

You need not incubate the newly inoculated tubes since this is simply an exercise in manipulation. Therefore, discard all tubes when you are satisfied that you have mastered this activity.

Activity 4: Aseptic Transfer of a Measured Amount of Broth Culture Using a 10 ml Pipet

Transfer 2 ml of broth culture to the flask containing sterile nutrient broth. *Precaution:* The same as in Activity 3. Asepsis must again be observed and contamination avoided.

Normally you would use a 5 ml pipet for transferring this amount, but since you did your practice activity with a 10 ml pipet, you may make this substitution if your instructor so indicates. If the flask is closed with a cotton plug, it is removed exactly the same as a capall, with the little finger of your pipetting hand. See Figures 16-6, 16-7, and 16-8 in Module 16. As before, discard the pipet and flask when you are satisfied with your aseptic technique. You need not incubate.

Activity 5: Transfer of Liquid Culture Medium to a Sterile Empty Petri Dish

Use a sterile, cotton-plugged, 1 ml pipet, and aseptically transfer 1 ml of broth culture from the same tube used in Activity 4 to the center of an empty petri dish. Remove the petri dish lid with your left hand. Keep the lid over the sterile plate as much as possible as you let the broth culture drain out of the pipet. The reason for protecting the bottom of the sterile petri dish with its lid is to prevent airborne bacteria from dropping into it.

Reminder: If the pipet has the ringed marks at the mouth end, you must blow out the remainder of the broth culture adhering inside the tip. (Refer back to the discussion section for a review of this.) If the rings are not present, you should touch the tip to the bottom of the petri dish to remove the drop clinging to the tip.

If you are not satisfied, you may repeat this activity by obtaining another petri dish. If you have mastered the aseptic manipulations, discard the petri dish, tubes, and pipets in the appropriate place, and take the post test.

POST TEST

The post test is a self-evaluation. It is not used for a grade. It is designed only to let you decide if you have successfully completed this module.

True or False

_____   1. The smallest graduation on a 1 ml pipet measures 0.1 of a milliliter.

_____   2. Rings around the mouth end of a pipet indicate that this is a blow-out pipet.

_____   3. The numbers "10 in 1/10" are found on the mouthpiece of a 1 ml pipet.

_____   4. The longer, numbered rings on a 10 ml pipet represent 1 ml.

_____   5. After taking a pipet out of a pipet can, you should put the pipet down in order to close the can.

_____   6. When pipetting from one tube to another, you should be holding both tubes in your left hand as you make the transfer.

_____   7. When making an aseptic tube transfer, you should flame the neck of the test tube after removing the capall and before putting it back on.

_____   8. Color code in reference to pipets is used to differentiate between the various sizes of pipets.

_____   9. In some instances it is acceptable to use your thumb as the pipetting finger.

_____ 10. When you have drawn the fluid up the pipet beyond the zero mark, you should press your tongue against the hole in the mouthpiece of the pipet to keep the fluid from running back out.

_____ 11. Sterile pipets used to transfer broth cultures should always have cotton plugs in the bore of the mouthpiece.

_____ 12. To transfer 2.5 ml, a 5 ml pipet is the most correct pipet to use.

1-F, 2-T, 3-F, 4-T, 5-F, 6-F, 7-T, 8-T, 9-F, 10-F, 11-T, 12-T.

KEY

MODULE 10

Pour Plates

PREREQUISITE SKILL

Successful completion of Module 8, "Aseptic Transfer of Microbes."

MATERIALS

test tube rack
Bunsen burner
burner striker
sterile petri dish (7 to 8)
tubes of melted agar medium (2) deeps*

screw-cap bottle or flask of melted nutrient agar, 100 ml*

*To be prepared by the student if the instructor so indicates; or the media from Module 2, "Preparing and Dispensing Media," can be used.

OVERALL OBJECTIVE

Make a pour plate by aseptically transferring sterile, melted agar from a test tube, flask, or screw-top bottle to a sterile petri dish.

Specific Objectives

1. Demonstrate aseptically the pour plate technique using a practice "dry run."
2. Demonstrate the technique developed in the practice dry run using sterile, melted agar and a sterile petri dish.
3. Demonstrate the pour plate technique by aseptically pouring several plates from a container with large amounts of medium.
4. Describe the various purposes for which pour plates are used in clinical laboratories.
5. Define the terms *holding temperature, holding water bath, pouring temperature, solidifying temperature,* and *agar.*

DISCUSSION

In order to grow bacteria to study colony morphology or to count numbers of colonies, you must grow the bacteria on or in a solid medium in a petri dish. The purpose of this module is to teach you only how to get the sterile, melted medium into a sterile petri dish without introducing any airborne contaminating bacteria.

You will find the figures in this module almost self-explanatory. However, following the figures, you will find a check list of the correct step-by-step procedure. Read the check list while studying the figures.

After reading the check list, studying the figures, and mastering Module 8, "Aseptic Transfer of Microbes," you are ready for the first activity in this module.

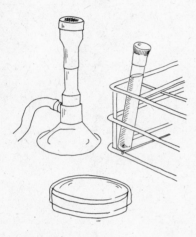

FIGURE 10-1
The necessary equipment.

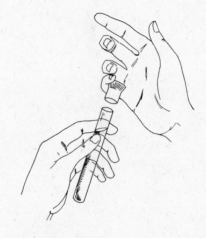

FIGURE 10-2
Removing the capall.

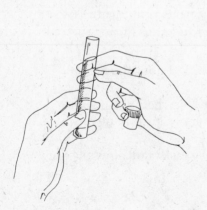

FIGURE 10-3
Transfer the tube to your right hand. Notice that you continue to hold the capall in the little finger of your right hand.

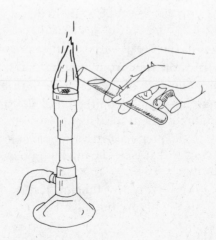

FIGURE 10-4
Flame the neck of the test tube.

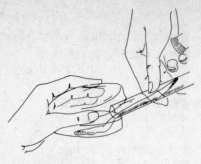

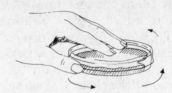

FIGURE 10-5
Simultaneously lift the lid of the sterile petri dish, and pour the melted medium into it. The lid should always be held over the plate to protect it from contamination by airborne bacteria.

FIGURE 10-6
If the melted medium does not cover the entire bottom of the petri dish, rotate the plate on the table top in a circle 6 to 8 inches in diameter. Once the medium covers the bottom, do not move the plate again until the medium has solidified.

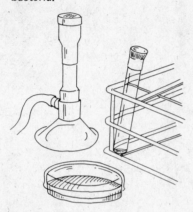

FIGURE 10-7
Replace the capall on the empty test tube, and return it to your test tube rack. Allow the plate to solidify.

Check List

1. Pick up your test tube of melted agar (or the empty dry run tube substituting for it), and place it in your left hand.*

2. Remove the capall with the little finger of your right hand as in Figure 10-2.

3. Transfer the tube from your left hand to your right hand as in Figure 10-3.

4. Flame the neck of the test tube as in Figure 10-4.

5. Simultaneously lift up the lid of the sterile petri dish, and pour the melted medium into it as in Figure 10-5.

6. Rotate the poured plate on your table in a circle 6 to 8 inches in diameter if necessary, that is, if the medium does not cover the entire bottom of the plate as in Figure 10-6.

7. Recap the empty test tube, return it to your test tube rack, and allow the pour plate to solidify as in Figure 10-7.

ACTIVITIES

Activity 1: Pour Plate Practice Activity

This is a practice dry run in which you use an empty test tube and a nonsterile petri dish. Pretend that you have melted medium in the empty test tube, and proceed through the check list, referring to the figures as you proceed.

*Everything is reversed if you are left handed.

Repeat this practice run as often as necessary to become so familiar with it that you can do it without your check list or the figures. If you wish, have a classmate check your ability to perform this activity.

Activity 2: Making a Pour Plate Using a Tube of Melted Agar

Successful completion of Activity 1 allows you to work with the real thing, that is, actually to pour sterile, melted medium aseptically into a sterile petri dish. The medium you use must first be autoclaved to sterilize it. If it already has been sterilized and allowed to cool, it must again be brought to boiling temperature to reliquefy it for pouring. After autoclaving, the liquid medium is placed in a holding water bath which is set at 50°C. This is called the holding temperature for media since agar, the solidifying agent in media, remains liquid at 50°C and becomes solid at 40°C. Sterile agar media can only be transferred in the liquid state. Therefore, once you remove your tubes of melted medium from the holding water bath and place them in your test tube rack, you must work rapidly to avoid solidification of the agar in the medium. You will note that the holding temperature is only 10° higher than the solidifying temperature of agar. It does not take long for media to cool 10° and solidify once placed at room temperature. Hence there is the need to work rapidly.

The average temperature between that of the holding water bath and that of solidification is called the pouring temperature, which is approximately 45°C. Media containing agar are poured at approximately 45°C for the following reasons:

1. Agar medium solidifies at 40°C and cannot be poured.
2. If the medium is just beginning to solidify while pouring, lumpy agar plates result.
3. Medium poured above 50°C causes too much moisture to condense on the petri dish lid. This moisture can drop onto the surface of the solidified medium and prevent the separation of bacterial cells and pure colonies.

Keeping this time factor in mind, get your two tubes of medium from the water bath, carrying them in your test tube rack. Wipe off any excess water on the outside of one tube, and aseptically pour the liquid medium into your sterile plate using the technique you mastered in Activity 1.

If the medium covers the bottom of the plate and no lumps of agar are present, you have indeed made a perfect pour plate. If the medium does not cover the bottom of the petri dish, place your fingers flat on the lid of your closed plate. Rotate the plate on the table top in a circle 6 to 8 inches in diameter, as shown in Figure 10-6. This moves the melted medium around so that it covers the entire bottom of the petri dish. Take care not to get the medium on the petri dish lid. Because of the limited amount of media a tube will hold, a thin media plate will result. You will want to pour thicker plates hereafter.

Precaution: As soon as the medium covers the bottom of the plate, do not move it again until it solidifies. The medium becomes opaque when it is solid.

Using your second tube of melted agar, pour another plate using the same technique. After your pour plates have solidified, label them on the bottom and incubate them for 24 hours in the 30°C incubator. Your plates must always be inverted (turned upside down) while they are incubating. This prevents any excessive condensation on the lid from dropping down onto the agar surface, which would cause the bacteria to float and the colonies to run together.

Observe your pour plates in your next lab period. If no colonies of bacteria or fungi have formed, then you have successfully mastered another basic microbiological technique. Congratulations!

Refrigerate the noncontaminated pour plates so that they can be used for a subsequent technique you will master in Module 14, "Streaking for Isolation." Put a label on the bottom of your pour plates, and refrigerate them upside down.

Perform the following variation of your pour plate technique now, using a bottle or flask containing large amounts of sterile medium and several sterile petri dishes.

Activity 3: Use of a Large Container to Pour Several Plates

A frequently used variation employs larger containers that hold enough agar to pour several plates. In this variation, you should put a larger amount of medium (20 ml per plate) into a flask or screw-top bottle to be sterilized and liquefied. Therefore, if you wish to make 10 pour plates, you should have approximately 200 ml of melted medium in a cotton-stoppered flask or screw-top bottle. You will find a bottle or flask of sterile, liquid agar medium in the holding water bath ready to pour. Always be sure the water in the bath covers the medium in the bottle.

   Reminder: Do not pour your agar when it is too hot or too cool.

   Line up five or six sterile petri dishes, and, using the same principles you learned in Activity 2, pour as many plates as you wish. Once you flame the neck of the container of medium, you can pour from four to six plates before reflaming the neck if you work quickly. After your plates have solidified, label them and incubate all your pour plates upside down for 24 hours to check on your asepsis. If colonies are present after incubation, you must repeat this module. If your pour plates are not contaminated, refrigerate them as you did earlier in Activity 2.

   Take the post test next.

POST TEST

The post test is a self-evaluation. It is not used for a grade. It is designed only to let you decide if you have successfully completed this module.

True or False

_____   1. In clinical laboratories pour plates are used for colony counts.

_____   2. You should be able to pour several plates from a container holding a large amount of melted medium before you reflame the neck of the container.

_____   3. Agar solidifies at 50°C.

_____   4. It is best to work slowly and deliberately when pouring melted agar from a test tube in order to avoid splashing the medium.

_____   5. After you have poured the melted agar, you may rotate the petri dish in a 6 to 8 inch circle to be sure that the whole bottom of the plate is covered.

_____   6. When you are sure the melted agar covers the bottom of the petri dishes, you can move your plates to the incubator before they solidify.

_____   7. Pour plates can be used to study colony morphology.

_____   8. When you incubate a pour plate, it should always be kept right side up.

_____   9. If you pour very hot agar, excessive moisture will condense on the lid of your petri dish.

_____ 10. The water level in the holding bath must always be just above the level of the media in your tubes or bottles.

_____ 11. The holding temperature of medium is approximately the same as the pouring temperature.

_____ 12. The solidifying temperature of agar media is close to room temperature.

KEY
1-T, 2-T, 3-F, 4-F, 5-T, 6-F, 7-T, 8-F, 9-T, 10-T, 11-T, 12-F.

68   Module 10

MODULE 11

Loop Inoculated Pour Plates

PREREQUISITE SKILL

Successful completion of Module 8, "Aseptic Transfer of Microbes," and Module 10, "Pour Plates."

MATERIALS

tubes of melted nutrient agar medium (2) deeps*

1:10,000,000 dilution of a 24 hr tube broth culture of *Escherichia coli* (Dilution 1)

1:100,000,000 dilution of a 24 hr tube broth culture of *Escherichia coli* (Dilution 2)

sterile petri dishes (2)

test tube rack

inoculating equipment:
  inoculating loop (diameter approx. 4 mm)
  Bunsen burner
  burner striker

*To be prepared by the student if the instructor so indicates, or deeps from Module 2, "Preparing and Dispensing Media," may be used.

OVERALL OBJECTIVE

Make a pour plate inoculated with bacteria so that after incubation the colonies are well distributed and are not touching each other.

Specific Objectives

1. Transfer aseptically a loopful of bacteria into a tube of melted medium.
2. Distribute these bacteria throughout the melted medium so that all bacterial cells are separated from each other.
3. Make a pour plate of this inoculated melted medium.
4. After incubation, colonies developing from these separated cells will not be touching each other. This objective is a measure of the correct performance of the first three objectives and is to be evaluated by your laboratory instructor.

DISCUSSION

You learned from Module 10, "Pour Plates," that bacteria are introduced into melted media for two reasons. The first is to study colony morphology. Remember, however, that this is only one means of studying colony morphology. Later you will learn several other ways. The second reason is to do a colony count, which is tantamount to doing a bacterial count. Remember also that there are many other ways to count bacteria in which you can determine the number of bacteria per milliliter.

This module is designed to show you how to make an inoculated pour plate, but it will also allow you to begin a study of colony morphology. For instance, from observation you will learn that a bacterium growing on the surface of the agar medium, where much oxygen is available, looks different from its sister cells growing embedded in the medium.

This module also teaches you the technique of mixing bacteria in melted agar medium so that the individual cells are so well distributed that after incubation the colonies are not touching each other. A colony is a macroscopic mass of daughter cells that arise from a single bacterial cell by asexual fission. (Asexual fission is the normal method of reproduction for bacteria—the mother cell splits into two daughter cells.) Such a plate could be used to count living bacterial numbers if a known amount of inoculum were used.

ACTIVITIES

Activity 1: Making an Inoculated Pour Plate Using Dilution 1

Gather and organize the materials listed at the beginning of this module, except for the tubes of melted medium. You will need to remove the melted medium from the holding (50°C) water bath *only* when you are ready to inoculate them since they solidify at room temperature.

Study Figures 11-1 and 11-2 to learn how to mix and distribute the inoculum in the tube of melted medium. You will notice that the tube is rotated vigorously back and forth between your hands approximately 50 times. The remainder of the procedure is identical to that of the aseptic tube transfer of bacteria you have already learned so well. A check list of this entire procedure follows the figures. If the check list is not a sufficient reminder of the procedure, refer to the figures in Module 10, "Pour Plates."

Before you begin, remember the importance of using the absolutely correct aseptic techniques listed as prerequisite skills. With the avoidance of contaminating bacteria in mind, now transfer one loopful of the 1:10 million dilution of bacterial culture to the melted medium. (*Reminder:* You should be holding both tubes in your left hand as you make this transfer.) Make the aseptic transfer *now.* Rotate the inoculated tube of medium back and forth in your hands as shown in the figures, and pour the plate. Work rapidly so that the agar does not solidify in the tube. Once you have poured the agar into the plate, allow the plate to solidify, label the plate, and incubate it for 48 hours at 30°C.

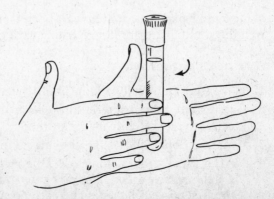

FIGURE 11-1
Rotate the tube (back). To disperse the inoculum in the melted medium, rotate the tube vigorously between your hands approximately 50 times.

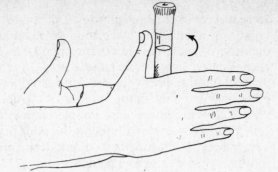

FIGURE 11-2
Rotate the tube (forth). Take care not to let any of the melted agar splash onto the capall. However, work rapidly so that the medium does not solidify in the tube before the plate is poured.

Check List

1. Distribute the cells in the dilution tube by gently shaking the tube from side to side.

2. Hold the dilution tube and melted medium in your left hand. Hold the inoculating loop in your right hand.

3. Flame the inoculating loop; allow it to cool.

4. Remove the capall from the dilution tube with the little finger of your right hand; then remove the capall from the melted medium with the next finger.

5. Flame the necks of the uncovered tubes.

6. Insert the inoculating loop into the dilution tube, and transfer a loopful to the melted medium tube. Loop diameter should be approximately 4 mm.

7. Reflame the necks of the tubes.

8. Recap the tubes, putting the caps back on the same tubes from which you removed them.

9. Flame the inoculating loop.

10. Put the inoculating loop and the dilution tube in the test tube rack.

11. You should still be holding the inoculated, melted medium tube in your left hand.

12. Mix the inoculum throughout the melted medium by rotating the tube between your hands as shown in Figures 11-1 and 11-2. Remember that you are dispersing thousands of cells so that they are separated from each other. Time is essential; rotate the tube and complete the rest of the procedure rapidly since the plate must be poured before the agar in the medium solidifies.

13. Remove the capall with the little finger of your right hand.

14. Transfer the tube to your right hand.

15. Flame the neck of the tube.

16. Lift the lid of the petri dish with your left hand, and pour the inoculated medium into it.

17. Rotate the poured plate in a circle 6 to 8 inches in diameter on your table top if the medium does not cover the bottom of the plate.

18. Let your pour plate remain unmoved until the medium solidfies. This should take 15 to 20 minutes.

19. Invert the plate and incubate it in the 30°C incubator for 48 hours.

Draw a few representative colony types, and submit the drawing to your file. Note that the difference in size of colonies is dependent upon their location and also that many colonies trapped in the medium are spindle shaped, while those growing on the surface are circular. For example, see Figure 11-3.

FIGURE 11-3
Different colony types.

Activity 2: Making an Inoculated Pour Plate Using Dilution 2

Repeat Activity 1 using Dilution 2 of the bacterial culture. The purpose of the repetition of activities using an increased dilution of the number of the original quantity of bacteria is to obtain a plate containing between 30 to 300 colonies, which is considered a countable plate. Plates containing lower or higher numbers than 30 or 300 colonies, respectively, are considered uncountable. Refrigerate the countable plates if you will be doing Module 12, "Quebec Colony Counter."

This module will be completed in your next lab period after the single mother cells you have just distributed in the pour plates have incubated, forming billions of daughter cells, resulting in a colony. Remember that after many generations, each mother cell gives rise to a colony.

Since good distribution of cells is so important to colony counts, show your incubated plates to your instructor in order to fulfill Specific Objective 4. Then take the post test.

POST TEST

The post test is a self-evaluation. It is not used for a grade. It is designed only to let you decide if you have successfully completed this module.

True or False

_____ 1. You cannot estimate the number of bacteria per milliliter of 24 hour broth culture from your loop inoculated pour plate when you use an unmeasured amount of inoculum.

_____ 2. A countable plate has between 30 and 300 colonies growing on it.

_____ 3. A 24 hour broth culture must be diluted to get a countable plate.

_____ 4. After inoculation of the melted agar, good distribution of the bacteria will occur by diffusion if allowed to set at room temperature long enough.

_____ 5. When you distribute the bacteria by rotation of the inoculated tube, you must be careful not to splash any agar on the capall.

_____ 6. Your inoculating loopful of Dilution 2 held more than 300 bacteria.

_____ 7. Aseptic technique is not important when inoculating the medium for a pour plate.

_____ 8. Colonies embedded in the agar of your pour plate are smaller because they are younger and little oxygen is available to them.

_____ 9. If the inoculum has been well mixed, most of the colonies will not be touching each other after incubation.

_____ 10. You should flame the neck of the inoculated melted agar tube before you pour it into the sterile petri dish.

MODULE 12

Quebec Colony Counter

PREREQUISITE SKILL

None.

MATERIALS

48 hour inoculated pour plates*
Quebec colony counter
tally register

*If you performed Module 11, "Loop Inoculated Pour Plates," and your plates are countable, then you may use these plates. If you did not perform Module 11, then plates will be prepared for you.

OVERALL OBJECTIVE

Learn how to use a Quebec colony counter.

Specific Objectives

1. Explain why a colony count is equal to a living bacterial cell count.
2. Define a countable plate.
3. Count the total number of colonies on a countable plate.
4. Repeat the results of Specific Objective 3 three times. The range of error in these three colony counts on the same plate will be no more than ±10 colonies. That is, there should not be a variance of more than 10 colonies per count.
5. Describe the parts and functions of a tally register.

DISCUSSION

The pour plates that you will be counting in this module have been inoculated with bacterial cells and incubated for 48 hours. Thus you will see that each separated invisible mother cell did, indeed, give rise to a colony of cells. Therefore, when you count all the colonies in the pour plate, you are, in essence, counting the number of living bacteria in the inoculum.

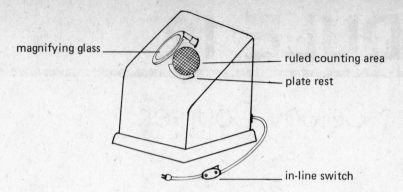

magnifying glass — ruled counting area — plate rest

in-line switch

Quebec colony counter

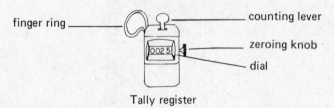

finger ring — counting lever — zeroing knob — dial

Tally register

FIGURE 12-1
Parts of the Quebec colony counter and
tally register.

It is important to realize that the colony count you will be doing in this module is not a quantitative one. To explain why it is not let us begin by defining "the quantitative determination of bacterial numbers." This means that you are able to determine the number of bacteria in a measured amount of liquid medium.

The plates you will be counting were inoculated by introducing a loopful of a diluted 24 hour bacterial culture into a tube of melted agar medium. Because the loops are not standardized the amount of liquid culture they can hold is variable, and you cannot estimate the number of bacteria transferred.

The colony count you will be doing in this module, therefore, is strictly to show you how to use a colony counter. The colony count you are about to make will not allow you to determine the number of bacteria in a known amount of liquid. This practice of colony counting, however, will be very important to you later if you do Module 13, "Quantitative Determination of Bacterial Numbers in Milk."

Let us begin with a description of the necessary equipment to do a colony count. First, consider the Quebec colony counter shown in Figure 12-1. The first part for you to note in this figure is the adjustable magnifying glass. It can be adjusted closer or farther away from the ruled area, depending on whether you want to reduce or increase the magnification. This magnifying glass also swings out of the way by moving it to the right or left. Next notice the in-line switch in the electrical cord. This switch controls a light that illuminates the ruled counting area. This light makes it easier to see the colonies to be counted. Figure 12-2 shows a close-up of the circular, ruled counting area. Study Figures 12-1 and 12-2 so that you understand the three principal parts of the Quebec colony counter that you will be using.

After reading this paragraph, inspect the Quebec colony counter in your laboratory. Become familiar with it by first turning on the in-line switch; next move the magnifying glass up and down and around, and take a good look at the circular, lighted, ruled area. Note well the half-moon protrusion at the bottom of the circular counting area. When you place your plate to be counted on the circular counting area, this half-moon protrusion will hold your plate. Beside the colony counter you should find an empty petri dish. Place it on the circular counting area so that you can see how the half-moon rest will hold your plate.

Now inspect the tally register shown in Figure 12-1. This register allows you to count colonies one at a time, adding the number of colonies to a final total. Get a

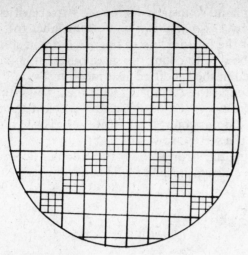

FIGURE 12-2
Ruled counting area in the Quebec colony counter.

tally register and inspect its working parts as they are described to you. The ring on the top goes over the forefinger of your right hand. Push the protrusion on the front with your thumb a few times. Your tally register is now adding for you one at a time. The knob on the side turns so that you can set your counting dial at zero. Try all the working parts until you become familiar with the tally register.

You are now ready to study Figure 12-3 which shows you how to cover systematically all areas of the plate. Notice how you use the horizontal lines for the direction in which you count the colonies. For this reason the vertical lines have been left out of this illustration of the counting circle. The arrows are added to this figure to show you that you begin counting at the uppermost left part of the ruled area and move to the right as you count between the first and second horizontal lines. Between the second and third horizontal lines, you move back to the left as you count. Follow the arrows left to right, right to left, left to right, and so forth, as you move toward the bottom of the plate and the finish point.

It is conventional to count all colonies between the horizontal lines as well as any colonies touching the top line, while avoiding colonies touching the bottom line of the two horizontal lines you are counting between. This means that in the return count the horizontal line on the top must be counted.

Begin count here.

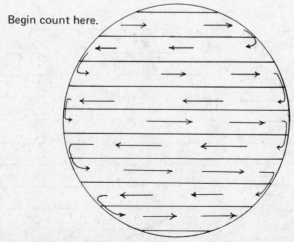

FIGURE 12-3
Direction of count.

This concept is difficult to understand so inspect Figure 12-4 for clarification. How many colonies are in Horizontal Row 1? *Count them now.* Remember to count all colonies touching the top line but not the bottom line of the row; so you should have counted seven colonies. You must be sure to count the small pinpoint colonies, too. They are younger colonies embedded in the solidified agar medium.

Now count the colonies in Horizontal Row 2 moving with the arrows, that is, from right to left. *Stop and count them now.* You must be sure to count the two colonies touching the top line since you did not count them in Row 1. Therefore, you should have counted 12 colonies in the second row.

Moving from left to right following the arrows, count the colonies in Row 3. Did you count 11 colonies? If you did, you have grasped the idea of doing a plate count systematically.

ACTIVITIES

Activity 1: Replica of Plate Count

For further practice of this conventional system for making a systematic plate count, you should count a replica of a pour plate. Turn to Figure 12-5, and note that it has been ruled horizontally to simulate the appearance of an actual pour plate on the Quebec colony counter. Use your tally register to click off the simulated colonies as you count them. Count the entire area, allowing your eyes to move back and forth from one horizontal line to the next lower one. Compare your count with the actual number. Did you count 163 colonies? Remember that you should be able to repeat your count within a range of error of ±10 colonies. Count it again, and see if you can do this. You should now be ready to count your pour plate using the same conventional system.

Activity 2: Colony Count

Count your pour plate, using your tally register to click off the colonies as you count them. Be sure to count the entire plate; let your eyes move back and forth from one horizontal row to the next lower one. Go to the Quebec colony counter, and *count your pour plate now.* Write down the total number of colonies on the entire plate.

Activity 3: Repeatability of Colony Count

Count the *same* pour plate two more times, and write down the total number of colonies each time. Submit the three plate counts to your file.

Did you satisfy Specific Objective 4? If you did, then you have learned to use the Quebec colony counter correctly.

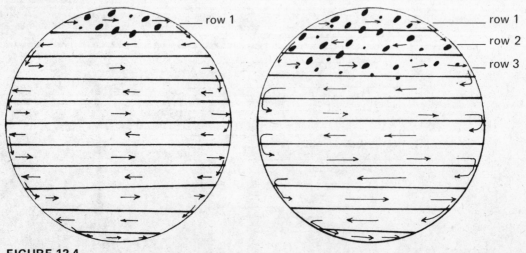

FIGURE 12-4
Counting colonies in horizontal rows.

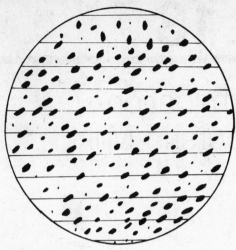

FIGURE 12-5
Replica of a pour plate.

POST TEST

The post test is a self-evaluation. It is not used for a grade. It is designed only to let you decide if you have successfully completed this module.

True or False

_____ 1. A colony count is equal to the number of living bacteria introduced into the pour plate.

_____ 2. Conventionally, a plate count is counted vertically.

_____ 3. Colonies embedded in the agar medium are smaller than those growing on the surface.

_____ 4. When counting between two horizontal lines, you should count the colonies on or touching the bottom line.

_____ 5. When counting the horizontal row beneath the row you just completed, you must count the colonies on or touching the top line.

_____ 6. You begin your count at the extreme upper left of the ruled area.

_____ 7. The extreme upper left of the ruled area at which you begin your count is a complete, large square.

_____ 8. When counting colonies, you should count from left to right only.

_____ 9. A pour plate in which the colonies are well distributed makes the final colony count more accurate.

_____ 10. When you place the plate to be counted on the ruled counting area, you must hold it in place until you finish your count.

_____ 11. The automatic counter used to total the number of colonies on a plate is called a tally register.

_____ 12. A countable plate must have 10 to 300 colonies.

If you did not get all the answers correct, repeat this module.

MODULE 13

Quantitative Determination of Bacterial Numbers in Milk

PREREQUISITE SKILL

Successful completion of Module 8, "Aseptic Transfer of Microbes," Module 9, "Aseptic Use of a Serological Pipet," Module 10, "Pour Plates," and Module 12, "Quebec Colony Counter."

MATERIALS

For Activity 1:

  pasteurized milk
  sterile 9 ml water blanks in screw-top
    (or rubber-stoppered) test tubes (2)*
  100 ml of sterile plate count agar
    (standard methods agar)*
  sterile petri dishes (4)
  sterile 1 ml pipets (3)

For Activity 2:

  raw milk
  sterile 99 ml water blanks in screw-top
    (or rubber-stoppered) bottles (2)*
  125 ml, sterile plate count agar*
  sterile petri dishes (6)
  sterile 1 ml pipets (3)

*To be prepared by the student if the instructor so indicates.

OVERALL OBJECTIVE

Determine the number of bacterial cells in a known quantity of milk, using serial dilutions and plate counts of the dilutions.

Specific Objectives

1. Explain the difference between the colony count in this module and the colony count you did in Module 12, "Quebec Colony Counter."

2. Given the number of colonies and the number of times a milk sample is diluted, be able to calculate the number of bacterial cells in 1 ml of milk and in 1 liter of milk.

3. Define the terms *quantitative determination of bacterial numbers, serial dilutions, plate count, dilution blanks, plating,* and *plated specimen.*

4. Describe why a plate count method is a determination of living bacterial numbers.
5. List seven bacterial diseases transmitted to man via ingestion of contaminated milk.
6. List five sources of milk contamination.
7. Explain why plate count agar is used instead of nutrient agar for milk colony counts.

DISCUSSION

Many methods are used to determine numbers of bacteria in liquids. Some methods determine the number of both living and dead cells, while others determine only the number of living cells. Once again you will be doing a plate count in this module. Therefore, you will actually be counting only living cells. As you know from Module 12, "Quebec Colony Counter," colony formation can result only from the growth and reproduction of the mother cell.

The plate count method is the standard method used by the American Public Health Association to determine the quality of milk. According to public health law, acceptable milk for market must have less than 15,000 bacteria per milliliter in pasteurized milk and less than 75,000 per milliliter in raw milk. Therefore, when you drink ¼ liter of milk (a little more than 8 oz), it is possible that you could be ingesting about 4 million living bacteria. Most pasteurized milk, however, has a much lower bacterial count, and the organisms that survive pasteurization are nonpathogenic. In fact, the survivors would probably be harmless thermophilic or thermoduric bacteria.

Technicians working in public health laboratories of every major city in the United States periodically check, by plate count, the number of bacteria in both pasteurized milk and raw milk. Therefore, the cleanliness of all producers and all dairies is under constant surveillance to prevent the sale of substandard milk. Interestingly, raw milk often has no more bacteria than pasteurized milk.

Many diseases are transmitted to man via ingestion of contaminated milk. The milk usually becomes contaminated from one or more of the following sources.

Diseased cows:
1. The cows themselves harbor the disease-producing microbes in their bodies. The microbes travel to the udders and thus to the milk.

Careless producers:
2. Inadequate cleansing of the udders before milking admits skin flora into the milk.
3. Dirty, dusty milking barns allow bacteria to enter raw milk via air currents.
4. Milking devices and milk cans that are not thoroughly cleansed, disfected, and rinsed can cause contamination of milk. Just a small amount of milk remaining in the milking equipment provides a culture medium in which bacteria will grow.

Careless dairy processors:
5. Lack of refrigeration as the raw milk is transported from the producer to the dairy can cause milk contamination. Milk has an initial bacterial flora at the time it is removed from the cow despite absolute cleanliness.
6. Insufficient pasteurization allows the initial flora to proliferate.
7. Contamination of pasteurized milk can result from unclean milk handlers or automated machinery.

Diseases transmitted to man by drinking milk from infected cows are bovine tuberculosis, brucellosis, Q-fever, and streptococcal infections. Cattle are given inoculations (vaccines) to prevent their contracting some diseases. Diseases of man introduced into milk by uncleanliness of the animals, their surroundings, the equipment, or milk handlers are human tuberculosis, Q-fever, diphtheria, cholera, typhoid, dysentery, scarlet fever, and other streptococcal infections.

Public health sanitation officers make periodic, unannounced inspections of cattle, buildings, and equipment to ensure careful handling of milk, cleanliness, and proper pasteurization.

Laboratory testing of milk is the final proof of carelessness on the part of producers or processors. Laboratory testing can pinpoint the exact source by which the milk becomes contaminated. If the plate count is high, that is, over 15,000 colonies/ml of pasteurized milk or over 75,000 colonies/ml of raw milk, a sanitation officer can point the finger of blame where it belongs. For pasteurized milk he can tell whether the farmer or the dairy is responsible for the high plate count and, hence, the unacceptable milk. Modifications of serial dilutions and plate counts can be used to determine bacterial numbers in or on any liquid, food, or utensil.

Definitions

Water blanks (dilution blanks)—Sterile tubes or bottles containing a measured amount of sterile water. In this module the tubes contain 9 ml of sterile water, and the bottles contain 99 ml of sterile water. Use screw cap or rubber stopper closures.

Plating—The act of pipetting from the dilutions to the empty, sterile petri dishes.

Serial dilutions—Dilutions of the original specimen (in this case, undiluted milk) in increments of 10 and/or 100 with sterile water.

Plating medium—Agar medium to be added to the plated specimens and dilutions of the specimens.

Plated specimens—Well-mixed plating medium and specimens or dilutions that are allowed to solidify.

PC agar—Plate count agar (standard methods agar).

Precautions:
1. Use aseptic technique when making serial dilutions.
2. Use aseptic technique when pipetting milk or dilutions into sterile petri dishes.
3. In Figure 13-1 note the lower case, circled letters. They mark the transfers that can be made with the same pipet, that is, all transfers marked "a" are done with the same pipet before discarding it. Use a clean, sterile pipet for all transfers marked "b" in Figure 13-1. Discard it, and use another clean, sterile pipet for "c" transfers. This is done to reduce the number of pipets needed, as well as to avoid carrying over an added amount of milk in the serial dilutions. Letters "d," "e," and "f" denote when the same pipet is used for diluting raw milk.
4. Do not immerse the pipet in the water blanks when delivering the milk or the previous dilution.
5. You may touch the tip of the pipet to the sterile petri dish when delivering into it.
6. Shake the milk up and down 25 times over a distance of 12 inches. This vigorous shaking disperses the bacteria evenly before transferring and plating.
7. Shake each serial dilution in the same fashion before transferring or plating. Be sure the screw cap (or rubber stopper) is on tightly before shaking.
8. After you have plated the serial dilutions, add approximtely 20 ml of plate count medium to each plate, using the technique you learned in Module 10, "Pour Plates."
9. Use the same aseptic precautions you used in Module 10 when pouring medium into petri dishes.
10. Pour only two plates at a time since the mixing of the milk dilutions takes place in the plate. Remember that you are dispersing bacteria, that is, separating cells, using only this table top rotation. Therefore, make 20 complete rotations of each plate.

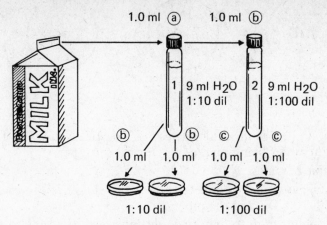

Activity 1: Pasteurized milk

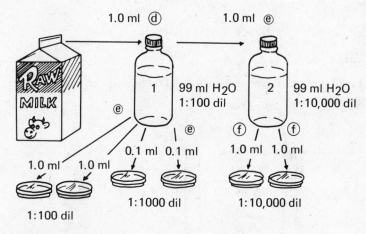

Activity 2: Raw milk

FIGURE 13-1
Serial dilutions and the amounts to be plated.

ACTIVITIES

Activity 1: Quantitative Determination of Bacterial Numbers in Pasteurized Milk

After sterilization, keep the medium in the 50°C holding bath so that it does not solidify. Next, gather together the rest of the materials needed. They are listed separately at the beginning of this module under Materials. Before proceeding, study the scheme for serial dilutions and the plating of these dilutions as shown in Figure 13-1.

Follow this procedure to perform Activity 1.

1. Label the two 9 ml water blanks as 1 and 2.
2. Label the sterile petri dishes as shown in Figure 13-2. Each dilution is plated in duplicate for greater accuracy of plate count. This is explained further in the calculations at the end of this module.
3. Shake the pasteurized milk thoroughly as follows: shake up and down 25 times over a 12 inch span. Do not allow the milk to settle out before performing Step 4.
4. Aseptically pipet 1 ml of milk to the dilution tube marked 1. Discard the pipet.
5. Shake the first dilution 25 times over a 12 inch span also. Perform Step 6 immediately, before the milk settles out.
6. Aseptically transfer 1 ml of this first dilution to each of the petri dishes labeled 1:10.

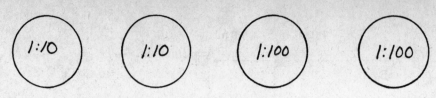

FIGURE 13-2

7. Using the same pipet, again before the milk settles out, aseptically transfer 1 ml of the first dilution to the water blank marked 2. Discard the pipet.
8. Shake the second dilution 25 times.
9. Immediately after shaking, aseptically pipet 1 ml of this final dilution to each of the petri dishes labeled 1:100. Discard the pipet.
10. Aseptically add the plating medium to only two dilution plates at a time since the coolness of the specimen and the glass dish will cause the medium to solidify rapidly.

Activity 1: Pasteurized Milk

If the pasteurized milk in the carton contains about 5,000 bacterial cells per ml the expected results are as follows:

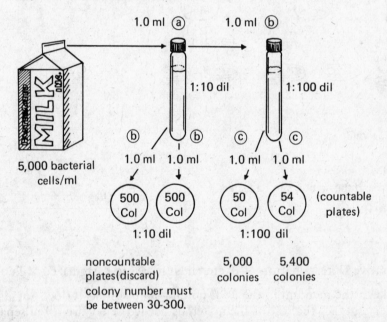

Average the two countable plates that are the 1:100 dilution; multiply by the dilution factor to get the number of microbes/ml of milk.

$$\begin{array}{r} 50 \text{ colonies} \\ 54 \text{ colonies} \\ \hline 104 \text{ colonies} \end{array} \quad \text{Divide by 2} = 52 \text{ colonies.}$$

Multiply by the dilution factor, 52 colonies x 100 = 5,200 colonies. This indeed shows that the original sample of milk you aseptically removed from the carton contains 5,200 bacteria per ml of milk. Does this milk with 5,200 organisms per ml meet the Public Health Standards?

FIGURE 13-3
Example for calculating number of bacteria in pasteurized milk.

Activity 2: Raw Milk

If the raw milk in the carton contains about 2,000,000 bacterial cells per ml the expected results are as follows:

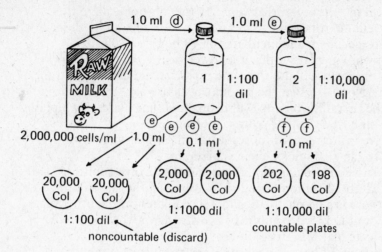

Average the two countable plates, which in this case are the 1:10,000 dilution. Multiply by the dilution factor and you will get the number of microbes/ml of raw milk:

202 colonies
198 colonies
_____
400 colonies   Divide by 2 = 200 colonies

Multiply by the number of times the original ml of milk has been diluted; 200 colonies x 10,000 = 2,000,000 colonies. This shows that the original sample of milk that you aseptically removed from the carton or bottle contains 2,000,000 microorganisms per ml of raw milk. Will the Public Health Department allow this milk to reach your local grocer?

FIGURE 13-4
Example for calculating number of bacteria per ml in raw milk.

11. Immediately rotate the plate 20 times on your table top to distribute the inoculum evenly throughout the medium. This table top rotation must be thorough since the bacterial cells must be separated from each other. (Review Module 10, "Pour Plates," and Module 11, "Loop Inoculated Pour Plates," if necessary.)

12. Add the plating medium to the other two plates, and rotate them on your table top as described above.

13. Allow the medium in all four plates to solidify.

14   After solidification, invert the plated specimens and place them in the 30°C incubator for 48 hours.

During your next lab period you will do a plate count using the Quebec colony counter on the dilution that has 30 to 300 colonies. The number of colonies from pasteurized milk is often so low that the plates are uncountable. If the count is 30 or more, calculate the number of bacteria present in 1 ml of milk. You will find the method of calculation described at the end of Activity 2, and example illustrations in Figures 13-3 and 13-4. Complete Activity 2 next since it would be impossible to do the calculations until after you count your plates in the next lab session. Review Module 12, "Quebec Colony Counter," before your next lab period if you feel it necessary.

Activity 2: Quantitative Determination of Bacterial Numbers in Raw Milk

Since the plate count from unpasteurized milk may be higher than the one you just did, you would expect to have to dilute the milk many more times. You will find this to be true by studying the dilution and plating scheme for raw milk shown in Figure 13-1. While following this scheme carefully, make and plate the serial dilutions employing the same procedure you used in Activity 1, that is, make the dilutions, plate the dilutions, add the plating medium, rotate the plates on the table top, allow the medium to solidify, invert the plated specimen, and incubate for 48 hours at 30°C.

Precautions: Note from Figure 13-1 that the lower case, encircled letters "d," "e," and "f" again indicate that you will use the same pipet for the dilution and the platings before discarding it. Note also that the amount of diluted specimen plated varies; for example, from the first dilution you will be making four plates. Two plates will have 1.0 ml, and two will have 0.1 ml of diluted specimen.

Review the procedure in Activity 1 if necessary, and *complete* Activity 2 now. Refer to Figure 13-4 for an example of the calculations.

Calculations to be derived from Quebec colony counts involve counting both plates of the dilution with 30 to 300 colonies and then averaging the two counts. For example, if you count 150 colonies on one plate and 160 colonies on the other plate, then your average colony count is 155. By public health standards, the acceptable degree of variation on duplicate plates is 10 colonies.

To determine the number of bacteria per milliliter of milk, multiply the average colony count by the dilution of the specimen that the count represents. For example, 155 colony average x 1:100 dilution = number of bacteria per milliliter of milk and 155 x 100 = 15,500 bacteria per milliliter of milk. If you plated 0.1 ml, you must multiply the dilution by 10 since you must determine the number of bacteria per milliliter.

Submit to your file the number of colonies on both countable plates and all your calculations for determining the number of bacteria per milliliter for both pasteurized and unpasteurized milk. Then take the post test.

POST TEST

The post test is a self-evaluation. It is not used for a grade. It is designed only to let you decide if you have successfully completed this module.

Part I

Submit a scheme of serial dilutions and plating similar to that shown in Figures 13-1, 13-3, and 13-4 for a broth culture with a predetermined bacterial cell count of 20,000,000. Indicate which dilution and which plate would be most likely to have an acceptable plate count.

Part II

List seven bacterial diseases transmitted to man via ingestion of contaminated milk.

1. _____
2. _____
3. _____
4. _____
5. _____
6. _____
7. _____

KEY

Part I: $\dfrac{1.0\ ml}{or} \longrightarrow$ 1:100 $\xrightarrow{1.0\ ml}$ 1:100 $\xrightarrow{1.0\ ml}$ 1:10 $\longrightarrow$ plate 1.0 ml

$\xrightarrow{1.0\ ml}$ 1:100 $\xrightarrow{1.0\ ml}$ 1:100 $\longrightarrow$ plate 0.1 ml

Part II: Seven diseases may be listed from the following: human or bovine tuberculosis, brucellosis, Q-fever, diphtheria, cholera, typhoid, dysentery, scarlet fever, and streptococcal infections.

Part III: Five sources of contamination may be chosen from the following: diseased cows, inadequately cleaned udders, dirty milking barns, improperly cleaned milking devices or cans, lack of refrigeration, insufficient pasteurization, contamination of pasteurized milk by handlers, and unclean machinery.

Part IV: 1. 22,500,000/ml.
2. 11,250,000,000/500 ml.

Part III

List five sources of milk contamination.

1. _____
2. _____
3. _____
4. _____
5. _____

Part IV

If a colony count is 225 from a specimen that has been diluted 100,000 times, how many bacteria are there per milliliter of specimen?

1. _____

How many bacteria are there in 500 ml of the specimen?

2. _____

MODULE 14

Streaking for Isolation

PREREQUISITE SKILL

Successful completion of Module 8, "Aseptic Transfer of Microbes."

MATERIALS

---

unlined paper

pencil or pen

empty petri dish

felt pen

wax glass-marking pencil

inoculating equipment:
  inoculating loop
  Bunsen burner
  burner striker

petri dish containing nutrient agar
  (6 to 8)*

beaker containing disinfectant (for swab
  discard)

slant culture of *Escherichia coli*

broth culture of *Escherichia coli*

slant culture of *Serratia marcescens* (for
  post test)

mixed broth culture of two different
  genera of bacteria (for post test):
  Serratia marcescens and *Escherichia coli*

sterile swab

*To be prepared by the student if the instructor
so indicates, or plates from Module 10, "Pour
Plates," may be used.

OVERALL OBJECTIVE

Streak a bacterial culture on a nutrient agar plate using a technique that will separate
the individual bacterial cells. When you have applied this technique successfully, each
isolated cell will develop into a pure colony after incubation.

Specific Objectives

1. Demonstrate the principle of streak dilution using paper and pencil.
2. Demonstrate the principle of streak dilution using an empty petri dish and a felt
pen.

3. Streak for isolation using living bacteria and a nutrient agar plate.
4. Define the terms *original inoculum, pure colony, mother cell, daughter cells, streak dilution,* and *bacterial isolation.*
5. Differentiate between bacterial growth and an increase in bacterial size.

DISCUSSION

Beginning microbiology students find it difficult to streak for isolated colonies for the following two reasons:

1. They do not utilize as much of the streaking surface as possible. This results in fewer dilutions.
2. They use too large an inoculum, which means that their streaking must dilute thousands and thousands of cells until the individual cells are separated from each other.

The practice activities in this module are designed so that you will use *all* the streaking surface and make the maximum number of dilutions. This module also strives to help you think small enough so that you will not have too many bacterial cells in your original inoculum.

It is difficult to realize just how very minute bacterial cells are, that is, that each individual bacterium is approximately 1/25,000 of an inch in size. Therefore, when you introduce an inoculating loop of bacteria onto the surface of a nutrient agar plate, you are actually placing tens of thousands of cells on the medium. When you learn to use a small amount of inoculum and learn how to separate these thousands of cells so that they are not touching each other, then you have correctly streaked for isolation.

Proper streaking for isolation results in pure colonies. A pure colony arises from a single mother cell. After your nutrient agar plate is streaked, separating the bacterial cells from each other, these separated single cells are then called *mother cells.*

Upon incubation of your streak plate, each mother cell divides by asexual binary fission, i.e., by a splitting in half of the cell, in 20 to 30 minutes, giving rise to two daughter cells. In the next 20 to 30 minutes, the daughter cells split in half, and another generation of four new daughter cells comes into existence. For example, see Figure 14-1.

The new cells continue to divide in exponential numbers, resulting in billions of daughter cells. These billions of cells pile up on top of and around each other, and a pure colony is born. You must remember, however, that a colony can be considered a *pure* colony *only* if it does not touch another colony.

In actuality for most bacteria after 24 hours of incubation, a pure colony consists of 50 to 72 generations of cells arising from a single mother cell. A colony is

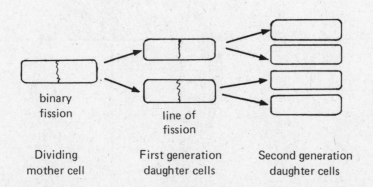

Dividing
mother cell

First generation
daughter cells

Second generation
daughter cells

binary
fission

line of
fission

FIGURE 14-1
Asexual reproduction (binary fission) of
bacterial cells.

therefore composed of billions and billions of daughter cells. Bacterial growth, then, means an increase in cell numbers, not an increase in cell size.

In summary, after completion of this module, you will have developed the correct technique for separating bacterial cells from each other, thus allowing them to develop into pure colonies.

When you have completed the practice activities in this module, you will be able to streak for isolation. This is a basic microbiological technique that is important to your success in the laboratory portion of this course. Therefore, do not skip any of the practice activities designed to help you become adept when using this technique.

ACTIVITIES

Activity 1: Simulation of Streaking for Isolation Using Paper and Pencil

On a plain sheet of paper, draw a circle about 3 inches in diameter. Line and label the circle exactly as shown in Figure 14-2. If you are left handed, the 0 sector should be on the right; look for other reversals as you proceed through the steps of this technique. (Notice that the 0 sector is *smaller* in relation to I, II, and III.) Sectors I, II, and III should be almost equal in surface area.

Using a pencil or pen, follow these steps in which you will be imitating the procedure you will use when streaking for isolation on a nutrient agar plate with living bacteria. Keep in mind that, when working with bacteria, you will be attempting to

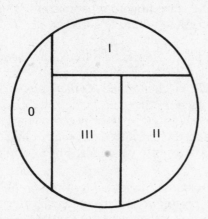

FIGURE 14-2
Dilution sectors as drawn on paper.

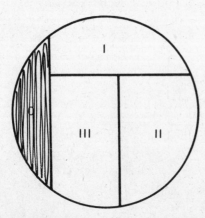

FIGURE 14-3
Step 1: Sector 0 represents the planting of a small amount of bacteria with your inoculating loop. This simulates the original inoculum.

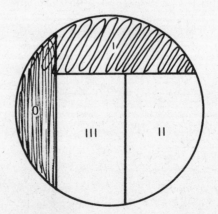

FIGURE 14-4
Step 2: Sector I represents the first dilution. Your lines in Sector I should be as uniformly apart as shown here.

dilute the number of cells in each sector, and that you will be *flaming* your inoculating loop between each sector. Flaming the inoculating loop kills the remaining cells on the loop, and thereby aids in diluting the cells. Each sector represents a dilution or a reduction in numbers of the thousands of bacterial cells in the original inoculum. Sector 0 is for the original inoculum and represents thousands of cells. The steps of this procedure are to be done in numerical order. Begin Step 1 now. Using your pencil, draw almost solid lines in sector 0. It should look like Figure 14-3, with your circle being larger. Note again that Sector 0 is smaller than the other sectors and that the inoculating lines are not carefully separated as they will be in the subsequent dilutions.

Study Figure 14-4, and then draw lines from Sector 0 into Sector I as shown in the figure. Take care to use as much of Sector I as possible. Also be careful not to let your lines in Sector I cross over each other. As you begin drawing your looping lines, be certain that only the first two or three lines enter Sector 0. In doing this, you are imitating the dilution of the number of cells in Sector 0. This is Step 2 of the procedure.

Rotate your piece of paper counterclockwise almost ¼ of a turn so that Sector I is on the left. Imitate the dilution or reduction of the numbers of cells again, as shown in Figure 14-5. This is Step 3 of the procedure.

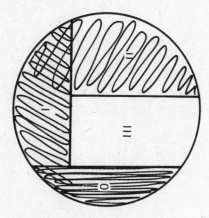

FIGURE 14-5
Step 3: Sector II represents the second dilution.

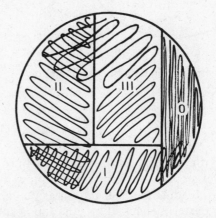

FIGURE 14-6
Step 4: Sector III represents an imitation of the final dilution of the numbers of organisms. Be sure to leave some space between each sector to allow for the expanding bacterial growth.

Rotate your piece of paper to the left again, and imitate the final dilution by streaking Sector III as shown in Figure 14-6, which is Step 4. When making your lines in each sector, note that *only* the first two or three lines enter the preceding sector.

Using another piece of paper, now draw another circle similar to the one you have just finished. Mark and label it in the same manner. Repeat this practice activity several times. Proceed very carefully through all four steps. When you have done this to your satisfaction, show the final sketch to your lab instructor. If your lab instructor approves your work, then you are ready to proceed to the next practice activity.

Activity 2: Dry Run Using Empty Petri Dish and Felt Pen

Gather together a petri dish, a felt pen, and a wax glass-marking pencil. Turn the petri dish upside down, and mark it *on the bottom* with the wax pencil as shown in Figure 14-7. Take careful note of the different position of the sectors as compared with their position in Figure 14-2. Note well, the position of Sector I is reversed while the petri dish is upside down, that is, Sector I is now on the bottom instead of on the top.

Next turn the unopened empty dish right side up. The marking now appears as shown in Figure 14-8, and the sector positions are now identical to Figure 14-2.

Now that you have learned how to make the four-step dilution, in this practice activity pretend that the felt pen is your inoculating loop and that your empty petri dish contains a solid medium. When you remove the petri dish lid, you will want to protect your medium from airborne bacteria. Therefore, the petri dish lid should be removed just far enough to let you peek in and see where you are placing your felt pen and making your lines. Figure 14-9 shows you how to handle the lid with your left hand. Now, with the petri dish lying on the table in front of you, carefully remove the lid as shown in Figure 14-9.

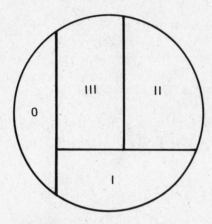

FIGURE 14-7
Dilution sector markings on bottom of empty petri dish.

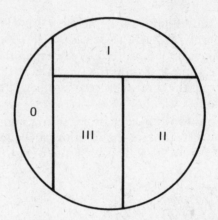

FIGURE 14-8
Dilution sector markings on bottom of plate as seen through top of empty petri dish. Notice that Sector I is again at the top when Sector 0 is at the left.

With Sector 0 on the left, draw lines on the inside of the glass bottom. Refer to the four steps in Activity 1, and follow them exactly. Proceed now through all four steps, reading the instructions as you go. Close the petri dish, and have your streaking patterns checked by your laboratory instructor.

Activity 3: Streaking for Isolation Using Living Bacteria

If your instructor has approved your felt pen dry run, you are now ready to work with the real thing, that is, with bacteria, an inoculating loop, and a nutrient agar plate. Label and line the bottom of the plate containing agar just as you did in Activity 2. Then follow the steps shown in Activity 1.

Precautions: The agar can be cut with the inoculating loop, so use a light but definite touch. The loop should be kept as flat, that is, as horizontal with the agar, as possible. The inoculum need not be so large that you can see it macroscopically; simply touch your loop to the bacterial growth. *Flame your loop, and let it cool between each dilution.*

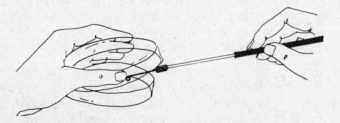

FIGURE 14-9
Proper handling of the lid of the petri dish. Open the petri dish slightly, but keep the lid over the dish. This will protect the sterile medium from airborne bacteria.

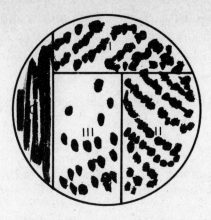

FIGURE 14-10
Incubated petri dish. After 48 hours of incubation, your plate should show isolated colonies.
Isolation may occur in any sector except Sector
0 depending upon the number of cells in the
inoculum.

Inoculate your nutrient agar plate from the slant culture of *Escherichia coli*
now, beginning once again with Sector 0. Proceed through your dilutions numerically
as you did before. When you have finished, invert your plate, label it, and incubate it
at 30°C for 48 hours. It is only after incubation that you will know if you have successfully streaked for isolated colonies. If you have mastered the technique in this module,
your streak plate should look similar to Figure 14-10. Isolated colonies often occur
before Sector III. The sector in which isolation takes place is not important as long
as you get separated colonies. Show your plate to your lab instructor for approval.

Precaution: If colonies appear on any area of the agar surface that you did
not streak, they are airborne contaminants and must not be used for subculturing. So
examine the location of the colonies carefully.

Activity 4: Use of a Cotton Swab for Original Inoculum

In medical laboratories, the specimen from which bacteria are to be grown and identified is often collected on a sterile swab. In the past you have probably had a sore
throat, and the doctor or nurse has taken a throat culture from you by using a sterile
swab to gather the bacteria from the back of your throat. This throat swab is teeming
with the bacteria causing your sore throat, and so it is used to plant the original inoculum on various types of agar plates. After the original inoculum is put on the medium
from the swab, the streak dilutions of this original inoculum are done with an inoculating loop exactly as in Activity 3.

Since using a swab for the original inoculum (Sector 0) is common practice
in hospital laboratories, this activity is designed to simulate hospital procedure. You
will use a broth culture of *Escherichia coli* instead of the organisms from a sore throat
swab.

Begin this activity by dividing the bottom of the agar plate into the same
sectors that you used in Activities 2 and 3. Next hold both the *Escherichia coli* broth
culture tube and the tube containing the sterile swab in your left hand. Remove the
capalls, using your best aseptic technique, that is, use your little finger and the one
next to it. Remove the sterile swab from the sterile tube by using the thumb and forefinger of your right hand. Flame the necks of the tubes. Dip the swab into the broth
culture tube, saturating it with *Escherichia coli*. Flame the necks of the tubes again,
recap them, and put them back in the test tube rack. You should still be holding the
swab in your right hand, so carefully lift the lid of the agar plate and streak Sector
0 heavily with the *Escherichia coli* saturated swab. Discard your swab in a container of
disinfectant or in another appropriate place, and streak for isolation with your inoculating loop using the same technique that you developed in the preceding activities.
Incubate this plate with the plate from Activity 3, and show it to your lab instructor.

Submit sketches of your isolation plates from Activities 3 and 4 to your file, and then take the post test.

POST TEST

The post test is a self-evaluation. It is not used for a grade. It is designed only to let you decide if you have successfully completed this module.

Part I: Activities

1. Using a pure culture slant of *Serratia marcescens,* streak three nutrient agar plates, and get isolated colonies on all three plates after incubation for 48 hours at 30°C.

2. Using a mixed culture containing two different genera of bacteria, streak for isolation, and after incubation get two different appearing, separated colonies. Incubate at 30°C for 48 hours.

To pass this part of your post test, your streak plates must be approved by your instructor. These streak plates can be used for Activity 1 in Module 15, "Cultural Characteristics of Bacteria," if your instructor so indicates.

Part II: True or False

_____ 1. A pure colony is made up of billions of daughter cells.

_____ 2. A pure colony may be touching another colony if they both look alike.

_____ 3. The original inoculum contains many more cells than does the first dilution.

_____ 4. The third dilution contains more cells than the second dilution.

_____ 5. Streak dilutions of bacterial cells indicate that the original inoculum has been diluted with sterile saline.

_____ 6. It is desirable to use as much as possible of the surface area of an agar plate when streaking for isolation.

_____ 7. A pure colony can arise from more than one mother cell.

_____ 8. Streaking for isolation requires that the original inoculum be diluted by means of an inoculating loop to single, separated cells.

_____ 9. Bacterial growth means an increase in the bacterial numbers, instead of in size.

_____ 10. Bacteria can be measured accurately with a ruler than has millimeter divisions.

If you did not get all the answers correct, then you should repeat this module.

SUMMARY OF STREAKING TECHNIQUES FOR ISOLATION

The following material can be used as a review in your future laboratory work. If you find that you do not get pure colonies when you streak for isolation, then this summary should be of great help to you. Use it until you become so adept at getting pure colonies that it is no longer necessary for you to divide the plate into sectors.

When you can make three streak dilutions of the original inoculum which result in isolated colonies, discontinue marking the bottom of the plate into sectors. You have now mastered the skill of obtaining pure colonies and no longer need this instructional aid.

Mark the bottom of your nutrient agar plate with a wax pencil, dividing it into sectors as indicated in Figure 14-11. If you are left handed, the 0 sector will be on the right.

Next turn the marked plate right side up. The markings will be reversed and will look like those in Figure 14-12.

Carefully hold the lid in your left hand,* using it to protect the nutrient from airborne bacteria. Using your inoculating loop and living bacteria, streak each sector in numerical order imitating the four steps shown in Figure 14-13. Use a light but definite touch so that the loop does not dig into the agar, but your streaks are uniform. Be sure to flame and cool your inoculating loop between each dilution.

Always keep the sector you are streaking from on the left.* Do this by rotating the plate almost ¼ of a turn counterclockwise.* Note how the arrow follows the correct position of Sector 0. Also note that Sector 0 (for the original inoculum) is small in comparison to the other sectors (I, II, and III). This allows you to use the major portion of the plate for dilution of the numbers of bacteria in the original inoculum.

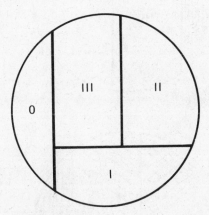

FIGURE 14-11
Sector markings on bottom of petri dish.

*Reverse this if you are left handed.

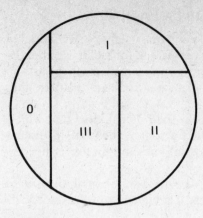

FIGURE 14-12
Sector markings on petri dish turned right side up.

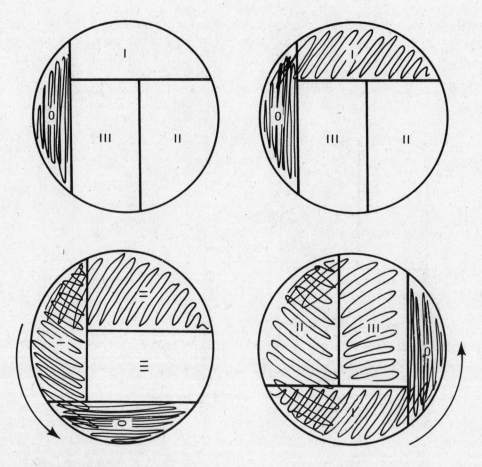

FIGURE 14-13
Summary of the steps used to streak for isolation.

MODULE 15

Cultural Characteristics of Bacteria

PREREQUISITE SKILL

Successful completion of Module 2, "Preparing and Dispensing Media," Module 8, "Aseptic Transfer of Microbes," and Module 14, "Streaking for Isolation."

MATERIALS

48 hr streak plates*
hand lens or dissecting microscope
sterile nutrient agar (1.5%) slants (3)†
sterile nutrient broth tubes (4)†
sterile thioglycollate broth tubes (2)†
sterile nutrient agar plates (3)†

slant cultures of *Escherichia coli, Bacillus subtilis, Proteus vulgaris*
nutrient broth cultures of *Streptococcus pyogenes, Staphylococcus aureus, Pseudomonas aeruginosa*

*You may use the streak plates from the post test activity in Module 14, or plates will be prepared for you. The *Bacillus subtilis* plates will be prepared for you.
†To be prepared by the student if the instructor so indicates.

OVERALL OBJECTIVE

Recognize, name, and describe the various growth patterns of different bacteria using several media preparations.

Specific Objectives

1. Draw and name the colonial morphology of two different-appearing colonies from your streak plates.
2. Make three nutrient agar stroke inoculations using three different genera of bacteria. Draw and name the stroke morphology of each.
3. Inoculate two different types of broth with two genera of bacteria. Draw, name, and use a descriptive term or phrase for the differing patterns of growth of the two bacteria in each type of broth.

4. Draw and describe the difference between a soluble pigment and a nonsoluble pigment.
5. Fill in Table 15-1 with descriptive names for the features of the various types of growth patterns that you will observe in all four activities of this module.
6. Define stroke slant inoculation.

DISCUSSION

Some microbes have characteristic growth patterns. Only if these growth patterns are distinctive for a single species can they aid in the identification of that species. Although some bacteria do grow in distinctive patterns, many others look very much alike. Since many different types of bacteria have similar growth characteristics, too much emphasis can be placed on the study of cultural characteristics for a beginning microbiology student. Therefore, to avoid confusion, your exposure will be limited to those organisms with distinctly different growth patterns. If you compare Figures 15-1 and 15-2 with figures in other microbiology lab manuals, you will see that only the most common cultural types have been included. Figures 15-1 and 15-2 give you enough information to complete all the activities in this module, and they give an adequate introduction to the study of the cultural characteristics that you will encounter in this lab course. Study Figures 15-1 and 15-2 now, and refer to them as you perform Activities 1, 2, and 3 of this module.

After studying Figures 15-1 and 15-2, you can see that many characteristics are used to attempt to categorize bacteria by their growth patterns. Four types of growth patterns are classically used:

1. An isolated colony on the surface of a nutrient agar plate.
2. A nutrient agar stroke culture.
3. A nutrient broth culture.
4. A gelatin stab.

Bacterial growth on these four different media preparations has become standard for the study of cultural characteristics. The gelatin digestion patterns have been omitted here because it is difficult to distinguish among the various patterns and also because it adds further confusion to the already confusing study of cultural characteristics. A later module discusses the gelatinase activity of microorganisms. It is not necessary to concern yourself with this now.

Begin now to study the morphology of an isolated colony by examining Figure 15-1. From Figure 15-1 you will learn that there are three aspects of a single colony used to study the characteristic growth features of that colony. They are:

1. Colony shape.
2. The margin of the colony.
3. The elevation of the colony.

Activity 1 tells exactly how to study and name these three colonial features.

Pigment production and different patterns of bacterial growth in broths are included when studying growth characteristics. Further explanation of pigment production by bacteria is included in Activity 4.

A few stroke slant patterns are depicted at the top of Figure 15-2. These stroke slant patterns will be sufficient to allow you to complete the activity related to them. The characteristic spreading growth of *Proteus* is confined to separate, distinct colonies on salt-free media. This is why you use NA (1.5%), which contains salt, instead of plain NA, which is salt-free, for Activity 2.

You will be able to complete Activity 1 during your present lab session. However, you will only be able to make the inoculations for Activities 2, 3, and 4; so they will be completed during the next lab session after they have had a chance to grow. Turn in all your work from the activities in this module together. You will find that completion of Activities 1 to 4 will allow you to do Activity 5 with ease.

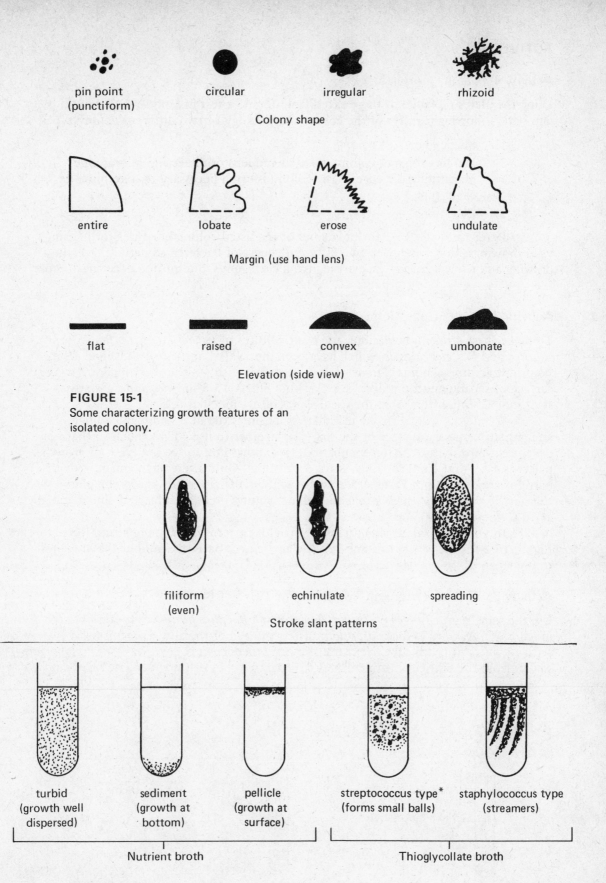

pin point
(punctiform)   circular   irregular   rhizoid

Colony shape

entire   lobate   erose   undulate

Margin (use hand lens)

flat   raised   convex   umbonate

Elevation (side view)

FIGURE 15-1
Some characterizing growth features of an
isolated colony.

filiform
(even)   echinulate   spreading

Stroke slant patterns

turbid
(growth well
dispersed)   sediment
(growth at
bottom)   pellicle
(growth at
surface)   streptococcus type*
(forms small balls)   staphylococcus type
(streamers)

Nutrient broth   Thioglycollate broth

FIGURE 15-2
Some growth patterns on stroke slants and in
different broths. *Thioglycollate broth formulated with resazurin indicator is preferred since
it is less inhibitory to streptococci than
methylene blue indicator.

ACTIVITIES

Activity 1: Colony Morphology

Using the plates provided or those saved from the post test in Module 14, draw and name the following features of the colony morphology of two different colony types:

1. Shape—Form of colony.
2. Margin—Edge of colony. (Use a hand lens or dissecting microscope. Magnifying the edge of the colony is often necessary to determine its exact type.)
3. Elevation.

Refer to Figure 15-1 for the names of the listed colonial features for labeling your drawings. You were given two different genera of bacteria, so be sure to make a drawing and name a colony type arising from each genus. Submit the drawings to your file.

Activity 2: Slant Stroke Morphology

To make these stroke inoculations, use slant cultures of *Escherichia coli, Bacillus subtilis,* and *Proteus vulgaris,* which have been inoculated in the usual fashion. Using your best aseptic technique, transfer each organism with your inoculating loop to the surface of separate nutrient agar (1.5%) slants. Figure 15-3 shows you how a stroke slant inoculation differs from the usual slant culture inoculations.

When you make a stroke inoculation, begin at the butt and make a single straight stroke up the surface of the slant. Remember to use a light touch so that you do not cut into the agar. After incubation, this allows you to see the types of growth patterns extending from the single-stroke inoculum. Only a few of the many growth patterns are depicted in Figure 15-2, which will be sufficient for you to complete this activity. Make the three stroke inoculations now. Incubate all stroke inoculations at 30°C for 48 hours.

In your next lab session, draw the growth patterns of each organism. Refer to Figure 15-2 for the names of some stroke slant growth patterns. Submit the drawings to your file.

Activity 3: Growth Patterns in Two Different Types of Broth

Use the same slant cultures of *Escherichia coli* and *Bacillus subtilis* that you used for inoculum in Activity 2. Make *aseptic* transfers of each of the above organisms to a separate nutrient broth tube. Next transfer *Streptococcus pyogenes* and *Staphylococcus aureus* from the nutrient broth cultures to separate tubes of thioglycollate broth and

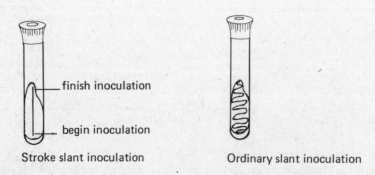

Stroke slant inoculation   Ordinary slant inoculation

FIGURE 15-3
The differences between a stroke slant inoculation and an ordinary slant inoculation.

nutrient broth. Use your best aseptic technique since *strep* and *staph* are possible pathogens. You should now have six newly inoculated broth tubes (four nutrient broths and two thioglycollate broths). Incubate *Bacillus subtilis* at 30°C and all other broth cultures at 37°C for 48 hours. Since *Bacillus subtilis* is not usually normal or transient flora of the human body, its optimum growth temperature is lower than body temperature (37°C).

After incubation, you will notice that the growth of *Staphylococcus aureus* and *Streptococcus pyogenes* in thioglycollate broth is quite different from their growth in nutrient broth. Make drawings of both organisms from both types of broth in your next lab session after incubating for 48 hours. Compare the growth patterns in the two different broth cultures. *Strep* and *staph* have characteristic growth patterns in thioglycollate broth that you can use as a clue to their identification in future exercises (see Figure 15-2). This is not absolute proof, however, because organisms other than staph will form streamers.

After incubation, make drawings of the nutrient broth growth patterns of *B. subtilis* and *E. coli*. Refer to Figure 15-2 for names and descriptions of growth types to accompany all six of your drawings. The pellicle of *B. subtilis* may take as long as a week to appear; therefore, reincubate and reexamine as necessary. Submit your drawings to your file.

Activity 4: Production of Pigment by Bacteria

Pigment production is best studied from isolated colonies since a separated colony will allow you to see if the pigment is soluble or nonsoluble. When we say a pigment is soluble, we mean that it is water soluble and therefore can diffuse out of the bacterial cells making up the colony into the surrounding medium. This soluble pigment turns the medium surrounding the colony the color of that pigment. If the pigment is nonsoluble, it remains confined inside the bacterial cells, and therefore only the colony itself is colored.

Most bacteria are not chromogenic, that is, they do not produce pigments; hence their colonies are white or gray. The few bacteria that are chromogenic produce most colors of the spectrum, for example, green to blue, yellow to orange, red to violet, and many shades in between. A few other microbes produce a black pigment. Most pigments are not water soluble.

As you proceed through this course, it will become clear to you that the pigment produced by an organism can be used as another clue leading to the identification of that organism. Therefore, as you study characteristic growth features of a colony, you should always be observant of the various kinds of pigment production that can occur.

This activity is designed to demonstrate the difference between a soluble pigment and a nonsoluble pigment. Begin work by using nutrient broth cultures of *Pseudomonas aeruginosa* and *Staphylococcus aureus*. Streak each organism for isolation on separate nutrient agar plates. Incubate the *Staphylococcus aureus* plate at 37°C for 48 hours and *Pseudomonas aeruginosa* at room temperature overnight and then refrigerate. Use the third nutrient agar plate as a comparative control to detect pigment production after incubation. In your next lab session, draw and describe the growth on both plates. *Pseudomonas aeruginosa* should produce a greenish water soluble pigment.

Activity 5: Summary of Significant Cultural Characteristics

Satisfy Specific Objective 5 by filling in Table 15-1. Use the descriptive names for the various growth patterns as you did in all the activities in this module. Write these descriptive names in the appropriate blanks. This chart reinforces the fact that the same organism must be grown several different ways in order to study its cultural characteristics as described in this module.

You have now completed all the activities in this module. Turn in all work that was required of you in each activity, along with Table 15-1.

All types of cultural characteristics have not been studied on all the organisms you used because of similarities among organisms. Only organisms with distinctive characteristics have been chosen for purposes of emphasis. That is why some blocks have been shaded in Table 15-1.

Take the post test now.

TABLE 15-1 Summary of Some Significant Cultural Characteristics*

Microbe	Shape of colony	Margin of colony	Elevation of colony	Stroke types	Broth growth Nutrient	Thio.	Pigment color
Activity 1							
B. subtilis							
S. marcescens							
E. coli							
Activity 2							
B. subtilis							
E. coli							
P. vulgaris							
Activity 3							
B. subtilis							
E. coli							
S. aureus							
S. pyogenes							
Activity 4							
S. aureus							
P. aeruginosa							

*Fill in all unshaded blanks.

POST TEST

The post test is a self-evaluation. It is not used for a grade. It is designed only to let you decide if you have successfully completed this module.

Part I: True or False

_____ 1. Most pigments produced by bacteria are water soluble.

_____ 2. Most bacteria do not produce pigments.

_____ 3. Water soluble pigments produced by bacteria remain confined inside the cell.

_____ 4. *Streptococcus pyogenes* and *Staphylococcus aureus* grow differently in nutrient broth than they do in thioglycollate broth.

_____ 5. Turbidity throughout a bacterial broth culture means that the organisms are growing in a pellicle.

_____ 6. When you make a stroke inoculation, you begin at the top of the agar slant, and move your inoculating loop toward the butt.

_____ 7. In order to study cultural characteristics, the same organism must be inoculated several different ways in different media.

_____ 8. Microbes produce most colors of the spectrum except blue-green.

_____ 9. *Streptococcus pyogenes* forms small balls of growth in thioglycollate broth.

_____ 10. A magnifying glass is useful in determining the margin type of a colony.

Part II

List four features of an isolated colony that are growth characteristics of the organism forming the colony. Use only the information you learned in this module.

1. _____

2. _____

3. _____

4. _____

PART THREE

Chemical and Physical Effects on Bacterial Growth

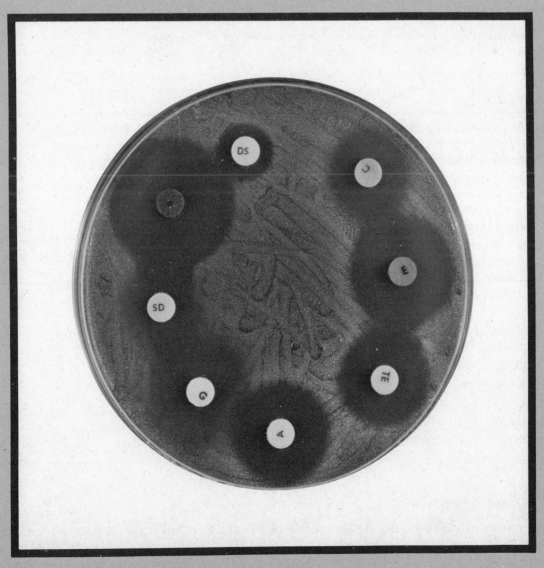

Antibiotic sensitivity plate.

MODULE 16

Cultivation of Anaerobes

PREREQUISITE SKILL

Successful completion of Module 8, "Aseptic Transfer of Microbes."

MATERIALS

sterile thioglycollate broth tubes, 10 ml in screw-cap tube (store at room temperature until inoculated) (3)*

sterile, cotton-plugged, Brewer anaerobic agar short slants, 10 ml/tube, slanted at a more acute angle for shorter agar surface on the slant (3)*†

broth cultures of
Escherichia coli
Micrococcus luteus
Clostridium sporogenes

pyrogallol crystals

10% sodium carbonate solution, dropper bottle

weighing spoon

rubber stoppers, #0

scissors

masking tape (¾ to 1 inch wide)

coffee can or wire basket

Brewer anaerobic jar (demonstration)
Brewer anaerobic agar plate inoculated with Clostridium sporogenes
Brewer anaerobic agar plate inoculated with Micrococcus luteus

candle jar (demonstration)
TSA plate inoculated with Streptococcus pyogenes and incubated in a candle jar at 37°C for 48 hours
TSA plate inoculated with Streptococcus pyogenes and incubated aerobically at 37°C for 48 hours

*To be prepared by the student if the instructor so indicates.
†To make short slants, increase the slope of slant board by placing a long piece of glass tubing across slant board under neck of tubes containing melted media. See Figures 16-6, 16-7, and 16-8 for "How to Make Cotton Plugs."

OVERALL OBJECTIVE

Demonstrate the oxygen requirements of three selected microorganisms and some techniques for obtaining anaerobic and microaerophilic conditions.

Specific Objectives

1. List the four major categories of bacteria based on their oxygen requirements.
2. List the three principal, significant ingredients of thioglycollate broth and the reason they are included in the formulation.
3. Demonstrate the variety of oxygen tension states that can be produced using thioglycollate broth.
4. Demonstrate your ability to produce anaerobic conditions for organisms on solid media by the use of the Wright's tube method.
5. Explain why Brewer's anaerobic agar is used in the Wright's tube preparation.
6. Summarize the theory and use of the self-contained Brewer anaerobic jar.
7. Describe the use of a candle jar.
8. Give the genus and species names of four pathological anaerobes and the diseases they cause.
9. Define the terms *obligate anaerobe, facultative anaerobe, microaerophile, aerobe, transport medium, exotoxin, crepitant tissue, tetany,* and *proteolytic.*

DISCUSSION

Microorganisms are ubiquitous, that is, they are everywhere in your environment. They are found at the depths of the seas, in hot sulfur springs, and on snow-capped mountain ranges. It is logical, therefore, to suppose that microorganisms have varying growth requirements since they are able to flourish in such diverse environments.

One physical condition that is extremely important for microbial growth is the oxygen requirement. Some microorganisms cannot live in the absence of oxygen; we call these organisms *strict aerobes.* A strict aerobe must have oxygen present in order to live and reproduce. At the opposite extreme are those organisms that cannot live if oxygen is present; these are the *strict* or *obligate anaerobes.* Oxygen is actually toxic to obligate anaerobes. Between these two extremes, we find the majority of organisms; these are the *facultative anaerobes.* They grow either in the presence or absence of oxygen, but they grow better if oxygen is available.

Still another bacterial category based on oxygen requirements is that of the *microaerophiles.* "Micro-aero-philic" literally means "small or little air loving." This is a well-chosen name since these organisms flourish in reduced oxygen conditions, although they still require some oxygen. Usually microaerophiles grow best with increased carbon dioxide (10%) and reduced oxygen. These conditions are commonly created in a candle jar. This is discussed completely in Activity 4.

There are many other methods of obtaining anaerobic conditions besides those considered in this module. For other methods you should consult a microbiology textbook or other references.

One of the most common methods of producing an anaerobic environment for microbes is the incorporation of a reducing agent into the medium. A reducing agent, such as sodium thioglycollate, removes oxygen from the bacterial environment by combining with the oxygen in a chemical reaction. You will use thioglycollate broth in Activity 1. Either methylene blue or resazurin, which are both oxygen indicating dyes, are incorporated into thioglycollate broth. Resazurin is less inhibitory to delicate organisms than methylene blue. These particular dyes are colorless in a reduced state (when free oxygen is not present). When the dye molecules become oxidized, their respective colors reappear. In the upper levels of your thioglycollate broth, where oxygen has diffused into the medium, you will be able to see a pink tinge if the broth contains resazurin, or blue if the indicator is methylene blue. A small amount of agar (0.5%) is also added to the medium. This tends to localize bacterial growth and encourages anaerobiosis.

Strict aerobes grow in the surface region of thioglycollate broth where more oxygen is present, while obligate anaerobes thrive throughout the depths of the broth. As you might expect, facultative organisms grow throughout the medium. Microaerophiles grow primarily in the regions between the oxygenated surface and the anerobic depths, as shown in Figure 16-1. Thioglycollate broth is probably the most commonly used liquid medium in clinical laboratories because it can provide proper oxygen requirements for such a variety of organisms.

When a specimen is collected for the diagnostic laboratory from a deep wound or abscess, gangrenous tissue, or other conditions where anaerobic organisms are suspect, it is essential to deposit the organisms immediately into a transport medium such as Stuart's or Amies'. Both are semisolid media. Stuart's transport medium is recommended as the transport medium in public health laboratory practice. The medium is highly reductive and nonnutritive. For this same reason, anaerobes are protected from the lethal effects of oxygen. Rapid delivery of specimens to the laboratory, however, is still extremely important, and it is recommended that the transported specimens be cultured without delay after their arrival in the laboratory. Stuart's medium is used primarily for swab-collected specimens.

A few of the most famous pathogenic, anaerobic bacteria are also spore-formers. A spore is a resistant structure formed by some bacteria. *Clostridium botulinum,* which causes fatal food poisoning (botulism), is most often found in inadequately processed, canned foods. The spores of this organism are resistant to heat and can survive the food canning process. The closed can creates ideal anaerobic conditions for botulism spores to grow, reproduce, and form their deadly toxin. A large amount of gas is also produced by *Clostridium botulinum* as it grows in the closed container, which causes cans to bulge or mason jars to explode. This is why you have often heard that you should never buy a can that bulges out. Such a can of food could indeed contain *Clostridium botulinum* and its fatal exotoxin (a toxin that is secreted out of the bacterial cell into the environment). Botulism is more accurately called an intoxication than an infection.

Another pathogenic, anaerobic bacterium belonging to the same genus is *Clostridium perfringens,* which causes gas gangrene. The spores of this organism enter a deep wound that contains much dead tissue. The lack of oxygen in the dead tissue again sets up anaerobic conditions. The spores germinate, the cells reproduce, and an exotoxin is formed that destroys more tissue. This bacterium also produces large amounts of gas, and for this reason the tissues that harbor the organism become crepitant; that is, the tissues contain so much gas that crackling sounds can be heard as the trapped gas bubbles break.

The other pathogenic member of this genus is *Clostridium tetani,* which causes tetanus (lockjaw). The spores of *Clostridium tetani* are often introduced via a puncture wound where anaerobic conditions already exist. As the tetanus organism grows and reproduces, it forms a neurotoxin that has a devastating effect on nerves and causes muscles to go into the state of tetany (rigid contraction of most of the voluntary muscles). A patient with advanced tetanus and muscle contraction has characteristic arching of the back, and his jaws are impossible to pry open, hence the common name of the disease, "lockjaw." Death is caused by respiratory failure as the diaphragm becomes involved.

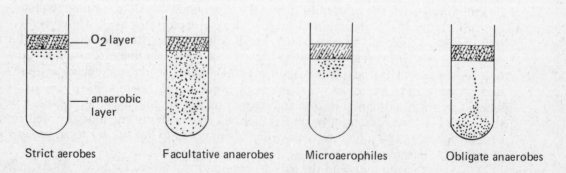

FIGURE 16-1
Slanted lines denote oxygen indicator.

Many other species of clostridia are implicated in pathological conditions, especially in gas gangrene. Gas gangrene is often a mixed infection of more than one species of *Clostridium*. For example, the anaerobe you will be working with in this module, *Clostridium sporogenes,* is proteolytic (breaks down protein) and promotes the dissemination of the more highly toxic *Clostridium perfringens* throughout the tissues.

Actinomyces, not a spore-former, is another anaerobic genus of medical importance. *Actinomyces israelii* is a normal inhabitant of the human mouth that can cause formation of tumorlike masses and damage to the jawbone. The anaerobic condition that allows this organism to proliferate usually follows tooth extraction or oral surgery.

ACTIVITIES

Activity 1: Various Oxygen Tension States in Thioglycollate Broth

Obtain three sterile tubes of thioglycollate broth and stock broth cultures of *Escherichia coli, Micrococcus luteus,* and *Clostridium sporogenes.* Examine your uninoculated thioglycollate broth tubes, and notice how far the color of the indicator extends down from the surface of the broth. If more than the upper ½ to ¾ inch is pink or blue (depending upon the indicator present), then you must place the tubes in a container of boiling water for about 5 minutes to drive off the absorbed oxygen. Remove the tubes from the boiling water bath, and carefully place them in your test tube rack without unnecessary agitation. Shaking causes more atmospheric oxygen to diffuse into the medium. Allow your tubes to cool before inoculating them. Always store thioglycollate broth at room temperature with the caps on tightly, and they will keep for a week or two without having to be reheated. Usually, if you tighten the screw caps after autoclaving, you will not have to heat them.

Label your thioglycollate tubes carefully and completely for the inoculations you will make. Now, using your best aseptic technique, inoculate a separate tube from each of the three broth cultures, that is, one tube with *Escherichia coli*, one tube with *Micrococcus luteus,* and one tube with *Clostridium sporogenes.* Be sure the screw caps are on tightly so that oxygen does not continue to enter the tube. Incubate *Escherichia coli* and *Clostridium sporogenes* at 37°C for 48 hours. Incubate *Micrococcus luteus* at 30°C since it has a lower optimum temperature.

In your next lab period, you will see that, although thioglycollate broth (thio) is intended primarily for the cultivation of anaerobes, aerobes can grow in the region near the surface where more oxygen is available. Facultative anaerobes can grow throughout the medium, and microaerophiles can grow in the middle regions, but obligate anaerobes are found throughout the depths of the tube where true anaerobic conditions prevail. Therefore, thio can be considered an all-purpose medium with respect to oxygen tension. Examine Figure 16-1 to see how this looks.

In your next lab period, examine the thio tubes that you inoculated, disturbing them as little as possible. Examine the growth pattern of each organism carefully, comparing them again to Figure 16-1, and make drawings of what you see. Label your drawings with the names of the organisms. Name the types of growth pattern with respect to their oxygen requirement, and submit these drawings to your file.

Activity 2: Wright's Tube Method

The Wright's tube method allows you to grow anaerobes on an agar slant. You must, of course, exclude the oxygen from the culture environment for the anaerobes to grow. In Wright's tube method, this is accomplished by activating pyrogallol with sodium carbonate, which then acts as a strong reducing agent and removes the free oxygen from a tightly stoppered tube. Figure 16-2 shows this type of preparation.

The Wright's tube method can be tricky, and a number of precautions should be observed to ensure your success.

1. Pyrogallol and sodium carbonate are caustic so you must work quickly

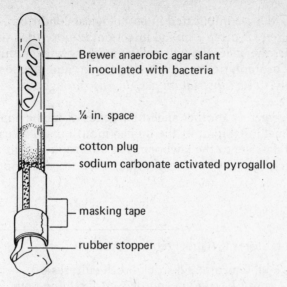

— Brewer anaerobic agar slant
   inoculated with bacteria

— ¼ in. space

— cotton plug
— sodium carbonate activated pyrogallol

— masking tape

— rubber stopper

FIGURE 16-2
Completed Wright's tube preparation. Place one
strip of masking tape over the top of the stopper
so that it attaches to both sides of the glass
tube. Wrap a second strip of masking tape
several times around the glass tube-rubber
stopper juncture (see Figure 16-3).

     to avoid seepage of these chemicals through the cotton onto the agar.
Invert the tube as soon as possible.

2. The cotton plug must be tight enough to keep the chemicals from running
onto the agar until you can invert the tube, but it must allow free passage
of gases from one area to the other so that the oxygen can be chemically
removed.

3. The outer plug must be a rubber stopper, and it *must be inserted and
twisted well into the tube* to form an airtight seal.

4. As soon as you have added the sodium carbonate, you must insert the
rubber stopper since this small amount of sodium carbonate and pyrogallol can only reduce a limited amount of oxygen. *You must work
rapidly.* You should practice these steps with an empty tube, if you
feel you need to, until you are able to begin inverting the tube as you
twist the stopper into place.

    The main cause of failure with Wright's tube method of anaerobiosis is the
inability to achieve an airtight seal. To help ensure proper sealing use masking tape,
as described in Figure 16-3, that is, one strip wrapped over the top of the stopper and
another strip wrapped several times around the tube-stopper juncture.

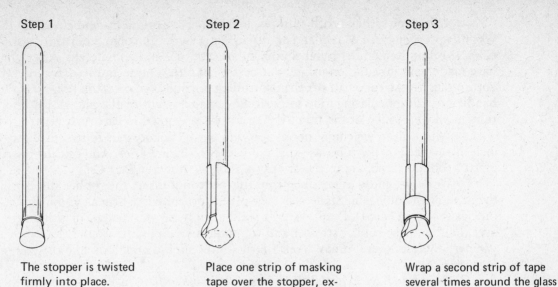

Step 1

The stopper is twisted
firmly into place.

Step 2

Place one strip of masking
tape over the stopper, extending onto the glass tube.

Step 3

Wrap a second strip of tape
several times around the glass
tube-rubber stopper juncture.

FIGURE 16-3
Achieving an airtight seal for Wright's tube
method of anaerobiosis.

Figure 16-4 shows you the steps that you must perform after you have inoculated your slant with an organism. Because oxygen is toxic to strict anaerobes, you
should work rapidly after you have made your inoculation. Study Figure 16-4 now,
and read these directions as you follow the steps labeled in the figure. The need to
work rapidly becomes more urgent after you have added sodium carbonate to the
pyrogallol.

Obtain three Brewer anaerobic agar slants, and inoculate a separate slant for
each of the organisms: *Escherichia coli, Micrococcus luteus* and *Clostridium sporogenes.*
Note that the color of the medium is pink (or purplish) because it contains resazurin
as an oxygen indicator. Brewer anaerobic agar is an excellent medium to use in your
Wright's tube because it allows you to actually see that anaerobic conditions have been
created.

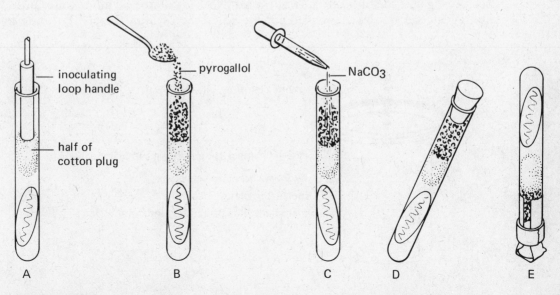

inoculating
loop handle

half of
cotton plug

pyrogallol

NaCO3

A   B   C   D   E

FIGURE 16-4
Preparation of Wright's tube.

Using your stock broth cultures, make your inoculations, and complete the Wright's procedure for *E. coli* and *M. luteus* before you inoculate your third slant with *C. sporogenes.* Remember to label your tubes carefully and completely. After you have made your inoculation using your best aseptic technique and have returned the cotton plug to the tube, cut off the protruding portion of the cotton plug. Use the handle of your inoculating loop to push the remaining half of the plug down into the tube to within ¼ inch of the tip of the agar slant, as shown in Part A of Figure 16-4.

With a clean weighing spoon, now add pyrogallol crystals to within about ½ inch of the neck of the tube, as shown in Part B of Figure 16-4. Add one dropperful of 10% sodium carbonate to the dry pyrogallol, as shown in Part C.

Work as rapidly as possible from this point on. Insert the rubber stopper into the neck of the tube, simultaneously twist it to form an airtight seal, and invert the tube as shown in Part D. Ensure a tight seal by applying one strip of masking tape over the top of the rubber stopper and wrapping a second strip around the tube-stopper juncture several times. Your finished tube should now look like the one in Figure 16-2.

Once you have inverted the tubes, they must remain inverted. So place them inverted in a coffee can or wire basket, and incubate them at the appropriate temperature for 48 hours.

In your next lab period, examine your Wright's tube preparations carefully. Notice the color change in the medium. Write a brief description of the amount and type of growth for each organism, and submit this to your file.

Precaution: Use extreme care when removing masking tape and rubber stopper. Sometimes the organisms produce so much gas that the stopper and contents are forcibly expelled.

Activity 3: Demonstration of Self-Contained Brewer Anaerobic Jar

The Brewer anaerobic jar is the most common method for creating anaerobic conditions in solid medium cultures in clinical laboratories. The atmospheric oxygen is removed from a large, sealed container by catalyzing the chemical combination of oxygen with hydrogen to form water. In Figure 16-5, examine the various parts of the self-contained Brewer jar (BBL Gas Pak).

To cultivate anaerobes in the Brewer jar, Brewer anaerobic agar plates are inoculated with an organism. The plates are inverted as usual and placed into the jar. Working rapidly, use a scissors to cut the corner off a foil-packaged hydrogen generator, and place this opened generator package into the jar between the petri dishes and the side of the jar. Next open a methylene blue indicator packet, and place it in a similar position against the opposite side of the jar. Now introduce 10 ml of water into the hydrogen generator envelope, and put the lid on the jar immediately. Put the clamp

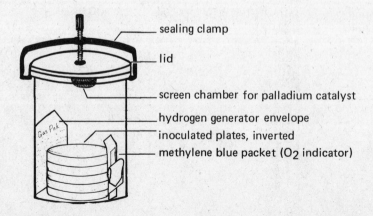

FIGURE 16-5
Self-contained Brewer anaerobic jar.

in place over the lid, and turn the thumb screw for a tight seal. As you noticed in Figure 16-5, a small, screened chamber that holds the palladium catalyst is attached to the inner surface of the lid.

Within 30 minutes, droplets of moisture will begin to condense on the inside of the jar, indicating that hydrogen has been generated and chemically combined with the atmospheric oxygen, aided by the catalyst, to form water.

$$2 H_2 + O_2 \xrightarrow{\text{catalyst}} 2 H_2O$$

The entire Brewer jar is placed in an incubator at the optimum temperature for the organism with which the plates are inoculated for 24 to 48 hours.

After incubation, examine the Brewer jar. Before opening the jar, you can see that the methylene blue indicator is quite colorless, which shows an absence of free oxygen inside the sealed jar. Open the jar and examine the plates. One has been inoculated with *M. luteus* and the other with *C. sporogenes*. Write a short description of the amount and type of growth for both organisms. Submit this to your file.

Activity 4: Demonstration of a Candle Jar

A candle jar is used primarily for the cultivation of microaerophiles some of which are pathogens, notably *Neisseria gonorrheae* and *Neisseria meningitidis,* which thrive best in conditions of increased carbon dioxide (10%) and decreased oxygen. Micro-aerophilic conditions for growth on solid media are usually created in a candle jar.

Any container that holds several petri dishes and can be tightly sealed can be used as a candle jar (a Brewer jar will do). Simply inoculate ordinary nutrient agar plates, invert them, and place them into the jar. Now affix a candle to the inside of a petri dish lid, and set it in the jar on top of your inoculated plates. Light the candle, and seal the jar tightly. The oxygen inside the jar will be used up by the burning candle and replaced by carbon dioxide from the combustion. When the oxygen has been replaced by carbon dioxide, the candle will go out, and a 10% carbon dioxide atmosphere (microaerophilic conditions) will then be created.

To demonstrate microaerophilism, two plates are inoculated with *Streptococcus pyogenes.* One is placed in the candle jar and incubated at 37°C for 48 hours. The other is incubated aerobically at 37°C for 48 hours. Compare the amount of growth on the two plates. Is the growth more luxuriant on one than on the other?

Make a drawing of the growth differences, along with a written description of the amount of growth that occurred compared with the amount of oxygen present. Also make a rough sketch of the candle jar setup. Label the sketch, and submit it to your file. When you are satisfied with your mastery of the information in this module, take the post test.

HOW TO MAKE COTTON PLUGS

Plugs are made by stuffing an appropriate amount of cotton into each tube or flask with the handle of an inoculating loop. A mechanical device is available for this job in many large laboratories, but the procedure is simple and easy to learn.

1. Tear a strip of cotton about 1 inch wide and 3 inches long for each tube (see Figure 16-6). To plug an Erlenmeyer flask you will need a piece of cotton about 2 inches wide and 4 inches long.
2. Center the cotton strip over the mouth of the tube or flask. Push the cotton into the tube with the handle of your loop (see Figure 16-7).
3. Half of the cotton should be inside the mouth of the tube and half of it protruding from the tube (see Figure 16-8). It is the protruding portion you will grasp with your little finger to remove the cotton plug, just as you removed the capall in Module 8. When removing or replacing the cotton plug, twist the test tube. This allows the plug to slip in and out of the tube more easily.

FIGURE 16-6

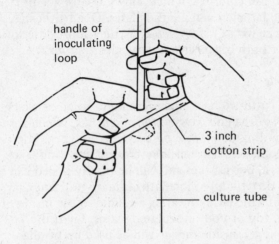

FIGURE 16-7

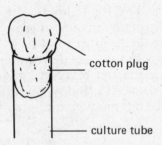

FIGURE 16-8

POST TEST

The post test is a self-evaluation. It is not used for a grade. It is designed only to let you decide if you have successfully completed this module.

Part I

Match the definition and the organism you used to demonstrate the four major conditions of oxygen tension. Fill in the blanks with the letters denoting the correct answers.

Definitions:

      A.   Cannot grow in the presence of oxygen.

      B.   Grows best in 10% CO_2.

      C.   Grows in the presence or absence of oxygen.

      D.   Will not grow in the absence of oxygen.

Organisms:

      a.   *Escherichia coli*

      b.   *Micrococcus luteus*

      c.   *Clostridium sporogenes*

      d.   *Streptococcus pyogenes*

		Definition	Organism
1.	Microaerophile	_____	_____
2.	Aerobe	_____	_____
3.	Anaerobe	_____	_____
4.	Facultative anaerobe	_____	_____

Part II

List the significant ingredients of thioglycollate broth and their functions.

	Ingredient	Function
1.	_____	_____
2.	_____	_____
3.	_____	_____

Part III

A.   List the components of the Brewer jar necessary to create and indicate anaerobic conditions.

      1. _____

      2. _____

      3. _____

B.   Write an equation showing how oxygen is removed from the environment in the Brewer jar.

_____

Part IV: True or False

_____   1.   Crepitant tissue means damaged, nonoxygenated tissue.

_____   2.   *Clostridium sporogenes* aids the invasion of gas gangrene organisms by breaking down tissue protein.

_____   3.   A candle jar removes all the oxygen from the environment within it.

_____   4.   The clostridia produce powerful endotoxins.

_____   5.   Tetany is violent contraction of the voluntary muscles.

_____   6.   The pathological conditions caused by the clostridia are more correctly called infections instead of intoxications.

Part V

Match the name of the anaerobic organism with the pathological condition it causes.

	Organism		Condition
_____ 1.	*Clostridium tetani*	a.	Fatal food poisoning
_____ 2.	*Actinomyces israelii*	b.	Gas gangrene
_____ 3.	*Clostridium botulinum*	c.	Lockjaw
_____ 4.	*Clostridium perfringens*	d.	Tumorlike mass in jaw

Part VI

1. Name two types of transport media.

   a. _____

   b. _____

2. Why do anaerobes survive in Stuart's transport medium?

   a. _____

MODULE 17

Effects of Temperature

PREREQUISITE SKILL

Successful completion of Module 8, "Aseptic Transfer of Microbes," and Module 9, "Aseptic Use of a Serological Pipet."

MATERIALS

nutrient broth culture of *Serratia marcescens* in tube (do not use a lyphollized culture)

nutrient broth cultures in flasks:
Bacillus subtilis (48 to 72 hour culture)
Escherichia coli

For Activity 1:
sterile nutrient agar slants (2)*

For Activity 2:
sterile nutrient agar plates (2)*
sterile empty tubes with capalls (10)*
sterile cotton-plugged 10 ml pipets(2)*
culture tube with about 10 ml tap water
thermometer, Celsius
wax glass-marking pencil
water bath setup:
hot plate stirrer (or burner and ring stand or tripod)
600 ml beaker
300 ml tap water

*To be prepared by the student if the instructor so indicates.

OVERALL OBJECTIVE

Demonstrate the principal effects of temperature on microbial growth.

Specific Objectives

1. Demonstrate pigment variation in a bacterium based on incubation temperatures and be able to explain it in terms of enzyme activity.

2. Record variations in the growth of two microorganisms following heating to various specified temperatures for a constant time.
3. Define the terms *thermal death point, thermal death time, lethal temperature, mesophilic, psychrophilic, thermophilic,* and *thermoduric.*
4. Contrast the effects of elevated temperatures on a spore-forming bacterium as opposed to a non-spore-forming organism.
5. Determine the thermal death point (if any) of both organisms used in Activity 2.
6. Discuss the importance of thermoduric bacteria to the food-processing industry and the dairy industry.

DISCUSSION

Many environmental and physical factors affect the activity of bacterial enzymes. These factors include temperature, osmotic pressure, hydrogen ion concentration (pH), radiation energy such as ultraviolet radiation and X rays, oxygen tension states, and chemical toxins.

Temperature is one of the most important physical factors affecting microorganisms. The simple bacterial cell lacks homeostatic mechanisms found in higher plants and animals to regulate heat generated by metabolism. Because these mechanisms are lacking, the bacterial enzyme systems are directly and readily affected by environmental factors, including temperatures.

Enzymatic reactions proceed at maximum speed and efficiency at an *optimum* temperature that varies with the bacterium. Beyond the maximum and minimum extremes of temperature for the microorganism, the enzymes become inactive. Enzymes are composed primarily of proteins. Low temperatures, merely inactivating the enzymes, are generally less damaging than high temperatures, which denature proteins, causing irreversible changes and total enzyme destruction, resulting in the death of the cell. This lethal effect is commonly considered in discussions of temperature effects on microorganisms since it is especially important in the food-processing industry.

For some microorganisms the optimum temperature for growth (and therefore for enzymatic reactions) is between 0° and 20°C; these organisms are called *psychrophilic. Mesophilic* microorganisms grow best between 20°and 40°C, and those that flourish between 40° and 80°C are called *thermophilic.* When we talk about destructive temperatures, we must bear in mind these natural variations. Temperatures in the range between 50° and 100°C are normally in the lethal range for bacterial cells and spores. Generally, most bacteria are adversely affected by temperatures above 50°C, but there may be considerable variation as to the length of time required to kill them. Some microorganisms are extremely resistant to destruction when exposed to temperatures that are in the lethal range for most bacteria. These *thermoduric* bacteria are resistant to destruction by lethal temperatures even with relatively long exposure times. For example, the common thermoduric bacterium *Microbacterium lacticum,* which is isolated from pasteurized milk and persists on dairy utensils, can survive temperatures of 72°C for 30 minutes in skim milk. Milk is normally pasteurized at 62.8°C (145°F) for 30 minutes. The newer flash pasteurization process heats milk to 71.7°C (161°F) for 16 seconds.

Time of exposure, therefore, is a vital factor in assessing the lethal effect of high temperatures on bacterial cells. We must have some measure when attempting to compare the susceptibility of various organisms to elevated temperatures. For this purpose, two methods are useful: the *thermal death point* (TDP), which is the temperature at which an organism is killed in 10 minutes of exposure, and the *thermal death time* (TDT), which is the time required to kill a suspension of cells or spores at a given temperature.

In your first activity, you will demonstrate the effect of incubation temperature on the pigment production of a bacterial culture. Pigment production is also controlled by enzyme activity that can be inactivated by temperature extremes.

ACTIVITIES

Activity 1: Effects of Incubation Temperature on Bacterial Pigment Production

Gather together a nutrient broth culture of *Serratia marcescens* and two sterile nutrient agar slants. Carefully and completely label your nutrient agar slant tubes for inoculation with *Serratia marcescens*, but mark one for incubation at 25°C and the other at 40°C. Using your best aseptic technique, inoculate the two sterile nutrient agar slants with organisms from the broth culture of *Serratia marcescens*. Incubate one tube at 25°C (room temperature) and the other at 40°C for 24 to 48 hours.

In your next lab period, examine your two tubes for the presence of pigment. Write and submit to your file a brief description of the growth and pigment in each tube, and explain the results in terms of enzyme activity.

Activity 2: Lethal Effects of Temperature on a Spore-forming Bacterium and a Non-spore-forming Bacterium

With a wax glass-marking pencil divide the bottoms of two sterile nutrient agar plates into six segments and label them as shown in Figure 17-1. Now, with tape, label one plate to be inoculated with *Bacillus subtilis* and the other with *Escherichia coli*. Remember to label the *bottom* of your plates.

Inoculate the sector marked C on your plates from your stock culture flasks. Make all inoculations as a *single straight stroke in the center of the sector*. Take care that your inoculum does not run into another sector, and stays at least ½ inch from the center of the plate where the sectors converge. An example of a single straight stroke inoculation is shown in Sector C of the *B. subtilis* plate in Figure 17-1.

Line up five sterile test tubes for each organism. Label five tubes *E. coli* and five tubes *B. subtilis*. Now add to the labels of the *E. coli* tubes the temperatures shown in Figure 17-2. Do the same for your *B. subtilis* tubes. After you have labeled the tubes, aseptically make the 2 ml transfers, using a sterile 10 ml pipet as shown in Figure 17-2.

Set up a water bath on a hot plate stirrer or ring stand at your work table, using a 600 ml beaker for your water container. Place your thermometer in a clean test tube into which you have put 10 ml of tap water, and place this in your water bath container, as shown in Figure 17-3.

When the temperature of your water bath has reached 40°C, place the tube of *E. coli* labeled 40°C and the tube of *B. subtilis* labeled 40°C in the water bath. These tubes are to be heated at 40°C for 10 minutes. To do this, adjust the controls or flame to maintain an even 40°C for 10 minutes, or remove the beaker from the heat source for a short time. Do what is necessary to keep the temperature constant. After 10 minutes at a constant 40°C, remove both tubes of bacterial culture from your water bath, and inoculate your 40°C sectors on each plate with a straight stroke before the broth cools. Discard your 40°C broth tubes.

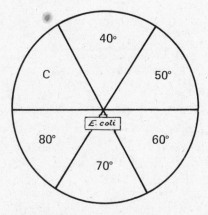 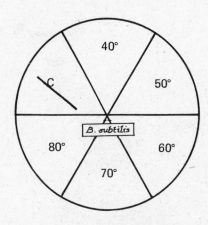

FIGURE 17-1
Sterile nutrient agar plates.

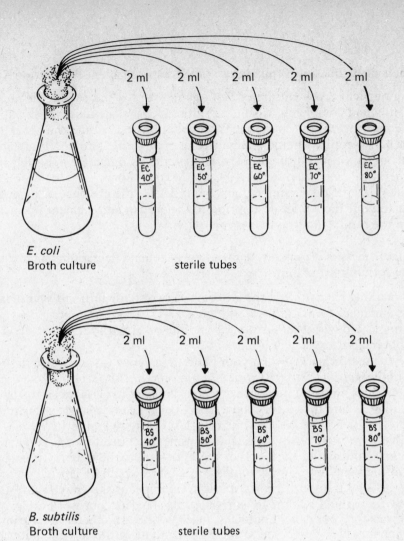

E. coli
Broth culture

sterile tubes

B. subtilis
Broth culture

sterile tubes

FIGURE 17-2
Pipetting and labeling scheme for Activity 2.

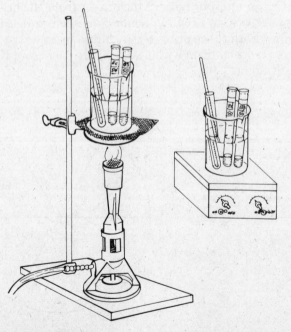

FIGURE 17-3
Two types of water bath setups

Raise the temperature of your water bath to 50°C. Put your second set of culture tubes (labeled 50°C) into the water bath, and begin your timing. Keep the tubes at 50°C for 10 minutes. Remove the tubes from the water bath, and inoculate the sector labeled 50°C on your plates with the separate broth cultures before they cool. Discard your 50°C tubes. Continue to raise the temperature, time the 10 minute exposure, and make your inoculations as before for the remaining temperatures (60°C, 70°C, 80°C). When you have inoculated all sectors of both plates, invert and incubate them at 30°C for 24 to 48 hours.

In your next laboratory session, examine your plates, gather your data, and arrange them in tabular form as shown in Table 17-1. Record the presence of bacterial growth as + and the absence of growth as - .

TABLE 17-1 Collection of Data to Determine TDP

Organism	40°C	50°C	60°C	70°C	80°C	Control growth
Escherichia coli (Gram - non-spore-forming rod)						
Bacillus subtilis (Gram + spore-forming rod)						

When you have tabulated your data, you should be able to determine TDP for each of these organisms under these conditions. Remember that these determinations hold true only for our laboratory stock cultures and do not necessarily hold true for other strains of *Bacillus subtilis* and *Escherichia coli.* Circle TDP on your table for each organism if you determined the TDP.

Submit a duplicate copy of your completed table to your file. Also submit a brief discussion of the differing effects of elevated temperature on spore-forming bacteria in contrast to non-spore-forming bacteria based on your laboratory observations only.

When you are satisfied that you have mastered the material in this module, take the post test.

Related Experience: TDT

Using the same two bacterial cultures, set up an experiment to determine the TDT. The length of time that microbes are exposed to heat contributes to the lethal effect, as shown in regular pasteurization and flash pasteurization. Devise your scheme for determining TDT, and discuss it with your instructor before you proceed.

POST TEST

The post test is a self-evaluation. It is not used for a grade. It is designed only to let you decide if you have successfully completed this module.

Part I

1. List five physical or environmental conditions that affect bacterial growth.

    a. _____

    b. _____

    c. _____

d. _____

e. _____

2. Define thermal death time. _____

_____

3. Define lethal temperature. _____

_____

4. Define thermal death point. _____

_____

Part II: True or False

_____ 1. Extremes of temperature above the maximum for an organism are usually less damaging than extremes below the minimum temperature.

_____ 2. Below minimum temperatures act by destroying enzyme proteins within the bacterial cell.

_____ 3. Organisms that grow best between 20° and 40°C are called psychrophilic.

_____ 4. Enzyme systems affecting pigment production in some microorganisms are directly controlled by temperature and other environmental factors.

_____ 5. Thermal death time for an organism varies with the temperature.

_____ 6. The lethal effect on bacteria is due to irreversible changes of enzyme protein.

_____ 7. Thermoduric bacteria such as *Microbacterium lacticum* are killed by being exposed to a temperature of 65°C for 30 minutes.

_____ 8. Thermoduric organisms are a problem to the food processing industry.

MODULE 18

Effects of
Ultraviolet Radiation

PREREQUISITE SKILL

Successful completion of Module 8, "Aseptic Transfer of Microbes."

MATERIALS

broth culture of
   Serratia marcescens
sterile cotton swab*
test tube racks (2)
ultraviolet lamp

½ sheets of notebook paper (2)
sterile nutrient agar plates (4)*

*To be prepared by the student if the instructor
so indicates.

OVERALL OBJECTIVE

Study the effect of ultraviolet radiation on bacterial growth.

Specific Objectives

1. Correlate by description the amount of bacterial growth with the amount of exposure to ultraviolet light.
2. Sketch patterns of growth at different lengths of exposure time to ultraviolet light.
3. Define the terms *bactericidal, ultraviolet radiation, confluent growth, bacteriostatic, hereditary molecule, taxon, mutant,* and *DNA*.
4. List other physical factors that influence the growth and survival of micro-organisms in nature.
5. Describe the medical importance of radiation.

DISCUSSION

Some of the physical factors affecting the growth of bacteria are oxygen tension, temperature, pH, osmotic pressure, and radiation. The helpful and harmful effects of oxygen tension and increased temperature on microbial growth have been discussed in other modules. In this module you will study the effects of another physical factor: ultraviolet radiation.

In nature, the growth and survival of microorganisms are profoundly influenced by the physical factors in their environment. Once again, different bacterial taxa (classification groups) demonstrate the individuality of bacteria. Members of one particular genus and species may exhibit requirements different from those of another genus and species; for example, the relationship of aerobic and anaerobic bacteria to oxygen and the relationship of mesophiles and thermophiles to heat. This specificity of physical requirements can be used as an aid in the identification of bacteria.

You can control physical conditions in a limited environment to kill, inhibit, or remove microorganisms. An example of this is the use of ultraviolet lamps as sterilizing devices in an attempt to reduce the number of bacteria in operating rooms or meat storage lockers. Unfortunately, sterilization by ultraviolet radiation is not complete for the following reasons:

1. The bacterial cell must be in the direct path of the radiation and cannot be shielded by water, glass, or objects. (Ultraviolet rays are converted to heat as they strike glass or water.)
2. The killing efficiency is related to both the distance between the organism and the ultraviolet light, as well as to the time of exposure.

The killing of bacteria (bactericidal effect) and the inhibition of bacterial growth (bacteriostatic effect) by radiation are not limited to ultraviolet light. Other radiations such as X rays, beta rays, and gamma rays have a similar effect. These radiations have wavelengths much shorter than white light and are readily absorbed by the hereditary molecule of the cell. Prolonged radiation causes irreversible damage to the hereditary molecule. Hence the cell cannot reproduce, and cell death ensues.

Microbial geneticists use very small dosages of ultraviolet light to make inheritable changes in the hereditary molecule, deoxyribonucleic acid (DNA). These slight changes in inheritable characteristics are often irreversible, but not lethal, to the cell. In this way mutant offspring are produced instead of cell death. A mutant is an organism that has different genetic characteristics from its parent.

Radiation is used medically to change or destroy malignant cells. Malignant cells reproduce much more rapidly than normal cells. Cobalt radiation, which is shorter than ultraviolet light, therefore penetrating deeper tissues, is commonly used to alter the DNA of malignant cells in an attempt to halt their uncontrolled reproduction; that is, the radiation alters the gene (piece of DNA) that controls the speed of cell reproduction. You will want to think about these various aspects of the effect of radiation on the growth and reproduction of bacteria and human cells as you perform Activity 1.

Precaution: Be very careful not to look at the rays coming from the ultraviolet lamp. Ultraviolet rays can damage your eyes by sunburning them with relatively short exposure times.

ACTIVITY

Activity 1: Ultraviolet Irradiation of Bacterial Cells

The ultraviolet lamp should be placed about 1 foot from the cells you wish to irradiate. To do this, stand two test tube racks on end, and place them at each end of the ultraviolet lamp to support it, see Figure 18-1.

Label four nutrient agar plates 1, 2, 3, and 4. Disperse the cells of the broth culture of *Serratia marcescens* by gentle, side-to-side shaking of the tube. Now, using a sterile swab, aseptically dip it into the broth culture, and streak for confluent growth on all four plates. To streak for confluent growth you should use a heavy inoculum that covers the entire surface of the plate. To do this brush the swab, laden with bacteria from the broth culture, over the *entire* surface of the agar plate so that the

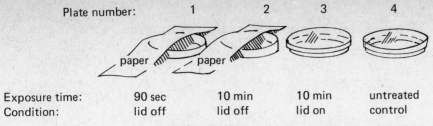

Plate number: 1 2 3 4

| Exposure time: | 90 sec | 10 min | 10 min | untreated |
| Condition: | lid off | lid off | lid on | control |

Paper covers one-half of inoculated plates 1 and 2.

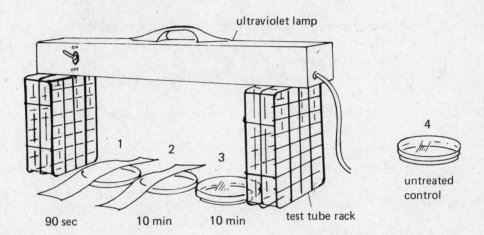

FIGURE 18-1
Ultraviolet irradiation of bacterial cells.

bacteria are evenly and thickly dispersed. Rotate the plate on your table top as you do this to facilitate an even distribution of the inoculum with the swab. This also ensures that you will cover the *entire* surface of the plate. Redip the swab in the broth culture before aseptically streaking each of the remaining three plates. Redipping the swab will give you a very heavy inoculum that will make it easier for you to obtain confluent growth. *Inoculate all four plates now.*

Next study Figure 18-1. Treat the numbered, inoculated plates with ultra-violet light by placing them on the table top under the ultraviolet lamp according to the times and conditions shown in Figure 18-1. Begin timing as you turn the ultra-violet lamp switch on. *Expose Plates 1, 2, and 3 now.*

Replace the covers on Plates 1 and 2, invert all four plates, and label them with all pertinent information as usual. Incubate them at 30°C for 48 hours. Submit sketches of the growth patterns on all four plates along with short, written descriptions to your file. Take the post test now.

POST TEST

The post test is a self-evaluation. It is not used for a grade. It is designed only to let you decide if you have successfully completed this module.

Part I

List five physical factors that influence the growth of microorganisms.

1. _____

2. _____

3. _____

4. _____

5. _____

Part II

List two reasons that sterilization by ultraviolet radiation is not always effective.

1. _____

2. _____

Part III: True or False

_____ 1. The killing of bacterial cells is called bacteriostasis.

_____ 2. Ultraviolet light kills cells by causing irreversible changes in the hereditary molecule.

_____ 3. Small doses of ultraviolet light are always lethal to the bacterial cell.

_____ 4. The ultraviolet lamp should be placed about 1 foot from the cells you wish to irradiate.

_____ 5. You will get eye damage if you look at the ultraviolet lamp with your glasses on.

_____ 6. Cobalt radiation is often used to make the DNA of malignant cells revert to normal.

_____ 7. Ultraviolet lights completely sterilize hospital operating rooms.

_____ 8. Geneticists use small doses of ultraviolet light to study mutants.

Part IV

Define the following terms.

1. DNA _____
   _____

2. mutant _____
   _____

3. taxon _____
   _____

4. confluent growth _____
   _____

Part V

Try to explain why ultraviolet radiation is not used to alter deep-seated tumors.

_____

_____

MODULE 19

Effects of Disinfectants and Antiseptics

PREREQUISITE SKILL

Successful completion of Module 8, "Aseptic Transfer of Microbes."

MATERIALS

---

4 disinfectants or antiseptics (bring from home or work)*

broth cultures of
    Escherichia coli
    Staphylococcus epidermidis
sterile cotton swabs (2)*
sterile nutrient agar plates (2)*
sterile filter paper disks (paper punch tailings)

For related experience:
sterile wide-mouth containers (6) (such as screw-cap specimen jars)
thermometers (4)
Zephiran (100 ml)

For related experience *(Continued):*
70% isopropyl alcohol (100 ml)
sterile water (300 to 400 ml in flask or screw-cap bottle)*
broth cultures of *Escherichia coli* and *Bacillus subtilis* (48 hour culture)
sterile nutrient agar plates (4)*
sterile swabs (4)*
forceps (optional)

*To be prepared by the student if the instructor so indicates.

OVERALL OBJECTIVE

Determine the effect of chemical agents on bacterial growth.

Specific Objectives

1. Describe the difference between an antiseptic and a disinfectant.
2. Describe the phenol coefficient of an unknown disinfectant.
3. Discuss why the filter paper disk method of evaluating disinfectants is not accurate.
4. Discuss the medical importance of disinfectants used as sterilizing agents.

5. Relate by tabulation the effect of four disinfectants or antiseptics on bacterial growth.

6. Tabulate the effect of two disinfectants on a spore-forming bacterium and on a non-spore-forming bacterium.*

7. Do a plate count so that you can describe the effect of two disinfectants on a spore-forming bacterium and on a non-spore-forming bacterium.*

DISCUSSION

In this module you will study how certain chemicals affect the proliferation of bacteria. These antimicrobial chemical agents have been given the names of antiseptics and disinfectants.

Are disinfectants and antiseptics sterilizing agents? You know that sterilization is bactericidal, that is, that sterilization kills all living organisms in or on an object or substance. However, disinfectants and antiseptics do not always sterilize if all organisms and spores are not killed. Disinfectants are described as antimicrobial agents used on inanimate objects such as instruments or structural surfaces. If a disinfectant frees an object of all infectious agents, sterilization has occurred. The term *antiseptic* is usually applied to antimicrobial agents that are used on living tissue such as skin and throat mucosa. The presence of organic material interferes with the antimicrobial action of antiseptics and disinfectants, while sterilization by autoclaving is not affected by organic materials.

Joseph Lister first introduced aseptic surgery in 1867 when he used carbolic acid as a germicide. Carbolic acid is phenol and is used as a standard for determining the effectiveness of other disinfectants by comparing the bactericidal effect of other disinfectants to it. Standard dilutions are made of phenol in order to determine which dilution will kill the test organism in 10 minutes but not in 5 minutes. In like manner, dilutions and killing effect are determined for the disinfectant. The dilution of the disinfectant is divided by the dilution of phenol, and the resulting number is called the phenol coefficient of the disinfectant. For example, if the dilution of phenol that kills at 10 minutes but not at 5 minutes is 1:90 dilution, and the dilution of disinfectant that kills at 10 minutes but not at 5 minutes is 1:450 dilution, then the phenol coefficient is arrived at thus:

$$\frac{450}{90} = 5 \text{ (phenol coefficient of disinfectant)}$$

The more a disinfectant has been diluted, the higher the numerator will be, which results in a higher phenol coefficient. Therefore, the higher the phenol coefficient is, the stronger the disinfectant is because it has been diluted more to kill at 10 minutes. The determination of the phenol coefficient must be done on standard test organisms. These test organisms are *Salmonella typhosa, Staphylococcus aureus,* and *Bacillus subtilis.* If any other bacteria are used to determine the phenol coefficient, the comparative measurement has no validity.

In the field of medicine, it is important to know the phenol coefficient of a disinfectant or an antiseptic. However, for practical reasons you will not be doing a phenol coefficient test in the laboratory.

Sterilization by disinfection at room temperature is often necessary, especially for items that would be damaged by heat. Disinfectants are often used for instruments, thermometers, and plastics, since the heat and moisture of autoclaving would dull cutting instruments, destroy thermometers, and change the shape of many plastic materials; 70% alcohol is widely used in hospitals as a disinfectant and an antiseptic, for example, for thermometers and venepuncture, respectively.

A few of the many types of disinfectants are:

*Specific Objectives 6 and 7 are to be fulfilled if you perform the related experience.

1. Phenolic compounds
   a. Concentrated cresol
   b. Diluted and saponified
      (1) Lysol
      (2) pHisohex
2. Alcohols (70%)
   a. Ethyl
   b. Isopropyl
3. Synthetic detergents = QAC (quaternary ammonium compound)
   a. Roccal
   b. Zephiran
4. Sterilizing gases
   a. Formaldehyde
   b. Ethylene oxide mixtures
      (1) Carboxide
      (2) Cryoxide

No single disinfectant is ideal. Each has its advantages and disadvantages. For example, phenols sterilize well but are quite corrosive. Detergents and 70% alcohol have some microbicidal effect but are not sporicidal.

ACTIVITY

Activity 1: Nonquantitative Demonstration of Antibiosis

Using a sterile swab dipped into the broth culture of *Escherichia coli,* inoculate the surface of one agar plate heavily to obtain confluent growth after incubation. Do the same with the other agar plate, using the broth culture of *Staphylococcus epidermidis* and a second sterile swab. Label the bottom of each plate with the name of the organism. Number the four antiseptic or disinfectant bottles that you will be testing 1, 2, 3, and 4. On the *bottom* of the agar plates, mark four sectors as shown in Figure 19-1, using a wax marking pencil. Label the sectors 1, 2, 3, and 4. These numbered sectors correspond to the numbers that you put on the antiseptic or disinfectant containers.

Sterilize the tip of your forceps by passing it through the flame of your Bunsen burner two or three times. Aseptically pick up a sterile filter paper disk with your sterile forceps. Dip the disk into the disinfectant or antiseptic numbered 1. Be sure that the excess disinfectant has drained off; then place the disk in Sector 1 of the *Staphylococcus epidermidis* inoculated plate. Using the same disinfectant, place another disk in Sector 1 of the *Escherichia coli* inoculated plate. You are comparing the

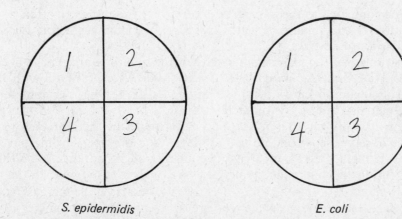

S. epidermidis   E. coli

FIGURE 19-1
Sector markings on bottoms of plates.

effectiveness of each disinfectant or antiseptic on both organisms, so repeat this procedure, placing the other disinfectants in the other sectors, that is, Disinfectant 2 goes into Sector 2 of both plates, and Disinfectant 3 goes into Sector 3, and so forth. Gently press down the disks with the tip of your flamed forceps to ensure contact with the agar. When all four disinfectant-soaked disks have been placed in all four sectors of both plates, invert the plates, label them properly, and incubate them at 30°C for 48 hours.

In your next lab session, observe and compare the zone of no growth (inhibition), if any, around the disk for each disinfectant or antiseptic for both organisms. On a separate sheet of paper, prepare and submit your evaluation of the antiseptics and disinfectants in tabular form using the following key:

3+ = most effective
2+ = moderately effective
1+ = slightly effective
0 = not effective

Set up your table as shown in Table 19-1.

TABLE 19-1 Evaluation of the Effectiveness of Various Antiseptics and Disinfectants

Antiseptic number	Name of antiseptic or disinfectant	Inhibitory effect	
		S. epidermidis	E. coli
1	for example, Lysol	3+	0
2			
3			
4			

Submit a duplicate of this completed table to your file. Include a short conclusion of the effectiveness of the disinfectants or antiseptics that you used.

Do not be disappointed if your favorite antiseptics or disinfectants do not appear to be as bactericidal as advertised. The amount of disinfectant (or antiseptic) that the filter paper disk will absorb and the amount of diffusion into the medium surrounding the disk are dependent upon the molecular weight of the disinfectant. The rate of diffusion is directly proportional to molecular size. So do not discard your favorite mouthwash on the basis of this activity.

Related Experience

This activity can be compared to the immediate treatment a thermometer often receives as it is removed from a patient's mouth. This was especially true before the advent of disposable thermometers.

Now gather together the materials. Label the wide-mouth containers as 1, 2, 3, and 4. Fill (¾ full) Jar 1 with Zephiran, fill Jar 2 with sterile water, fill Jar 3 with 70% isopropyl alcohol, and fill Jar 4 with sterile water. Label the four thermometers 1 to 4 also. Dip the mercury end of Thermometer 1 into the broth culture of *Escherichia coli,* and then allow it to dry for 10 minutes. You may handle the thermometer with your fingers if you observe asepsis and do not admit the skin organisms from your fingers into the experiment. (Use your forceps if you feel you are not able to do this.) Place only the intentionally contaminated mercury end of the thermometer in the Zephiran jar for 10 minutes of disinfection. After 10 minutes, rinse the mercury end of the thermometer in Jar 2 (sterile water). You are finished with Thermometer 1. Put it in an appropriate container for proper cleaning by the laboratory assistant. Now take a sterile swab, and inoculate the entire surface (streak for confluent growth) of a nutrient agar plate with the rinse water in Jar 2. Stir the contents of Jar 2 well before

Thermometers 1 and 2 have been dipped in *Escherichia coli* broth.

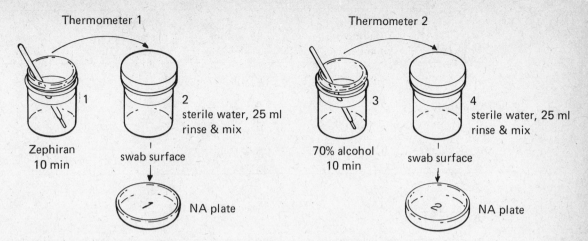

Replace jars 2 and 4 with clean, sterile jars & water, & repeat the procedure. Thermometers 3 and 4 have been dipped in *Bacillus subtilis* broth.

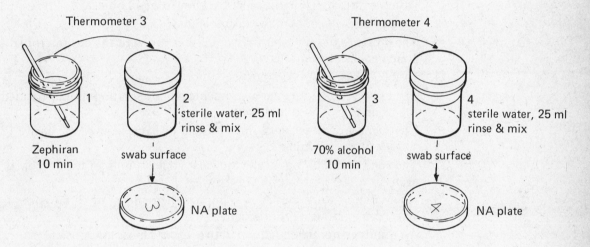

FIGURE 19-2
Effect of disinfectants on a spore-forming
organism and on a non-spore-forming organism.

streaking the plate. You can use the swab as a stirrer; then make the inoculation. Discard the bacteria-laden swab in an appropriate container for sterilization by autoclaving. Label the plate appropriately. After the surface appears dry, invert and incubate at 30°C for 48 hours. Repeat this procedure using Thermometer 2 and Jars 3 and 4. Here you will be evaluating the effect of 70% alcohol on *Escherichia coli.*

Before testing the effect of the same disinfectants on *Bacillus subtilis,* replace Jars 2 and 4 with two other jars of clean, sterile water. Label these replacement jars of water 2 and 4. Now repeat the same procedure using Thermometer 3 and Jars 1 and 2 to evaluate the effect of Zephiran on the spore-forming *Bacillus subtilis.* Finally, repeat the procedure using thermometer 4 and Jars 3 and 4 to evaluate the effect of 70% alcohol on *Bacillus subtilis.* Study Figure 19-2 for a schematic explanation of this activity.

After incubating all four plates at 30°C for 48 hours, collect your data in tabular form. Estimate the number of colonies per plate if possible. Design your table as shown in Table 19-2, using a separate sheet of paper. Submit a duplicate of the completed table to your file.

TABLE 19-2 Collection of Data

Disinfectant	Thermometer number	Bacterium used	Effective sterilization yes	Effective sterilization no	Colony count
Zephiran	1	E. coli			
70% alcohol	2	E. coli			
Zephiran	3	B. subtilis			
70% alcohol	4	B. subtilis			

POST TEST

The post test is a self-evaluation. It is not used for a grade. It is designed only to let you decide if you have successfully completed this module.

Part I: True or False

_____ 1. Disinfectants always remove all bacteria from an object.

_____ 2. Antiseptic is a synonym for disinfectant.

_____ 3. The presence of blood serum would interfere with the antimicrobial action of disinfectants.

_____ 4. Disinfectants are used as gargles.

_____ 5. The use of carbolic acid as a germicide in an operating room provided the beginning of aseptic surgery.

_____ 6. Confluent growth means a thick growth of bacteria over the entire surface of the medium.

_____ 7. The zone of inhibition means the clear zone around an antimicrobial agent in which no bacteria are growing.

_____ 8. Room temperature sterilization implies the removal of all microbes from an object by using a disinfectant.

_____ 9. All disinfectants are equally effective against most microbes.

_____ 10. Cutting instruments, thermometers, and plastic materials are best sterilized by autoclaving.

_____ 11. The lower the phenol coefficient is, the stronger the disinfectant is.

_____ 12. Louis Pasteur first introduced aseptic surgery by using a disinfectant.

Part II

List the test organisms used to determine the phenol coefficient of an unknown disinfectant.

1. _____

2. _____

3. _____

MODULE 20

Effects of Antibiotics

PREREQUISITE SKILL

Successful completion of Module 8, "Aseptic Transfer of Microbes."

MATERIALS

broth cultures of
  Escherichia coli
  Staphylococcus aureus
sterile swabs (2)*
sterile nutrient agar plates (2)*

antibiotic disk dispenser with sensitivity
  disks
millimeter ruler

*To be prepared by the student if the instructor so indicates.

OVERALL OBJECTIVE

Determine the susceptibility of bacteria to selected antibiotics.

Specific Objectives

1. Tabulate the inhibitory effect of eight different antibiotics on *Staphylococcus aureus* and *Escherichia coli*.
2. List the genera of microbes that produce the largest number of antibiotics useful to humans.
3. Describe the difference between an antibiotic and a chemotherapeutic agent.
4. Describe the differences between antibiotics and antiseptics or disinfectants.
5. Define the terms *biosynthesis, by-product of metabolism, drug-resistant mutant, in vivo, in vitro,* and *Kirby-Bauer sensitivity test.*
6. Describe the medical importance of antibiotic sensitivity testing.

DISCUSSION

In this module you will be studying the antimicrobial effects of chemicals in the form of antibiotics. Antibiotics differ from disinfectants and antiseptics in that they are chemicals that are biosynthesized. Biosynthesis refers to the production of substances by living organisms. These antibiotic substances are of no use to the microbes that produce them so we call them by-products of microbial metabolism. The genera of microbes that produce the largest number of antibiotics useful to humans are *Bacillus, Penicillium,* and *Streptomyces.* Antibiotics also differ from disinfectants and antiseptics in that they are administered internally and travel via the bloodstream to all parts of the body. This means that the antibiotic must have a low toxicity to body cells, while being toxic or destructive to the parasitic invader.

Other antimicrobial chemicals used therapeutically are artificially synthesized in the laboratory. Because of their artificial synthesis they are often called chemotherapeutic agents. More commonly, however, you hear them referred to as broad spectrum antibiotics.

Often a physician not only wishes to know the organism causing an infection, but the antibiotic that he must use to control the infection. The antibiotics most effective in inhibiting the growth of the causative bacterium *in vitro* are determined by a simple laboratory test. *In vitro* is the Latin term for "in glass," as opposed to *in vivo* meaning "in life" or in a living animal.

The Kirby-Bauer method for determining bacterial sensitivity to antibiotics is employed in diagnostic laboratories because it is a standardized test and therefore gives more accurate results. It involves growing pure colonies of the etiological agent in a special broth, diluting the broth culture to match a comparative standard, and inoculating a specific medium before the sensitivity disks are added. A certain size zone of inhibition for each different antibiotic has been empirically determined to be effective therapeutically for each bacterial type, and an organism is not considered to be sensitive to the antibiotic unless its zone of inhibition is as large as or larger than the predetermined zone for that antibiotic.

In summary, the Kirby-Bauer standardized sensitivity test as performed in the clinical laboratory *(in vitro)* allows a physician to determine the most effective antibiotics to use in treating his patient *(in vivo)* for a particular bacterial infection. You, however, will do a simplified antibiotic sensitivity test that is not standardized but which will demonstrate the principle of bacterial antibiosis nicely. It will also allow you to see that certain types of bacteria, depending on their Gram reaction, are more sensitive to most antibiotics.

The mutant resistant strains of bacteria often arise because of indiscriminate use of antibiotics or from undertreatment with the proper antibiotic. Therefore, it is important for the patient to take *all* the correctly prescribed dosage. In the case of antibiotic therapy, undertreatment can be more harmful than overtreatment. In low dosage, it is possible that the antibiotic can act as a selective agent by killing the normally susceptible bacteria, thereby selecting out the drug resistant mutant cells which grow into a difficult to treat population.

Antibiotics kill bacteria in several ways. A few examples are the following. Penicillin inhibits the synthesis of bacterial cell walls. Streptomycin competes wtih PABA (p-aminobenzoic acid) as a substrate for an enzyme reaction. Streptomycin actually enters the reaction in place of PABA, thereby blocking the synthesis of an essential cellular component. Some broad spectrum antibiotics interfere with enzyme synthesis. All the above result in cell death. The death of bacteria in the antibiotic sensitivity test you are about to perform is manifested by a zone of inhibition around the antibiotic impregnated disk. Since Activity 1 is not a standardized test, the largest zones of inhibition are not indicative of the drugs of choice, but they do allow you to see bacterial antibiosis.

ACTIVITY

Activity 1: Simplified Antibiotic Sensitivity Testing

Dip separate sterile swabs in the well-dispersed broth cultures, and inoculate two agar plates for confluent growth using *Staphylococcus aureus* on one plate and *Escherichia coli* on the other. Label the plates carefully and completely. Next, using a measure of aseptic precaution, admit the eight different antibiotics to the surface of each plate with the antibiotic disk dispenser. Gently press down the disks with the corner of your flamed forceps to ensure contact with agar. Invert the plates and incubate them at 30°C or 37°C for 48 hours.

On a separate sheet of paper, make a table like Table 20-1 to show the collection of your data in your next lab session. Measure the diameter of the zone of inhibition, and record it in millimeters. After you have completed your table, submit a copy of it to your file. Include a short, written description comparing the sensitivity to antibiotics of the two organisms that you used.

TABLE 20-1   Measurement of Inhibition Zones

| Name of antibiotic | Zones of inhibition in millimeters | |
	S. aureus	*E. coli*
1. _____		
2. _____		
3. _____		
4. _____		
5. _____		
6. _____		
7. _____		
8. _____		

Related Experience

Do research reading and perform the Kirby-Bauer test on the organisms used in this module.

POST TEST

The post test is a self-evaluation. It is not used for a grade. It is designed only to let you decide if you have successfully completed this module.

Part I

A. List the three genera that produce the largest number of antibiotics.

1. _____

2. _____

3. _____

B. List the two major differences between an antibiotic and a disinfectant.

1. _____

2. _____

C. How does an antibiotic differ from a chemotherapeutic agent?

_____

_____

D. List three ways that antibiotics are bactericidal.

1. _____

2. _____

3. _____

E. Name a standardized sensitivity test.

_____

Part II: True or False

_____ 1. By-products of metabolism produced by microbes are useless to all cells.

_____ 2. If an antibiotic is effective in sensitivity test against the causative organism of an infection, the same antibiotic will always stop the infection in the patient if it is caused by the same organism.

_____ 3. Chemotherapeutic agents are synthesized *in vitro*.

_____ 4. There are 100 mm in a centimeter.

_____ 5. A good antibiotic is toxic to all cells.

_____ 6. Zones of inhibition are measured and recorded in centimeters.

_____ 7. Gram positive bacteria are normally more sensitive to antibiotics than Gram negative bacteria.

_____ 8. Drug resistant mutants often arise because the patient does not take all the proper antibiotic prescribed for him.

_____ 9. To determine the size of the zone of inhibition, it is customary to measure its radius.

PART FOUR

Bacterial Stains

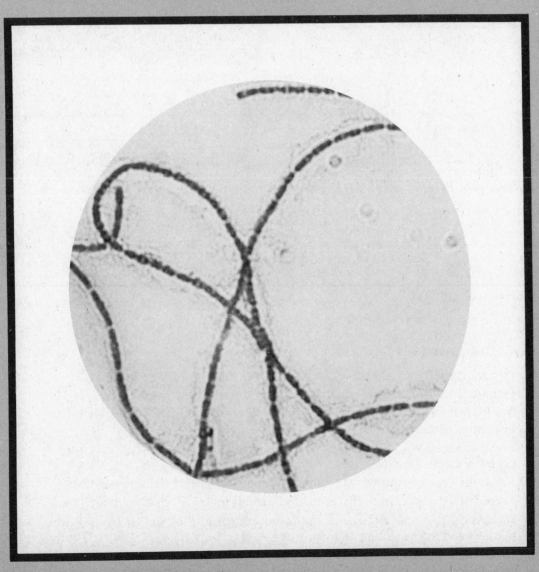

Simple stain of chained bacilli
(1000 x).

MODULE 21

Preparing a Bacterial Smear

PREREQUISITE SKILL

Mastery of Module 5, "Cleaning Microscope Slides," and Module 8, "Aseptic Transfer of Microbes."

MATERIALS

microscope slides (6)
broth culture of
   Bacillus subtilis
slant culture of
   Escherichia coli

inoculating loop (diameter of loop about
   4 mm)
Bunsen burner
burner striker
glass-marking pencil

OVERALL OBJECTIVE

Prepare bacterial cells to be stained for microscopic observation.

Specific Objectives

1. Prepare three bacterial smears from a broth culture of *Bacillus subtilis*.
2. Prepare three bacterial smears from a slant culture of *Escherichia coli*.
3. Name two microbial preparations commonly used to observe bacteria with a light microscope.
4. List three reasons why dirty microscope slides are unsatisfactory to use for bacterial smears.
5. Name the type of culture that is likely to result in smears that are too thin.
6. Name the type of culture that is likely to result in smears that are too thick.
7. Discuss the principal differences in procedures for making smears from broth cultures and from solid-medium cultures.
8. Name the type of smear preparation that requires a longer period of air drying.

9. Describe the heat-fixing process of a smear, why heat fixing is necessary, and how to determine if the smear is heat fixed.
10. Describe a "good" smear.

DISCUSSION

This module deals only with the preliminary procedure (preparation of a smear) that must be done before bacteria can be stained. Your next module tells you how to stain microbial smears.

Since bacterial cells are so small, special problems arise in the preparation of a good smear. Three principal precautions that you must take to avoid problems are:

1. Do not use scratched microscope slides since scratches in the glass slide can be confused with microorganisms.
2. Be sure that you are working with a very clean slide.
3. Avoid making smears that are too thick or too thin.

A dirty slide may be greasy or may be laden with dirt and dust. Such a slide will result in an unsatisfactory smear for three reasons:

1. The smear containing the desired microbes will wash off the slide during the staining process.
2. When you deposit the bacterial suspension on the microscope slide, it will coalesce, that is, it will not remain spread out.
3. Dirt, dust, and other debris can be mistaken for microbes.

Another difficulty in making a good smear is to get the right *amount* of bacteria on the slide. A good smear should be neither too thick nor too thin. If a smear is too thick, this means that too many cells have been put on the slide. A thick smear does not allow the proper penetration of light through the smear, which lessens the microscopic visibility of the microbes. Also, in a thick smear the bacterial cells are often packed too closely together and are piled on top of each other to the extent that you cannot determine the shape of the individual cells. You are most likely to make thick smears when your cells are obtained with a loop from solid culture media, such as slant cultures and agar plate cultures.

The opposite extreme of this is a smear that is too thin. As you know, bacterial cells are extremely small so if only a few cells are deposited on the slide, they are very difficult to find. You may search your slide for a long time with your microscope and not find a bacterium. Thin smears usually result when the smears are made from broth cultures.

The ideal smear, then, is one that is neither too thick nor too thin. That is, there are enough cells to be seen immediately, but not so many that cells cannot be visualized individually. The following activities are designed to *let you decide* how many organisms must be deposited on a slide in order to make a good smear.

ACTIVITIES

Activity 1: Smears from Broth Cultures

Labeling is as important in making smears for stained slides as it is for cultures. Therefore, number three clean slides 1, 2, and 3 with a glass-marking pencil. This labeling should be at one end of the slide where it is less likely to come in contact with stain. Later you may wish to use a different code for identifying your smears. Number your slides now.

Disperse the bacteria that have settled to the bottom of the broth culture tube by tapping the culture tube against your left hand or by gently shaking the tube back and forth. Usually you will see the settled out bacteria rise from the bottom of the tube in a swirl of cells. Take care not to do this so vigorously that the broth culture splashes onto the capall or other closure.

Be sure to use aseptic technique as you handle the broth culture tubes throughout the transferring of broth culture to the slides. Do not contaminate the broth culture, yourself, or the table top. Before you begin making the following smears, it is important that you adjust the diameter of your loop to approximately 4 mm. Keep the size of all three smears the same as in Slide 1.

Slide 1: Transfer two loops of broth from the culture tube to the slide, and spread the broth out on the slide to about the size of a quarter.

Slide 2: Transfer four loops of broth culture, and spread as above.

Slide 3: Repeat again using six loops of broth culture.

Allow the slides to air dry in a flat position on your table top for at least one-half hour after they appear dry. After this prolonged air drying, heat fix the smears. Heat fixing kills the bacteria and makes them adhere to the slide. To heat fix the smear, simply pass the slide back and forth three times through the flame of your Bunsen burner until the slide feels hot when touched quickly to the back of your hand. You should be holding the slide with your slide forceps or slide holder as you do this. Heat fix one slide at a time now.

You have now learned how to make bacterial smears from broth cultures. You used varying amounts of broth culture to determine how many loopfuls of broth culture are necessary to make a "good" smear. You will make this judgment when you stain the bacteria in your next module. So be sure to save your heat-fixed broth smears for staining. Also keep a written record of the amount of broth deposited on each of the three slides. You will need to refer to this data after completion of Module 22, "Simple Stain."

At the end of this module, you will find a check list on smear preparation from a broth culture. This check list may be a handy reference for future smear preparation.

Activity 2: Smears from Slant Cultures

Smears made from growth on solid media differ from those made from a broth culture. The bacteria in a broth culture are suspended in liquid so the smears can be made by spreading the broth culture onto the microscope slide. However, the bacteria from a solid agar must be put into a liquid before they can be spread out on the slide. This is accomplished in the following manner:

Continue numbering your three remaining clean slides. Number them 4, 5, and 6.

Slide 4: Add a small amount of tap water (two or three inoculating loopfuls or one small drop) to the center of a clean slide. A large drop is undesirable because it takes much longer for the slide to air dry. Now, using your inoculating loop, aseptically remove about half a loopful of bacteria from the slant culture. Mix the large amount of organisms on the loop into the drop of water on the slide. Next spread the mixture out on the slide to about the size of a quarter. Use your loop for the mixing and spreading. Be sure to flame your loop to prevent contamination of your pure culture slant, yourself, and your work table. A smear from a solid medium does not need to air dry as long as a smear made from broth. Allow the smear to air dry until it looks like a white film on the slide, and then heat fix it as you did in Activity 1.

Slide 5: Repeat the same procedure as followed for Slide 4, using an amount of bacterial growth about the size of a pin point. Mix these organisms into the small amount of water on the slide, and then air dry and heat fix them.

Slide 6: Repeat the same procedure as followed for Slide 4, using an invisible amount of growth. Do this by barely touching your loop to the growth. Some organisms will be adhering to your loop even though you cannot see them. Mix these organisms into the small amount of water, and then air dry and heat fix them.

Here again you are using varying amounts of cells to make the three smears from slant cultures to determine how much growth is necessary to make a "good" bacterial smear. Keep a written record of the amount of growth added to each of the slides. Save these smears as you did in Activity 1. They, too, will be used for your next module on the staining of bacteria. It is only after you stain these slides that you will be able to judge which of the smears is a "good" smear.

Now that you have completed Activity 2, you have mastered the skill of making smears from solid media also. At the end of this module, you will find a check list on smear preparation from a slant culture. It, too, may be a handy reference for future smear preparations.

You may want to review the following check lists before taking the post test.

Check List: Smear Preparation from a Broth Culture

1. Label slides.
2. Disperse settled-out bacteria throughout the broth by gently shaking the tube back and forth.
3. Aseptically transfer loops of broth to the center of a clean slide. Be sure to flame and cool your loop each time you introduce it into the culture tube.
4. Spread the broth culture out with your loop to about the size of a quarter.
5. Reflame the loop to kill bacteria still clinging to it.
6. Allow the smear to air dry for at least one-half hour after it looks dry.
7. Heat fix.

Check List: Smear Preparation from a Slant Culture

1. Label slides.
2. Place a small drop of tap water on the center of a clean slide.
3. Aseptically transfer a very small amount of slant growth to the drop of water with your inoculating loop.
4. Mix the organisms on your inoculating loop into the drop of water.
5. Spread the water-bacteria mixture out to the size of a quarter.
6. Allow the smear to air dry.
7. Heat fix.

POST TEST

The post test is a self-evaluation. It is not used for a grade. It is designed only to let you decide if you have successfully completed this module.

Part I

A. List the three reasons why a dirty, greasy slide will result in an unsatisfactory smear.

1. _____

2. _____

3. _____

B. List the following steps in the correct order of procedure for making a smear from a broth culture. Be careful; some steps *not* in this particular procedure have been included.

    a.   Label slides.

    b.   Allow smear to air dry.

    c.   With your inoculating loop, aseptically transfer a just-visible amount of growth to the drop of water on the slide.

    d.   Disperse settled-out bacteria throughout the broth.

e.   Heat fix.

f.   Mix the organisms into the drop of water with your inoculating loop.

g.   Aseptically transfer four to six loops of broth to the center of a clean slide.

h.   Reflame the loop to kill bacteria still clinging to it.

i.   Spread the water-bacteria mixture out on the slide to about the size of a quarter.

j.   Place a small drop of water on the center of a clean slide.

k.   Spread the broth culture out on the slide to about the size of a quarter.

1. _____   5. _____

2. _____   6. _____

3. _____   7. _____

4. _____

C.   From the list of steps above, put *in order* the four steps that apply only to the procedure for preparation of a smear from a slant culture.

1. _____   3. _____

2. _____   4. _____

Part II: True or False

_____ 1.   A smear made from broth must be air dried for at least one-half hour after it appears dry.

_____ 2.   All smears are heat fixed unless contra-indicated by the procedure.

_____ 3.   A heat fixed slide feels just slightly warm when you rest it on the back of your hand.

_____ 4.   Since the microscope slide is not sterile, asepsis need not be observed while making a smear.

_____ 5.   The amount of inoculum is most important in making a "good" smear.

MODULE 22

Simple Stain

PREREQUISITE SKILL

Successful completion of Module 21, "Preparing a Bacterial Smear," and Module 4, "Compound Microscope for the Study of Microbes."

MATERIALS

staining equipment:
  stainless steel pan or other receptacle
  staining rack
  wash bottle
  bibulous paper pad

6 smears from Module 21
methylene blue stain*
*To be prepared by the student if the instructor so indicates. The formula appears at the end of this module.

OVERALL OBJECTIVE

Perform a simple stain and describe the chemical reaction involved in the staining process.

Specific Objectives

1. Explain how a basic stain, such as methylene blue, colors the surface of a bacterium.
2. Describe the steps of the procedure involved in performing a simple stain.
3. Name an acid dye and a basic dye for staining bacteria.
4. Name that part of the cell stained by an acidic dye and by a basic dye.
5. Determine the correct amount of cells needed to make a "good" smear, by microscopic observation of simple stains.

DISCUSSION

The chemical compounds used to stain bacteria are called dyes. We stain bacteria to make them more readily visible, since unstained cells are practically transparent.

Therefore, for microscopic observations, stained bacteria are most often used.

Dyes can be acidic or basic. Acidic dyes such as acid fuchsin and eosin have a strong affinity for basic portions of the cell, that is, for the cytoplasmic components of the cell that are more alkaline in nature. Basic dyes such as crystal violet, methylene blue, and safranin have a strong affinity for the acid portions of the cell. The surface of a bacterial cell has an overall acidic characteristic because of a larger number of carboxyl groups located at the cell surface. The carboxyl groups (COOH) are the acid radicals of amino acids, and thousands of amino acids are combined in the cell wall. Therefore, when ionization of the carboxyl groups takes place, the surface of the cell has negative charges. For example,

$$COOH \xrightarrow{\text{ionization}} COO^- + H^+$$

In nature, however, the hydrogen ion is replaced by another positive ion such as Na^+ or K^+ and the H^+ bonds with oxygen to form water. The surface of a bacterial cell could be represented as shown in Figure 22-1.

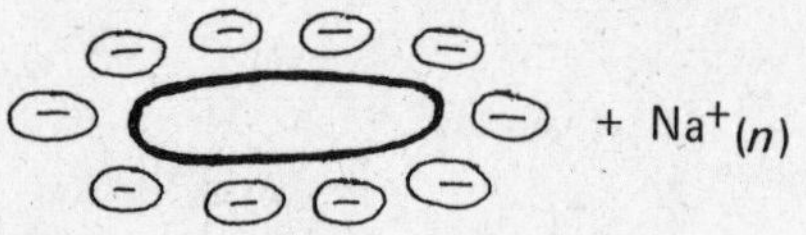

FIGURE 22-1
The predominance of negative charges on cell surface.

Basic dyes are commercially prepared as salts. For example, when you purchase methlene blue, it is actually methylene blue chloride. When rehydrated, the methylene blue chloride ionizes to have a positive charge on the colored part of the molecule, that is, MB^+. It is this cation that allows us to say methylene blue is basic because in electrolysis MB^+ will move to the negative electrode. It is a law of chemistry that unlike charges attract each other so the MB^+ molecules are ionically bonded to the negative charges at the surface of the bacterium. When this happens, the cell is stained. A word equation to show the staining process can be written as follows:

$$\text{Bacterial cell surface}^- Na^+ + MB^+Cl^- \longrightarrow \text{Bacterial cell surface MB} + NaCl$$

The staining of bacteria with a simple stain is therefore, an exchange of positive and negative charges between molecules to form an ionic bond. If only a single dye is used to stain bacteria, we call this a simple stain. In a later module, you will be doing a differential stain that employs two different colored dyes.

ACTIVITY

Activity 1: Staining Bacterial Smears with a Simple Stain

Place Slide 1 (smear 1) that you saved from your last module on your staining rack. Flood the slide with methylene blue. Allow the methylene blue to react with the smear for 1 minute. Pick up the stained smear with your slide forceps. Tilt the slide so that the stain runs off into the staining pan or staining sink as shown in Figure 22-2. While the slide is still held tilted over the staining receptacle, wash all the excess stain off immediately with water from your squeeze bottle. Blot the smear gently with bibulous paper to remove the water. Do not rub. Rubbing will remove the stain.

Repeat the staining procedure with the remaining 5 smears. If you stain more than one smear at a time, do not allow the stain to dry on the smear as the drying stain will crystallize and obscure the bacteria. You can avoid crystallization of the

FIGURE 22-2
Draining a bacterial smear before rinsing.

stain by using an excess of stain to flood the slide and by rinsing the stain off immediately with water, that is, while the stain is draining off the smear.

After you have stained all six smears, examine the stains under the oil immersion objective. *Apply the direct use of the oil-immersion objective as described in Activity 4, Module 4.* Using Table 22-1, attempt to determine the correct quantity of cells needed to make a "good" smear.

TABLE 22-1   Determining the Number of Cells Needed to Make a Good Smear

Smear number	Amount of cells deposited	Amount of cells present in stained smear
1	2 loops of broth culture	
2	4 loops of broth culture	
3	6 loops of broth culture	
4	½ loop of slant culture	
5	pinpoint amount	
6	not visible	

For the amount of cells present in the stained smear, use these categories: "too many," "good," and "too few."

Make a mental note of the number of loops from broth cultures and the amount of growth from a slant culture that yielded a "good" smear in your judgment. You will be making smears throughout the entire course so this is important to you.

Submit a copy of the table to your file, and then take the post test.

POST TEST

The post test is a self-evaluation. It is not used for a grade. It is designed only to let you decide if you have successfully completed this module.

True or False

_____   1.   A cation is negatively charged.

_____   2.   A basic dye is negatively charged when ionized.

_____   3.   A simple stain involves the use of two different colored dyes.

_____ 4. The surface of a bacterial cell has an overall negative charge.

_____ 5. The COOH groups on the bacterial cell ionize so that an ionic bond forms between the charged surface of the cell and the methylene blue ion of the ionized methylene blue salt.

_____ 6. The cytoplasm of a bacterial cell has an overall positive charge.

_____ 7. When a simple stain is being done, the slide must be rinsed with water immediately after the stain has been drained off.

_____ 8. The surface of a bacterial cell has a strong affinity for acid dyes.

_____ 9. Eosin is an acid dye.

_____ 10. Crystal violet is an acid dye.

_____ 11. Unlike charges repel each other.

_____ 12. Smears must be stained before they can be evaluated as "good" or "poor."

FORMULA FOR REAGENT

0.5% METHYLENE BLUE

| methylene blue chloride | 0.5 gm |
| distilled water | 100.0 ml |

Dissolve the methylene blue in the distilled water.

MODULE 23

Gram Stain

PREREQUISITE SKILL

Successful completion of Module 4, "Compound Microscope for the Study of Microbes," Module 8, "Aseptic Transfer of Microbes," Module 21, "Preparing a Bacterial Smear," and Module 22, "Simple Stain."

MATERIALS

clean microscope slides (5 to 15)
inoculating equipment:
  Bunsen burner
  burner striker
  inoculating loop
staining equipment:
  stainless steel pan or other receptacle
    for staining process
  staining rack
  slide forceps
  wash bottle
  bibulous paper pad
Gram stain reagent set* consisting of:
  crystal violet
  Gram's iodine
  acetone-alcohol
  Gram's safranin

slant cultures of:
  Bacillus subtilis
  Escherichia coli
  Staphylococcus aureus
  Neisseria sicca
  unknown organisms

For related experiences:
  1. sterile trypticase soy broth tubes (4)*
  2. hay infusion (4 to 7 days)*

*To be prepared in advance by the student if the instructor so indicates. See the end of this module for formulae for Gram reagents.

OVERALL OBJECTIVE

Become adept at Gram staining bacterial smears with consistent Gram reactions and gain a general understanding of some theoretical explanations for differing Gram reactions.

Specific Objectives

1. Discuss some probable reasons for the characteristic Gram reactions.
2. List five factors that could cause variability in the Gram reaction of an organism.
3. Gram stain both positive and negative organisms and obtain their characteristic, classical reactions.
4. Gram stain several unknown organisms and obtain the accepted Gram reaction for each organism.
5. Define the terms *Gram reaction, mordant, G+,* and *G-.*
6. List three important physiological or cytological characteristics that are correlated to the Gram reaction of a bacterium.
7. Describe the Gram reaction, shape, and arrangement of five or six different organisms.

DISCUSSION

Microbiologists find the Gram stain a most useful aid in the identification of bacteria. The Gram stain is a differential stain requiring a primary stain and a counterstain. The primary stain is crystal violet, which is followed by an iodine solution. The iodine is called the *mordant* (a specialized term used in dyeing), which is a substance, often a metallic compound, that combines with a dye to form an insoluble colored compound. This insoluble precipitate is called the crystal violet-iodine complex in the Gram stain procedure. Gram positive organisms do not retain the primary dye (after decolorization) if the iodine mordant is omitted. After decolorizing, usually with 95% ethanol or acetone-alcohol, a safranin counterstain is applied to the smear. If the acetone-alcohol decolorizer step is omitted from the Gram stain procedure, all bacteria will appear Gram positive.

Organisms that resist decolorizing and retain the crystal violet-iodine complex *appear purple or deep blue* microscopically and are called *Gram positive* (G+). Conversely, those cells that decolorize or give up the crystal violet-iodine complex more rapidly will accept the safranin counterstain and *appear red.* These are the *Gram negative* (G-) organisms.

In 1883, Christian Gram accidentally discovered what eventually was called the Gram staining reaction. He was working on the etiology of respiratory diseases in the Municipal Hospital in Berlin. As he attempted to stain biopsy specimens to differentiate microorganisms from the surrounding tissues, he applied crystal violet and then Lugol's iodine as a mordant. Both these solutions were standard reagents used in laboratories at that time. The precipitate formed was so thick that Gram used 95% ethanol as a clearing agent. He found that the tissue cells decolorized much more rapidly than the bacteria present in them. Gram originally thought he had developed a differential stain for all bacteria in tissue, but by 1884 he had observed that some bacteria did not retain the primary stain but were decolorized with the tissue cells and accepted the counterstain. Hence, there was no differentiation between these bacterial cells and tissue cells. This presents no problem, however, when bacterial cells are stained by themselves, that is, when they are not in tissue cells.

Most living cells, including animal tissues, are Gram negative. It is the Gram positive characteristic that is distinctive. Some bacteria, yeasts, and a few molds possess the Gram positive characteristic.

Today there still is no single, universally accepted explanation for the differences in Gram reactions of certain cells. There are many theories to explain the differences, but none of them is completely satisfactory. The theory now considered most valid relates the Gram reaction to a difference of permeability of the cell wall based on structural differences. Gram negative cells have a greater lipid content in their cell walls than Gram positive cells. Lipids are soluble in alcohol and acetone, which are used as decolorizers in the Gram stain procedure. The removal of the lipid by the decolorizer is thought to increase the pore size of the cell wall, and this would account for the more rapid decolorization of Gram negative cells.

Another theory suggests that a crystal violet-iodine-ribonucleate complex may form in Gram positive cells but not in Gram negative cells. This theory implies that the ribonucleic acids of the Gram positive cell cytoplasm must be different and therefore bind more firmly with the crystal violet-iodine complex. The chemical bond formed with the G+ ribonucleate is not readily broken by the decolorizer.

You should remember that the differentiation of the Gram reaction is not an absolute, all-or-none phenomenon. It is based on the *rate* at which the cells release the crystal violet-iodine complex to the decolorizer. Even Gram positive organisms *can* show a Gram negative reaction if decolorized too much. A number of other factors can result in variable Gram reactions, such as the following:

1. Improper heat fixing of the smear. If a smear is heated too much, the cell walls can rupture, causing G+ cells to release the primary stain and accept the counterstain. This supports the theory that the Gram reaction is dependent on cell wall structure.

2. Cell density of smear. An extremely thick smear may not decolorize as rapidly as one of ordinary density.

3. Concentration and freshness of the Gram staining reagents.

4. Length and thoroughness of washing after crystal violet, and the amount of water remaining on the smear when iodine is added.

5. Nature, concentration, and the amount of decolorizer applied.

6. Age of bacterial culture. Gram reactions are only reliable for cultures 24 hours old or less. This, too, is related to cell wall permeability in old cultures resulting in loss of Gram positivity.

It is important to keep these variables as constant as possible to ensure reliable and consistent Gram differentiations. For this same reason, you should practice your Gram staining procedure repeatedly until you are confident that you are able to obtain consistent reactions. The time spent now in practice will serve you well since you will be using this differential stain constantly in your study of microbes.

The Gram stain is an indispensable aid to the identification of unknown bacteria. This simple procedure will allow you to place any bacterium into one of five broad areas and, at the same time, to eliminate the remaining four areas. That is, the organism will be either a Gram positive rod, a Gram negative rod, a Gram positive coccus, a Gram negative coccus, or a Gram nonreactive. The Gram nonreactives include some spirochetes. Acid-fast organisms such as those of the genus *Mycobacterium* are only weakly Gram positive and are better studied by other staining procedures. You can see that your search has been narrowed rapidly by eliminating several large groups of bacteria through the use of this most helpful staining technique.

Diagnosis and treatment of bacterial diseases are facilitated by determining the Gram reaction of the causative organism. The Gram positive bacteria include the causative organisms of anthrax, rheumatic fever, diphtheria, botulism, septic sore throat, and boils. A few representatives of the Gram negative group are the organisms causing cholera, typhoid, dysentery, whooping cough, certain types of food poisoning, bubonic plague, and many other diseases.

In later modules, you will be involved in activities that demonstrate the correlation of the Gram reaction with the following important physiological and cytological traits:

1. G+ organisms are generally more susceptible to penicillin and dye bacteriostasis (inhibition of bacterial growth) than are G- organisms.

2. G- organisms are more sensitive to lysis and digestion by strong alkali, acids, and lysozyme (a cell lysing enzyme).

3. G+ organisms tend to be more fastidious i.e., have complex nutritional requirements for growth.

4. G+ organisms usually produce exotoxins, while G- bacteria form endotoxins.

5. Gram positivity is a characteristic easily lost while Gram negativity is never lost. Therefore, you will find Gram negative cells in Gram positive stained slides but you should never find Gram positive cells on a Gram negative slide from a pure culture.

Because of the great usefulness of the Gram stain procedure it is well worth your time to master this technique as you perform the following activities.

ACTIVITIES

Activity 1: Performing the Gram Stain

Obtain stock slant cultures of *Bacillus subtilis, Escherichia coli, Staphylococcus aureus,* and *Neisseria sicca.* Prepare a separate smear of each organism. You may put two smears on a single slide if you wish. Refer to Module 21, "Preparing a Bacterial Smear," if you feel that you need to refresh your memory. Be sure to air dry and heat fix your smears, and label the slides carefully so that you can distinguish one organism from another.

The first or second time that you do a Gram stain you will probably find it easier to stain one slide at a time. After you are more familiar with the procedure, you will be able to put several slides on your staining rack and stain a whole series of slides together. Place your carefully marked slides with the heat-fixed smears on the staining rack over your stainless steel pan or other receptacle.

Reminder: Once you begin the staining procedure, you must *never* let a smear become dry until you have completed the procedure. By flooding the slide with an excess amount of stain each time, you will prevent drying and consequent precipitation of crystallized dye on the slide, which can obscure the bacterial cells.

Perform the Gram stain as follows:

1. Flood the slide with *crystal violet,* and allow it to react for *1 minute.*
2. Handling the slide with your slide forceps, tilt it to about a 45° angle to *drain* the dye off the slide into the pan or staining sink as shown in Figure 23-1.
3. Continue to hold the slide at a 45°angle, and immediately *rinse* it thoroughly with a gentle stream of water from your wash bottle.
4. Replace the slide on the staining rack, and flood it with *iodine.* Allow the iodine to react for *1 minute.*
5. With your slide forceps, tilt the slide, and allow it to *drain.*
6. Immediately *rinse* the slide thoroughly with water from your wash bottle.
7. With your slide still held at a 45° angle, *decolorize it quickly* by allowing the acetone-alcohol to run over and off the smear. *Do not decolorize it too much.*
8. *Rinse immediately* with water from your wash bottle. This will stop the decolorizing process.
9. Replace the slide on your staining rack, and flood it with *safranin* counterstain. Allow the counterstain to react for *30 to 60 seconds.*
10. *Drain* the slide.
11. *Rinse* the slide thoroughly with water from your wash bottle.
12. Carefully *blot* your stained slide in your booklet of bibulous paper. *Do not rub* as you could rub off a very thin smear.

See Figure 23-1 for a pictorial representation of this procedure. You may be using different staining bottles. The procedure, however, is the same.

Repeat this procedure until all four of your smears are stained. Examine each smear microscopically, and draw several representative cells from each smear as they appear under your oil-immersion objective. Label the four drawings with the name of

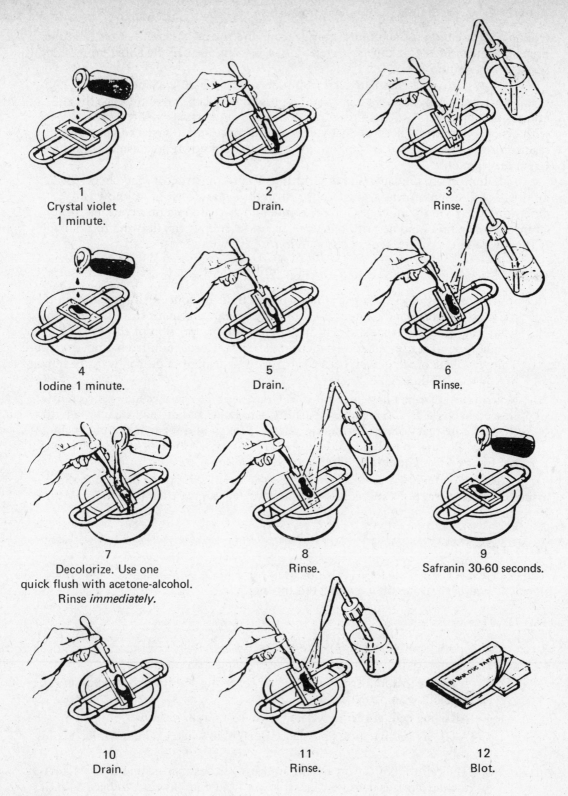

1
Crystal violet
1 minute.

2
Drain.

3
Rinse.

4
Iodine 1 minute.

5
Drain.

6
Rinse.

7
Decolorize. Use one
quick flush with acetone-alcohol.
Rinse *immediately*.

8
Rinse.

9
Safranin 30-60 seconds.

10
Drain.

11
Rinse.

12
Blot.

FIGURE 23-1
The Gram stain procedure.

the organism, the Gram reaction, cell shape, and arrangement, and then submit them to your file.

If you feel you need more practice to perform the Gram stain procedure with consistency, ease, and confidence, you should repeat this activity as time permits.

Activity 2: Gram Staining Unknown Organisms

Obtain a slant culture of an unknown organism, and record the number on the tube. Prepare a smear of the unknown bacterium; air dry and heat fix it. Label your slide for easy recognition.

Now perform a complete Gram stain on several unknown smears. Examine the smear microscopically, and sketch several representative cells as seen under the oil-immersion objective. List Gram reaction, cell shape, and arrangement, if any. Repeat with other unknowns. There are many genera and species that have the same Gram reaction and shape. Therefore, you *cannot* name the unknown organisms from their Gram reaction only.

Submit these drawings and descriptions to your instructor, and check the numbers of your unknowns against your instructor's master list of unknowns to determine the accuracy of your Gram reaction and microscopic observations. If your Gram stain reactions and descriptions were accurate, submit the drawing to your file. If they were not accurate, repeat this activity.

Related Experience: Optional Investigation

1. Inoculate trypticase soy broth tubes with each of the four organisms used in this package. Incubate and Gram stain. Compare cell arrangement from broth with what you observed from slant cultures in the activity portion of this module. Remember, to study cell arrangement it is best to Gram stain from a broth culture. Thioglycollate broth is even better for true arrangement. You will want to remember this for Module 53.

2. Make a smear from a hay infusion, and Gram stain it. Draw as many differently shaped bacteria as you are able to distinguish. Label the different bacteria with shape, Gram reaction, and grouping, if any. You should see many spiral shaped bacteria. Be sure to include a drawing of them. For a review of the spiral shapes, see Figure 6-1 in Module 6, "Preparing a Wet Mount."

When you feel that you have practiced the Gram stain procedure enough to assure consistent reactions and confidence in your technique, then take the post test.

POST TEST

The post test is a self-evaluation. It is not used for a grade. It is designed only to let you decide if you have successfully completed this module.

Part II: True or False

_____ 1. Gram positive organisms accept the counterstain.

_____ 2. Gram negativity is more widespread than Gram positivity.

_____ 3. It is best to add just enough dye to cover the smear on your slide in order to avoid wasting reagents.

_____ 4. All living cells are either Gram positive or Gram negative.

_____ 5. A G+ organism appears purple or deep blue when examined microscopically.

_____ 6. The cell wall of G+ organisms appear to be less permeable to the effect of the decolorizer because it has a lower lipid content than the cell wall of G- organisms.

_____ 7. The Gram reaction of a specific bacterial species will never vary if the staining procedure is consistent.

_____ 8. Gram negative organisms release the primary stain to the decolorizer, accept the counterstain, and appear red when examined microscopically.

_____ 9. G+ bacteria are usually more susceptible to antibiotics than are G- bacteria.

_____ 10. G- bacteria generally have more fastidious growth requirements than G+ organisms.

Part II

List five factors that can cause variability in the Gram reaction of an organism.

1. _____

2. _____

3. _____

4. _____

5. _____

Part III

Give the Gram reaction, shape, and arrangement of four different microbes.

Genus and species name

1. _____   _____

2. _____   _____

3. _____   _____

4. _____   _____

FORMULAE FOR REAGENTS

1. GRAM'S CRYSTAL VIOLET consists of two solutions, Solution A and Solution B.
    SOLUTION A

crystal violet	2.0 gm
ethanol, 95%	20.0 ml

    Dissolve the crystal violet in the ethanol.
    SOLUTION B

ammonium oxalate, C.P. (chemically pure)	0.8 gm
distilled water	80.0 ml

    Dissolve the ammonium oxalate in the distilled water.

    Make up solutions A and B separately. Then pour the two solutions together, and mix well.

2. GRAM'S IODINE

iodine, C.P.	1.0 gm
potassium iodide, C.P.	2.0 gm
distilled water	300.0 ml

    Grind the iodine and potassium iodide together in a mortar until finely divided. Add water in small quantities to wash the contents out of the mortar. Add the rest of the water, and mix thoroughly. This solution should be stored in a tightly closed amber bottle.

3. GRAM'S SAFRANIN

safranin O	0.25 gm
ethanol, 95%	10.0 ml
distilled water	100.0 ml

    Dissolve the safranin in the ethanol. Mix thoroughly. Add the distilled water, and mix well. Filter the solution through filter paper.

4. ACETONE-ALCOHOL

ethanol, 95%	70.0 ml
acetone	30.0 ml

    Mix the two liquids thoroughly.

Dispense these four Gram stain reagents in separate dropper bottles, labeling the dropper bottles appropriately. Store the remainder of each reagent in a large tightly stoppered or screw-cap bottle. As your working bottles (dropper bottles) become empty, replenish the reagents from the large storage bottles.

MODULE 24

Capsule Stain

PREREQUISITE SKILL

Successful completion of Module 4, "Compound Microscope for the Study of Microbes," Module 6, "Preparing a Wet Mount," Module 8, "Aseptic Transfer of Microbes," Module 21, "Preparing a Bacterial Smear," and Module 22, "Simple Stain."

MATERIALS

clean microscope slides (4)
cover slips (2)
inoculating equipment
staining equipment
fresh India ink
physiological saline*
Loeffler's methylene blue*

slant cultures on tryptose phosphate agar
   of *Klebsiella pneumoniae*
      Staphylococcus epidermidis

*To be prepared in advance by the student if the instructor so indicates. Formulae for reagents are at the end of this module.

OVERALL OBJECTIVE

Demonstrate proficiency in the performance of two types of capsule stain procedures and an understanding of the structure and functions of the capsule.

Specific Objectives

1. Discuss the bacterial capsule and two reasons why ordinary staining procedures cannot be used to visualize it.
2. Stain organisms demonstrating bacterial capsules by two different methods, and submit drawings of your results to your file.
3. Define the terms *negative stain, slime layer, phagocytosis,* and *ionic bond.*
4. Name the composition of most bacterial capsules.
5. Explain why a Gin's method capsule stain is considered a differential stain.

6. Describe the role of the capsule in disease.
7. Explain how Gin's method employs a negative stain and a positive stain.
8. Give the genus and species name of an organism whose virulence is entirely dependent on its capsule being present.

DISCUSSION

Most bacterial cells secrete a viscous substance that accumulates around the outside of the cell and "coats" the cell wall. This structure, depending on the thickness of the layer and its viscosity, is called the *capsule* or the *slime layer*. Most bacteria secrete at least some slime that is more soluble and less viscous than a capsule. The capsule appears as a more definite structure and, therefore, can be demonstrated by staining.

The size of the capsule is influenced by the environment in which the organism is cultured. For example, tryptose phosphate agar induces the production of larger capsules than does nutrient agar. It is also true that capsule producers, *in vivo,* produce large capsules if they are disease-causing bacteria.

The bacterial capsule has significance both for the bacteria and for humans. The capsule serves a protective function for the bacterium by acting as an osmotic barrier between the cell body and the environment. The capsule also appears to interfere with the phagocytic action of leucocytes when encapsulated bacteria invade the human body. There is speculation that the bacterial capsule may also be a reservoir of stored food or a disposal site of waste products.

For some disease-producing organisms, the virulence and infectivity of the organism are increased by, or entirely dependent upon, the presence of the capsule. For example, *Streptococcus pneumoniae* becomes avirulent when it loses the ability to produce capsules.

The bacterial capsule is usually composed of polysaccharides that are water soluble and nonionic. As you learned in Module 22, "Simple Stain," most staining techniques are based on chemical bonding between ionized particles of the dye molecules and ionized areas on the surface of the cell. That is, an ionic bond is formed by the attraction of unlike charges on the dye molecules and the cell surface, and the bacterium is stained. Because the bacterial capsule is nonionic it cannot be stained in the usual manner.

Since we cannot stain the capsules, techniques have been developed that allow us to stain the background and leave the capsule clear. This is called a negative stain. In essence, a negative stain dyes everything *except* the structure you wish to visualize.

The phase microscope actually provides the best method for visualizing bacterial capsules. However, phase equipment is expensive and not always readily available. A similar effect can be achieved by preparing a wet mount of the bacterium and adding carbon particles (India ink) to the suspension. When this preparation is examined microscopically with reduced light intensity, a phaselike effect is achieved. The capsules appear to be halos or clear rings around the bacterial cell, and the background looks dark, as shown in Figure 24-1. This method can be completed very rapidly but is a temporary preparation.

A variation of this technique is to prepare a suspension of bacteria, saline solution, and carbon particles on a slide just as for the wet mount. This suspension is then spread out on the slide and allowed to air dry, but it *is not heat fixed.* The dry smear is counterstained with methylene blue, which stains the cell body, and then is rinsed. This is called Gin's method for the staining of capsules. It differs from the wet mount demonstration of capsules in that it employs a negative stain and a positive stain. When examined microscopically, the background is dark, the capsules appear as clear unstained rings (negative stain), with the small blue cell body in the center of the rings (positive stain), as shown in Figure 24-1. This variation is not restricted by the shortcomings of the wet mount preparation. The counterstain makes the cell body more clearly visible and allows you to compare the size of the cell to that of the capsule. The oil-immersion objective can be used very easily with Gin's method but only with great difficulty with the wet mount technique.

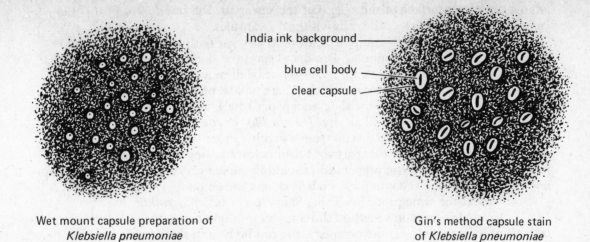

India ink background

blue cell body

clear capsule

Wet mount capsule preparation of
Klebsiella pneumoniae

Gin's method capsule stain
of *Klebsiella pneumoniae*

FIGURE 24-1
Two methods of demonstrating capsules.

Either of these capsule stain techniques can be readily adapted for use with a broth culture by simply omitting the saline solution when you prepare the suspension of organisms and carbon particles. Be sure that the organisms are well dispersed in the broth.

ACTIVITIES

Activity 1: Wet Mount Capsule Stain

Obtain a stock slant culture of *Klebsiella pneumoniae,* a stock slant culture of *Staphylococcus epidermidis,* and a bottle of new India ink. Using your medicine dropper, place one drop of physiological saline on a clear slide. Observing asepsis, secure the smallest amount of growth possible from the *Klebsiella pneumoniae* culture by just *touching* your inoculating loop to the growth. Gently but thoroughly mix this growth with the saline on the slide. When this suspension is thoroughly mixed, flame and cool your inoculating loop, and use it to obtain a very small amount of India ink. Add the ink to your bacterial suspension, and mix it thoroughly until it is a uniform dark gray (not black). Carefully add a cover slip, and examine your wet mount under your high dry objective. If you did not obtain the correct mixture of bacterial cells and India ink, you will not see the capsules. So repeat the procedure, varying the amounts of the mixture from the amounts that you just used. When you have the correct mixture, you *will* see the capsules. It is almost impossible to avoid trial and error as you learn to make this preparation correctly. Draw a representative sample of what you observe.

Follow the same procedure using *Staphylococcus epidermidis,* which does not form a capsule. This will allow you to compare it with *Klebsiella pneumoniae* which does form a large capsule. Submit both drawings to your file.

Do not dwell upon this technique because capsules and bacteria are difficult to distinguish from the India ink particles at 440x. You may wish to try observing your wet mount with a large drop of oil and your oil-immersion objective. Your organisms may become visible if you take care not to disturb the coverslip. In any case, in the next activity you will see capsules with less difficulty.

Activity 2: Gin's Method Capsule Stain

On a clean slide, mix equal amounts of India ink and saline. With your inoculating loop, add a small amount of organisms from the *Klebsiella pneumoniae* culture, and mix the suspension thoroughly and gently. As you mix, spread the smear out to cover almost the whole surface of the slide. Let the smear air dry. but *do not heat fix it.* Heat fixing tends to distort the cells within the capsules.

Using your staining equipment, flood the smear with Loeffler's methylene blue, and let it react for 3 minutes. Drain and rinse the smear. It is common for part of the smear to wash off as you rinse it—do not be dismayed! Remember that this smear is not heat fixed and is quite thick. There will be plenty of organisms left for you to study. Put one end of the slide on a paper towel, and prop the slide up at a 45° angle. Leave it to drain and air dry. *Do not blot or rub it.*

When making a capsule stain from a broth culture it is not necessary to use a drop of saline. Simply mix two parts of broth culture to one part of India ink and proceed as usual. The broth preparation should be mixed very gently to avoid disrupting any characteristic cell grouping, such as chains and so on.

Repeat the same procedure using *Staphylococcus epidermidis.*

Examine your Gin's method slides under oil immersion, and draw representative cells from each slide. Label the organism, the cell body, and the capsule, if present. Submit drawings of both organisms to your file. Then take the post test.

Related Experience

In this activity, you will observe encapsulated and nonencapsulated bacteria using a phase contrast microscope. A demonstration can be set up by the instructor if a phase microscope is available.

POST TEST

The post test is a self-evaluation. It is not used for a grade. It is designed only to let you decide if you have successfully completed this module.

True or False

_____ 1. The size of the capsule produced is solely dependent on the genetic make-up of the bacterium involved.

_____ 2. Slime layer is a synonym for capsule.

_____ 3. The capsule often serves a protective function for the bacterium that produces it.

_____ 4. *Streptococcus pneumoniae* lose their virulence when they lose the ability to produce capsules.

_____ 5. Most bacterial capsules are composed of polysaccharides.

_____ 6. The best method of visualizing bacterial capsules is with a phase microscope.

_____ 7. A capsule stain can be done only with bacterial growth from a solid culture medium.

_____ 8. The saline, bacteria, and carbon particle suspension should be mixed only enough to blend it for the Gin's method capsule stain.

_____ 9. You should use an excess of India ink when you prepare a capsule stain.

_____ 10. You should never heat fix a capsule stain preparation.

_____ 11. The positive stain in Gin's method is methylene blue.

_____ 12. Methylene blue is called a positive stain because it stains the body of the bacterium.

_____ 13. Gin's method is considered a differential stain because the India ink stains the background black.

_____ 14. In general, bacteria that produce capsules are more readily phagocytized than those that do not produce a capsule.

_____ 15. Capsules are produced just as readily on nutrient agar as on any other agar.

_____ 16. The India ink forms an ionic bond to stain the capsule.

_____ 17. The capsule is water soluble.

_____ 18. *Klebsiella pneumoniae* does not produce a capsule.

_____ 19. Using water instead of physiological saline in capsule preparations should have no effect on the observability of the capsule.

FORMULAE FOR REAGENTS

1. LOEFFLER'S METHYLENE BLUE

methylene blue	0.3 gm
ethanol, 95%	30.0 ml
distilled water	100.0 ml

    Dissolve the dye in the ethanol. Add the distilled water, and mix well. Filter the solution through filter paper in a funnel.

2. PHYSIOLOGICAL SALINE

NaCl	0.85 gm
distilled water	100.0 ml

    Mix to dissolve NaCl.

MODULE 25

Bacterial Endospores

PREREQUISITE SKILL

Successful completion of Module 4, "Compound Microscope for the Study of Microbes," Module 8, "Aseptic Transfer of Microbes," Module 21, "Preparing a Bacterial Smear," and Module 22, "Simple Stain."

MATERIALS

clean microscope slides (2)
inoculating equipment
staining equipment
Loeffler's methylene blue, dropper
  bottle*
Schaeffer-Fulton stains: malachite green*
  safranin (*not* Gram's safranin)*

28 hour slant culture of *Bacillus subtilis*

*To be prepared by the student if the instructor so indicates. Formulae are at the end of this module.

OVERALL OBJECTIVE

Recognize bacterial endospores, understand their functions, and become adept at the performance of the Schaeffer-Fulton staining method for visualizing the endospores.

Specific Objectives

1. Discuss sporulation and what is known about why it takes place in some bacteria.
2. Describe the functions of sporulation for bacteria.
3. Name four pathogenic, sporulating bacilli and the diseases they cause.
4. List the two genera comprising most spore-forming bacteria.
5. Describe how spores are visualized in a simple stain.
6. Describe the visualization of endospores by the Schaeffer-Fulton spore stain.
7. Define the terms *vegetative cell, endospore,* and *germination.*
8. Name the reagents and the sequence in which the reagents are used in the Schaeffer-Fulton stain.

DISCUSSION

Bacterial endospores are small oval or spherical structures that are very resistant to high temperatures, radiation, desiccation, and chemical agents such as disinfectants. Spores are produced intracellularly by *some* bacilli, which is the reason they are called *endospores.* The ordinary bacterial cell that gives rise to the spore is called the *vegetative cell.* Endospores are smaller than the parent cells and display different qualities, most notably their great resistance to adverse conditions.

The spore, however, is not formed as a response to adverse conditions. The nutritional and environmental conditions for sporulation are similar to those necessary for vegetative growth. What stimulates certain bacterial species to form spores is still unknown. Sporulation in bacteria is not a form of reproductive multiplication, as it is in some higher plants, because each cell produces only one spore, and each spore, in turn, germinates into one vegetative cell, as shown in Figure 25-1. Reproduction, then, is by binary fission of the vegetative cell in spore-forming species as it is in other species of bacteria.

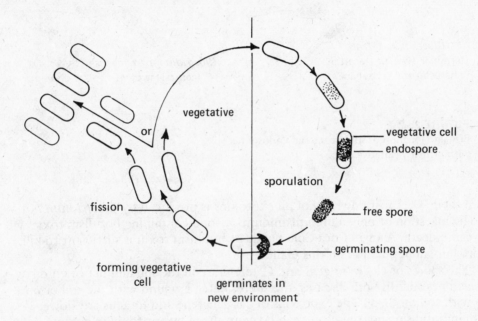

FIGURE 25-1
Life cycle of spore-forming bacteria.

It appears that the spore is just a part of the life cycle of some genera of bacteria. The spore is the dormant or resting phase of the bacterial cell and *in this respect only* is analogous to the seeds of higher plants or the cysts of protozoans. The spore, however, is *not* an agent of sexual reproduction as is the seed. The presence of spores in a culture is significant for identification and differentiation of bacteria since spore formation is primarily confined to the G+ rod-shaped organisms in two genera, *Bacillus* and *Clostridium.*

The size and location of the spore within the vegetative cell are also significant for differentiation of organisms. For example, spores can be centrally, subterminally, or terminally located, and they can be larger or smaller in diameter than the vegetative cell. When a spore is larger in diameter than the vegetative cell, a "swelling" or enlargement and distortion of the vegetative cell results, as shown in Figure 25-2. The sporulation characteristics of a species are constant each time sporulation occurs and so are a useful aid in the identification of the organism.

Several spore-forming bacilli are the causative agents of disease. The anaerobic clostridia are the most famous of these. *Clostridium botulinum* causes fatal food poisoning (botulism). *Clostridium perfringens* causes gas gangrene, and *Clostridium tetani* causes lockjaw (tetanus). All these spore-forming clostridia produce powerful exotoxins that

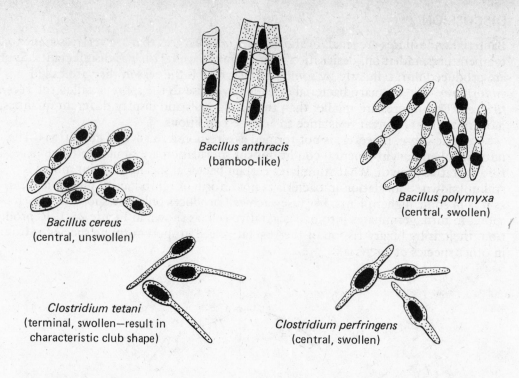

Bacillus anthracis
(bamboo-like)

Bacillus polymyxa
(central, swollen)

Bacillus cereus
(central, unswollen)

Clostridium tetani
(terminal, swollen—result in
characteristic club shape)

Clostridium perfringens
(central, swollen)

FIGURE 25-2
Size and location of some bacterial endospores
within their vegetative cells.

are often fatal. The most powerful of all exotoxins is produced by *Clostridium botulinum*. The ingestion of only a minute amount of food containing botulism toxin will usually cause death. A small Coca-Cola bottle of botulism toxin is sufficient to kill the entire human population on this planet.

The spores of *C. perfringens* and *C. tetani* are in the soil and hence on dirty objects and in food. In both diseases, the spores enter a wound with the soil or on objects with soil on them. The exotoxins of gas gangrene and tetanus are slower acting than botulism and, therefore, can be neutralized by antitoxin injections. Neutralization is followed by the use of antibiotics to kill the toxin-producing bacteria and excision of the damaged tissue to remove the anaerobic environment. Indeed, early treatment is life saving in *C. perfringens* and *C. tetani* intoxications.

Most species of the genus *Bacillus* are harmless saprophytes. *Bacillus anthracis* is the only *aerobic* spore-forming pathogen. The spores of this organism are also found in the soil where they remain viable for several decades. If the spores are ingested by sheep, goats, and other animals, then the disease, anthrax, is established. Although anthrax is primarily a disease of farm animals, it is transmissible to humans. It is an occupational hazard to farmers, veterinarians, and other handlers of infected animals because the organism can enter through a break in the skin. Workers handling animal products such as sheep's wool, goats' hair, and such can also contract the disease by inhalation of the spores. When the infection begins in the respiratory tract, it is called woolsorters' disease.

The anthrax bacillus is easily recognized microscopically. It is a large, Gram positive, spore-forming rod and forms characteristic chains. The ends of each bacillus are concave and this gives the chains a bamboo-like appearance, as shown in Figure 25-2.

Structures such as spores can often be visualized by taking advantage of certain peculiarities of the structure. For example, the bacterial endospore has very resistant spore coats. When a spore-forming organism is stained by ordinary staining methods, the spore resists the stain and is seen through the microscope as an unstained area within the vegetative cell. However, if only a few spores are present, or if they have been released from the vegetative cells and are free in the smear, then spores can often go undetected in a simple stain or a Gram stain.

The Schaeffer-Fulton stain is a differential stain developed to visualize both the endospore and the vegetative cell. Using this method, the spore itself is stained, and free spores are easily detected. In the Schaeffer-Fulton stain, heat is used to drive the primary dye (malachite green) into the spore coats. The same characteristics of the spore that make it difficult to stain cause it to retain the dye tenaciously once it has penetrated the spore coats. The malachite green is readily rinsed out of the vegetative cell because the cell wall has been disrupted by the heating process. Therefore, the vegetative cell accepts the counterstain, safranin. When examined microscopically, the spores appear as small, green ovals or spheres within the red vegetative cells.

ACTIVITIES

Activity 1: Visualization of Endospores by a Simple Stain

Obtain a slant culture of *Bacillus subtilis,* and prepare a smear in the usual manner. Air dry and heat fix it. Flood the slide with Loeffler's methylene blue, and allow it to react for 3 minutes. Drain the slide and rinse it. Blot the slide carefully. Examine it under your oil-immersion objective. The spores appear as clear, unstained areas within the vegetative cells. Draw several representative cells, label the sketch appropriately, and submit it to your file.

Activity 2: The Schaeffer-Fulton Spore Stain

Prepare a smear of *Bacillus subtilis,* and heat fix it as usual. Perform the Schaeffer-Fulton stain procedure as follows:

1. *Flood* the slide with *malachite green* stain.
2. Allow the malachite green to *react* at *room temperature* for about *1 minute.*
3. Next heat this stain-flooded slide to steaming by inverting your Bunsen burner and passing the flame over the stain periodically, as shown in Figure 25-3. When you observe steam rising from the slide, remove the burner. When the steaming stops, pass the flame over the stain again briefly. *Do not boil or allow the stain to dry. Steam* for at least 2 to 4 *minutes,* replacing the malachite green if it evaporates from the slide.
4. Allow the slide to *cool* about *5 minutes* to prevent breaking it. Continue to add stain as the slide cools, since the stain is still evaporating.

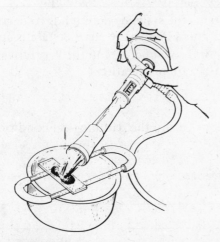

FIGURE 25-3
Heating process of the Schaeffer-Fulton stain procedure.

5. *Drain* the slide.
6. *Rinse* with water for *30 seconds.*
7. Replace the slide on your staining rack, and *flood* it with *safranin* counterstain. Allow the safranin to react for *1 minute.*
8. Drain the slide, and wash it thoroughly with water.
9. Blot the slide carefully, or allow it to air dry.

Examine your Schaeffer-Fulton stain with your oil-immersion objective. You should see oval or spherical green spores and red, rod-shaped vegetative cells. Draw several representative cells, label the spores and vegetative cells, and submit the sketch to your file

Practice the Schaeffer-Fulton stain procedure as often as you feel necessary, and then take the post test.

Related Experience

Do Gram stain of *Bacillus subtilis.* Spores can be detected using the Gram stain. Since the Gram stain will be the stain you will be using first with unknowns, it would be to your advantage to recognize spores from this preparation.

POST TEST

The post test is a self-evaluation. It is not used for a grade. It is designed only to let you decide if you have successfully completed this module.

Part I: True or False

_____ 1. The bacterial endospore is more resistant to adverse environmental conditions than is the vegetative cell.

_____ 2. Bacteria form spores as a survival mechanism in response to adverse environmental conditions.

_____ 3. The only bacteria capable of forming spores are those organisms of the genera *Bacillus* and *Clostridium.*

_____ 4. The spore is the dormant or resting phase of the bacterial cell.

_____ 5. The detection of endospores is a useful aid in the identification of bacteria.

_____ 6. The size and location of the spore within the vegetative cell are dependent upon physical and chemical conditions in the environment during sporulation.

_____ 7. Endospores can be visualized only by the Schaeffer-Fulton spore stain.

_____ 8. Safranin is the primary stain in the Schaeffer-Fulton technique.

_____ 9. Because the spore coat is so resistant, the stained spore tends to decolorize rapidly if the smear is washed too much before the counterstain is applied.

_____ 10. The spores appear as small, green ovals or spheres, and the vegetative cells are red, in a Schaeffer-Fulton spore stain preparation.

Part II

List the spore-forming bacilli of medical importance and the pathological conditions that they cause.

Genus and species   Disease

1. _____   _____

2. _____   _____

3. _____   _____

4. _____   _____

Part III

Define the following terms:

1. vegetative cell _____

    _____

2. endospore _____

    _____

3. germination _____

    _____

Part IV: Completion

1. Name an aerobic, spore-forming pathogen.

    _____

2. Give the Gram reaction and shape of all spore-forming bacilli.

    _____

3. Why is visualization of spores in a stained smear significant to a microbiologist?

    _____

FORMULAE FOR REAGENTS

1. LOEFFLER'S METHYLENE BLUE

methylene blue chloride	0.3 gm
ethanol, 95%	30.0 ml
distilled water	100.0 ml

    Dissolve the methylene blue in the ethanol. Add water and mix well. Filter through filter paper and funnel.

2. MALACHITE GREEN (5% aqueous)

    Dissolve 5.0 gm of malachite green in 100.0 ml of distilled water.

3. SAFRANIN COUNTERSTAIN (0.5% aqueous)

    Dissolve 0.5 gm safranin in 100.0 ml of distilled water.

MODULE 26

Metachromatic Granules

PREREQUISITE SKILL

Successful completion of Module 4, "Compound Microscope for the Study of Microbes," Module 8, "Aseptic Transfer of Microbes," Module 21, "Preparing a Bacterial Smear," Module 22, "Simple Stain," and Module 23, "Gram Stain."

MATERIALS

clean microscope slides (2)

inoculating equipment

staining equipment

Gram stain reagents*

Trypticase soy agar (TSA) slant culture of *Corynebacterium xerosis,* (72 hours growth)

Loeffler's *alkaline* methylene blue, dropper bottle*†

For related experience:

Albert's stain set consisting of:

  Albert's diptheria stain, dropper bottle*†

  Lugol's iodine solution, dropper bottle*†

*To be prepared by the student if the instructor so indicates.

†Formulae are at the end of this module.

OVERALL OBJECTIVE

Stain metachromatic granules in order to visualize them and gain an understanding of their role in the cell.

Specific Objectives

1. List three types of cytoplasmic inclusions that can be detected in bacterial cells.
2. List the different appearances that metachromatic granules have because of their location in different species of *Corynebacterium.*
3. Name a diphtheroid that normally does not contain metachromatic granules.
4. Discuss the significance of metachromatic granules to the clinical microbiologist.

5. Define the terms *volutin granule, bipolar staining, diphtheroid,* and *palisade arrangement.*

6. Give the genus and species name of a pathogenic organism that produces metachromatic granules.

DISCUSSION

Many types of cytoplasmic inclusions have been detected in bacterial cells. Cytoplasmic inclusions are usually contained in vacuolated areas in the cell or in concentrated deposits of some stored substance, for example, lipid droplets or polysaccharide granules composed of starch or glycogen.

Metachromatic granules, also called volutin granules, are another type of cytoplasmic inclusion found in many bacteria as well as in some fungi, algae, and protozoa. Metachromatic granules are composed mainly of polyphosphates, ribonucleic acid (RNA), and protein. Recent research indicates that a fairly extensive internal structure exists in the metachromatic granule. The function of metachromatic granules is unknown, although it appears that under starvation conditions the material stored in the granules becomes depleted. The metachromatic granules are most prominent in old cultures (at least 72 hours) before starvation occurs.

Metachromatic granules display a strong affinity for basic dyes such as methylene blue. When a basic dye is used to stain an organism that contains metachromatic granules, the granules stain much more intensely than the remainder of the cell. These cells typically look like a little string of beads or bands because of the more deeply stained granules. In some cells the granules occur at the ends of the cell, resulting in a bipolar staining appearance. This can be seen in Figure 26-1.

Although metachromatic granules occur in many types of bacteria, the clinical microbiologist is primarily interested in them as they occur in the genus *Corynebacterium.* This genus contains the causative organism of diphtheria, as well as the so-called diphtheroids. The diphtheroids are related to *Corynebacterium diphtheriae* and resemble it in many ways, but they are nonpathogenic normal flora of the human throat and skin. Diphtheroids display a characteristic arrangement of cells called palisading. The cells lie parallel to each other, similar to a log fence as seen in Figure 26-1. Most organisms of the genus *Corynebacterium* normally possess metachromatic granules. *Corynebacterium pseudodiphtheriticum,* however, usually lacks granules. The granules occur in characteristic parts of the cell in several species and so are a useful aid to species differentiation and identification. Figure 26-1 depicts metachromatic granules as you are likely to see them with your preparations.

Albert's differential stain was developed to stain the metachromatic granules of *Corynebacterium diphtheriae* in a contrasting color to that of the remainder of the cell body. This stain is also commonly called Albert's diphtheria stain.

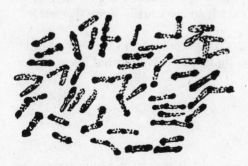

Metachromatic granule stain of diphtheroid
(Note some bipolar staining.)

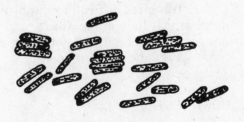

Gram stain of diphtheroid
(Note palisading.)

FIGURE 26-1
Diphtheroid stained to show metachromatic
granules and palisade arrangement.

ACTIVITIES

Activity 1: Loeffler's Alkaline Methylene Blue Stain for Metachromatic Granules

Obtain a slant culture of *Corynebacterium xerosis* (a diphtheroid), and prepare a smear in the usual manner. Air dry and heat fix the smear *very gently*. Flood the slide with Loeffler's alkaline methylene blue stain. Allow the dye to react for 5 minutes. Drain your slide and rinse it thoroughly. Place one end of the slide on a paper towel, prop the slide at a 45° angle, and allow it to air dry.

Examine your stained smear under your oil-immersion objective, and sketch several representative cells. The metachromatic granules will be stained an intense blue, while the cytoplasm will be a much paler blue. Label your drawing appropriately, and submit it to your file.

Activity 2: Gram Stain Showing Palisade Arrangement

As mentioned previously, the palisade arrangement is characteristic for diphtheroids. The pathogen, *Corynebacterium diphtheriae,* however, has a peculiar cell arrangement often likened to Chinese letters. Make a Gram stain from the *Corynebacterium xerosis* slant and look for palisades. Draw several representative cells, give their Gram reaction, and submit this to your file.

Related Experience: Albert's Diphtheria Stain

Prepare a smear of *Corynebacterium xerosis* as usual; air dry and heat fix it *very gently*. Place the slide on your staining rack, and complete Albert's differential stain as follows:

1. *Flood* the slide with *Albert's diphtheria stain,* and allow it to react for *5 minutes.*
2. *Drain* the slide of excess dye, but *do not rinse it.*
3. *Flood* the wet slide with *Lugol's iodine* solution, and allow it to react for *1 minute.*
4. *Drain* the slide.
5. *Rinse* thoroughly with water from your wash bottle.
6. Place one end of the slide on a paper towel, and prop the slide at a 45° angle to *air dry.*

Examine the stained smear under your oil-immersion objective, and sketch several representative cells. The metachromatic granules should appear as intense purple dots within a very pale green cytoplasm. The cytoplasm is often very faint and difficult to distinguish. In addition, purple droplets of stain are often deposited on the slide and must *not* be confused with the metachromatic granules. It is for this reason that critical focusing is necessary to see the entire cell. Look carefully *inside* the pale green cytoplasm to see the metachromatic granules, which are usually quite distinct. Label your sketch appropriately, and submit it to your file.

It is often difficult to distinguish metachromatic granules from the settled-out droplets that form in the Albert's stain. Therefore, do not be disappointed if you do not get clearcut results on your first try.

Take the post test now.

POST TEST

The post test is a self-evaluation. It is not used for a grade. It is designed only to let you decide if you have successfully completed this module.

Part I: True or False

_____ 1. Volutin granule is another term for metachromatic granule.

_____ 2. Metachromatic granules are found only in bacteria of the genus *Corynebacterium.*

_____ 3. Diphtheroids are the causative organisms of diphtheria.

_____ 4. Metachromatic granules have an affinity for basic dyes and will stain more intensely than the surrounding cytoplasm.

_____ 5. The location of metachromatic granules within the bacterial cell can be a useful aid to species differentiation.

_____ 6. Albert's differential stain is often called Albert's diphtheria stain.

_____ 7. *Corynebacterium xerosis* is a diphtheroid.

_____ 8. You should heat fix your smear intensively before you stain for metachromatic granules.

_____ 9. Bipolar staining results when metachromatic granules are located mainly at the ends of a bacterium.

_____ 10. *Corynebacterium diphtheriae* displays the characteristic palisading cell arrangement.

_____ 11. Lipid and glycogen can also be cytoplasmic inclusions.

_____ 12. Metachromatic granules become more distinct when all available nutrients are depleted but before starvation occurs.

_____ 13. *Corynebacterium pseudodiphtheriticum* does not form metachromatic granules.

_____ 14. The term diphtheroid means a diphtheria-like, nonpathogenic species of *Corynebacterium*.

Part II

1. Give the genus and species name of the metachromatic granule producing bacterium that causes diphtheria.

_____

2. Metachromatic granules give different appearances in different species of *Corynebacterium*. List these varying appearances.

a. _____

b. _____

c. _____

FORMULAE FOR REAGENTS

1. ALBERT'S DIPHTHERIA STAIN

toluidine blue	0.15 gm
methyl green	0.2 gm
acetic acid, glacial	1.0 ml
ethanol, 95%	2.0 ml
distilled water	100.0 ml

Dissolve toluidine blue and methyl green in water. Add acetic acid and ethanol, and mix well.

2. LUGOL'S IODINE SOLUTION

iodine, C.P.	5.0 gm
potassium iodide, C.P.	10.0 gm
distilled water	100.0 ml

Mix iodine and potassium iodide in a mortar, and grind with a pestle until finely divided. Add water in small amounts to wash contents of the mortar into a beaker. Add the remaining water, and stir until completely dissolved and well mixed. Store solution in a tightly closed amber bottle.

3. LOEFFLER'S ALKALINE METHYLENE BLUE

methylene blue chloride	0.3 gm
ethanol, 95%	30.0 ml
0.01% potassium hydroxide solution	100.0 ml

Dissolve the dye in the ethanol. Add the dye solution to the potassium hydroxide solution, and mix well.

MODULE 27 label module 25

Acid-Fast Stain

PREREQUISITE SKILL

Successful completion of Module 4, "Coumpound Microscope for the Study of Microbes," Module 8, "Aseptic Transfer of Microbes," Module 21, "Preparing a Bacterial Smear," and Module 22, "Simple Stain."

MATERIALS

clean microscope slides
slant cultures of
   Mycobacterium smegmatis (72 to 96 hours) (TSA slant)
   Staphylococcus epidermidis (24 hours)
inoculating equipment
staining equipment

Ziehl-Neelsen staining reagents, consisting of:
   carbolfuchsin*
   acid-alcohol*
   Loeffler's methylene blue*
*To be prepared by the student if the instructor so indicates. Formulae are at the end of this module.

OVERALL OBJECTIVE

Demonstrate an understanding of the theory and applications of the acid-fast stain and become adept in the performance of the Ziehl-Neelsen acid-fast staining procedure.

Specific Objectives

1. Define the terms *acid-fastness* and *mixed smear.*
2. List two organisms of the genus *Mycobacterium* that are pathogenic for humans, and list the diseases that they cause.
3. Discuss the probable reason that some bacteria display acid fastness in a Ziehl-Neelsen stain.
4. List two advantages to the bacterium of a relatively impermeable cell wall.
5. Describe the microscopic appearance of an acid-fast cell.
6. List in order the steps of a Ziehl-Neelsen acid-fast stain.
7. Give another term for Hansen's disease.

DISCUSSION

Paul Ehrlich developed the acid-fast stain in 1882 in his work with the etiological agent of tuberculosis, *Mycobacterium tuberculosis*. The Ziehl-Neelsen acid-fast stain procedure commonly used today is the result of changes in methodology to improve the original Ehrlich technique. The Ziehl-Neelsen reagents, in which carbolfuchsin is the primary dye and Loeffler's methylene blue is the counterstain, are much more stable than those of Ehrlich. The decolorizer is 3% hydrochloric acid (HCl) in 95% ethanol. This acid-alcohol is a very intensive decolorizer and should not be confused with acetone-alcohol used in the Gram stain procedure.

Most genera are not acid-fast, with the exception of the genus *Mycobacterium* and some species in the genus *Nocardia*. Both these genera contain species that are pathogenic. The two best known acid-fast human pathogens are *Mycobacterium tuberculosis*, the causative agent of tuberculosis and *Mycobacterium leprae*, the etiological agent of Hansen's disease or leprosy.

Human tuberculosis is diagnosed by clinical symptoms, X rays, and laboratory findings. The specimens of choice for bacteriological studies of pulmonary tuberculosis are sputum and bronchial secretions. The tubercle bacillus must be demonstrated in the specimen by acid-fast stain and culture and by X-ray examination of the patient before the diagnosis of TB is definitive.

Generally acid-fastness is very rare in cells. Organisms of the genus *Mycobacterium* have a high fat content, containing relatively large amounts of lipid materials including fatty acids, waxes, and complex lipids. They have waxlike cell walls that are relatively impermeable. The same relative impermeability that is associated with acid-fastness gives the genus *Mycobacterium* an above average resistance to disinfectants. They are also quite resistant to drying and can survive for long periods in dried sputum or other body fluids. These characteristics necessitate special precautions when caring for TB patients. However, TB bacilli are readily destroyed by pasteurization and ordinary sterilization by heat.

Because of the waxy, impermeable cell wall, special measures are necessary to allow the primary stain to penetrate. The primary dye (carbolfuchsin) is formulated with an aqueous 5% phenol (carbolic acid) solution as a chemical intensifier to assist penetration. Heat is also applied to the stain-covered bacterial smear as a penetrating agent. As in the Schaeffer-Fulton spore stain, the primary stain is driven into the cell by steaming.

Once the stain has penetrated the cell wall, the acid-fast cell retains it through very intensive decolorization. Therefore, acid-fastness means that once a cell is stained with carbolfuchsin, it resists decolorization with an acid decolorizer. Other genera of bacteria would lose the primary stain immediately upon exposure to an acid decolorizer. Decolorizing in the acid-fast stain is *not* the delicate procedure that it is in the Gram stain. Acid-fast cells retain the primary stain and appear red microscopically, whereas the non-acid-fast organisms will accept the counterstain and appear blue.

ACTIVITY

Activity 1: The Ziehl-Neelsen Acid-Fast Stain

Obtain stock cultures of *Mycobacterium smegmatis* and *Staphylococcus epidermidis*. Aseptically make a *mixed* smear of *M. smegmatis* and *S. epidermidis*. That is, take a small amount of organisms from the *M. smegmatis* slant and a small amount from the *S. epidermidis* slant and mix both organisms together in a drop of water to emulsify the suspension as well as possible. Allow the smear to air dry, and heat fix it as usual. Perform the Ziehl-Neelsen acid-fast stain as follows:

1. *Flood* the slide with *carbolfuchsin stain*.
2. *Heat* the stain-flooded slide *to steaming* by inverting your Bunsen burner and passing the flame over the pooled stain several times, as shown in Figure 27-1. When you see steam rising from the stain, remove the burner.

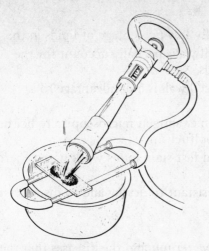

FIGURE 27-1
Heating the carbolfuchsin in the acid-fast stain.

When the steam stops rising, pass the flame over the stain again periodically as necessary to keep the stained smear just at steaming. *Do not boil or allow the smear to dry.* As stain evaporates from the slide, replenish with additional carbolfuchsin. *Steam for 5 minutes.*

3. Allow the slide to *cool for 5 minutes* to prevent breaking the slide when you rinse it with cool water.
4. Tilt the slide to *drain* it, and *rinse* it thoroughly with water.
5. *Flood* the slide with acid-alcohol and allow it to decolorize for 15 to 30 seconds. Then tilt the slide to a 45° angle, and add decolorizer drop by drop. If red color continues to come off in the decolorizer, repeat the flooding with acid-alcohol. When the red color no longer comes off in the decolorizer, perform the next step in this procedure. It is difficult to over-decolorize mycobacteria.
6. *Rinse* the slide with water.
7. Replace the slide on your rack, and *flood* it with Loeffler's methylene blue counterstain. Allow the stain to react for *1 minute.*
8. *Drain* and *rinse* the slide.
9. Blot the slide carefully in your pad of bibulous paper, or allow the slide to air dry in the tilted position.

Now examine your stained smear under your oil-immersion objective, and draw several representative acid-fast cells among the non-acid-fast *S. epidermidis.* Acid-fast bacilli are usually found in clusters and are long thin rods. Label your sketch appropriately, and submit it to your file. Remember that the acid-fast cells will retain the primary stain and appear red, and the non-acid-fast cells will appear blue because they have accepted the counterstain.

Repeat this procedure as often as you wish and time permits until you feel sure of your technique. Then take the post test.

POST TEST

The post test is a self-evaluation. It is not used for a grade. It is designed only to let you decide if you have successfully completed this module.

Part I: True or False

_____ 1. The acid-fast stain has very limited use medically.
_____ 2. Bacteria of the genus *Mycobacterium* are acid-fast.
_____ 3. The genus *Mycobacterium* contains some very important human pathogens.

_____ 4. Acid-fast cells are quite common.

_____ 5. Acid-fast bacteria have an abnormally low percentage of lipids in the cell.

_____ 6. Relatively impermeable, waxlike cell walls partially account for the acid-fastness of certain bacterial cells.

_____ 7. The relative impermeability of the cell wall is a disadvantage to a bacterium.

_____ 8. Acid-fast organisms appear red when examined microscopically because they retained the primary dye, carbolfuchsin.

_____ 9. Paul Ehrlich first developed the acid-fast stain in his work with tubercle bacilli.

_____ 10. Acid-fast organisms are relatively resistant to drying and to disinfectants.

Part II

List two pathogenic bacteria of the genus *Mycobacterium* and the diseases that they cause:

Organism   Disease

1. *Mycobacterium* _____   _____

2. *Mycobacterium* _____   _____

Part III

List in order the main steps of the Ziehl-Neelsen acid-fast stain procedure (omitting washes).

1. _____

2. _____

3. _____

4. _____

5. _____

FORMULAE FOR REAGENTS

1. ZIEHL-NEELSEN'S CARBOLFUCHSIN

basic fuchsin	0.3 gm
ethanol, 95%	10.0 ml
phenol crystals, C.P.	5.0 gm
distilled water	95.0 ml

    Dissolve basic fuchsin in the ethanol. In a separate container, dissolve phenol crystals in water. Mix the two solutions together thoroughly.

2. ACID-ALCOHOL DECOLORIZER

hydrochloric acid, concentrated (37%)	3.0 ml
ethanol, 95%	97.0 ml

    Add the acid to the ethanol, and mix well.

3. LOEFFLER'S METHYLENE BLUE

methylene blue chloride	0.3 gm
ethanol, 95%	30.0 ml
distilled water	100.0 ml

    Dissolve methylene blue in the ethanol. Add the distilled water, and mix well. Filter the solution through filter paper and funnel.

PART FIVE

Characteristics of
Other Selected Organisms

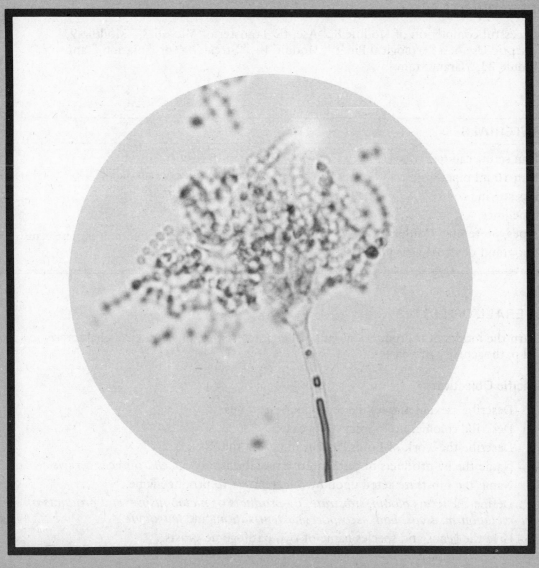

Fruiting head of *Aspergillus*
(440 x).

MODULE 28

Yeast Fungi

PREREQUISITE SKILL

Successful completion of Module 8, "Aseptic Transfer of Microbes," Module 9, "Aseptic Use of a Serological Pipet," Module 14, "Streaking for Isolation," and Module 23, "Gram Stain."

MATERIALS

clean screw-cap test tubes (5)
clean 10 ml pipet
glass stirring rod
grape juice
grapes, preferably Concord
Sabouraud dextrose agar plates (2)*

broth cultures of
   Saccharomyces cerevisiae
   Rhodotorula rubra

*To be prepared by the student if the instructor so indicates.

OVERALL OBJECTIVE

Learn the modes of reproduction, cultural characteristics, physiological characteristics, and pathogenicity of yeasts.

Specific Objectives

1. Describe asexual and sexual reproduction of yeasts.
2. Describe colonial morphology of yeasts.
3. Describe the work of Louis Pasteur in curing the "sick wine."
4. Name the by-products of fermentative metabolism of *Saccharomyces cerevisiae.*
5. Name the substrate acted upon by *S. cerevisiae* to produce wine.
6. Define the terms *bloom, substrate, by-products of metabolism = end products of metabolism, ascus, bud, ascospore, pasteurization,* and *leucocytes.*
7. Give the genus and species name of two pathogenic yeasts.

8. Describe the diseases caused by these two pathogenic yeasts.

9. Name a microscopic characteristic that differentiates these two pathogenic yeasts from *Saccharomyces*.

DISCUSSION

The fungi include yeast, molds, mushrooms, toadstools, smuts, and rusts. In nature their principal role is their power of degradation, that is, their ability to reduce organic material to inorganic molecules. In essence, they return dead organic material to the soil from which it came. Plants convert inorganic molecules, many obtained from the soil, into plant protoplasm (organic material). Animals eat the plants to make animal protoplasm. Fungi and other organisms return dead animal and plant protoplasm to the soil in the form of inorganic molecules so that these inorganic molecules can be used again in the formation of living protoplasm.

Fungi represent a large group of plants even though they are devoid of chlorophyll (photosynthetic pigment), roots, stems, and leaves. They range in size and complexity from one-celled yeast to filamentous molds to complex mushrooms.

The one-celled yeasts reproduce asexually by budding. The budding daughter cell begins as a small protrusion of the mother cell as shown in Figure 28-1. There is an equal amount of genetic material given up from the mother cell to the budding daughter cell since this is a mitotic division. Therefore, the inheritable characteristics of the daughter cell are the same as those of the mother cell. After complete separation from the mother cell, the daughter cell is smaller, indicating that there is not an equal division of cytoplasm as is true in bacterial fission.

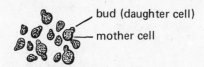

FIGURE 28-1
Saccharomyces cerevisiae — Gram stain.

Some yeasts and molds can participate in sexual reproduction also. In the yeasts (and some of the molds), this takes place by the conjugation and nuclear fusion of two fungal cells of different mating types. This is followed by meiotic divisions of the diploid nucleus, which results in more ascospores. The ascospores are retained in the parent cell, which is then called an ascus (sac). The ascospores with new genetic characteristics break out of the ascus and can then germinate to produce a yeast cell with altered inheritable characteristics. Yeasts and molds that undergo this particular type of sexual reproduction in the ascus are placed in the class *Ascomycetes*.

In this module you will be observing the more common mode of reproduction, that is, asexual reproduction. You will need only to make a Gram stain of a yeast culture to see the asexual buds arising from the mother cell. When you look at the Gram stain, take note of the size of yeast cells as compared to the bacterial cells that you have seen.

Yeasts, unlike the molds, culturally look more like bacteria when grown on artificial media. You will see this by simply observing your yeast streak plates after incubation. The physiological activities of yeast vary. For example, *Saccharomyces cerevisiae* (baker's yeast) converts simple sugars to ethyl alcohol and carbon dioxide when the condition of growth is anaerobic, while *Rhodotorula rubra* produces acid end products from the same substrate. One of the activities you will be doing in this module is designed to demonstrate the variability in the enzyme systems of these two different genera of yeasts. The yeasts that have the enzyme system to convert sugars to potable alcohol have been used by humans for centuries to make wine and beer. The same yeast *(Saccharomyces)* is used to cause bread dough to rise. We call this leavened bread.

In 1857, Louis Pasteur proved that a certain yeast *(Saccharomyces)* produced wine by fermentation of grape sugar. He was also hired by the French government to solve the problem of "sick wine." Until that time the bloom (organisms on the grape skins) was solely responsible for wine production. If organisms other than *Saccharomyces* predominated in the bloom, the wine would have a sour taste (sick wine), since acids instead of alcohol were probably being produced. So Pasteur heated the grape mash to kill *all* the organisms; then he inoculated the cooling mash with the alcohol producing yeast, *Saccharomyces.* This heating of wine mash was the birth of the process of pasteurization. Curing the "sick wine" made Pasteur a public hero to the wine producers and to all wine-loving Frenchmen. Activity 3 resembles this history-making work of Louis Pasteur.

Although most yeasts are beneficial to humans, a few are medically important. The two most famous of the disease-producing yeasts are *Candida albicans* and *Cryptococcus neoformans. Candida albicans* is a normal inhabitant of the intestinal tract, but it becomes a medical problem in patients on prolonged antibiotic therapy. Prolonged use of antibiotics kills the normal bacterial flora necessary to a healthy enteron and allows the yeast to predominate. If *Candida albicans* gets out of control in the vagina, it can cause a most uncomfortable vulvovaginitis with much irritation and discharge. For identification, the discharge is collected by means of a vaginal swab and transported to the lab in a tube containing sterile, normal saline. Gram stains and cultures are done in order to differentiate *Candida* from other organisms that cause vaginitis. *Candida* also causes the condition known as thrush in newborn infants. Identification of *C. albicans* can often be done from the Gram stain alone. Under the microscope *Candida albicans* looks very much like *Saccharomyces* since it is a budding yeast cell, but *Candida* differs in that the bud will often elongate to form a pseudo-hypha as shown in Figure 28-2. The observance of pseudohyphae is diagnostic.

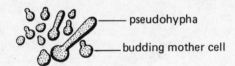

FIGURE 28-2
Candida albicans—Gram stain.

Cryptococcus neoformans is also a budding yeast cell and looks similar to *Saccharomyces* under the microscope. The major difference is that *Cryptococcus* has a large capsule, as shown in Figure 28-3. If this yeast gets into the spinal fluid of a human, it causes a fatal meningitis. The diagnosis is confirmed in the laboratory, primarily by doing a negative stain on the spinal fluid obtained from the patient. A negative stain would show encapsulated budding yeast cells if the patient has cryptococcal meningitis. Figure 28-3 shows typical findings from a positive spinal fluid. (Note the presence of leucocytes, that is, white blood cells.)

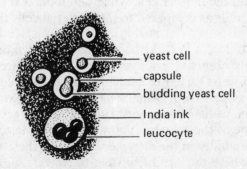

FIGURE 28-3
Cryptococcus neoformans — negative stain.

ACTIVITIES

Activity 1: Microscopic and Cultural Characteristics of Yeast Fungi

Using the broth cultures of *Saccharomyces cerevisiae* and *Rhodotorula rubra,* streak each yeast for isolation on a Sabouraud dextrose agar plate. Remember to shake the culture tubes gently to disperse the cells before you make your inoculations. Incubate the plates at 30°C for 48 hours. Examine the colony morphology, and make drawings of several isolated colonies of each organism. Accompany the drawings with short, written descriptions comparing the two yeasts to each other and to what you have already learned about bacterial colony morphology.

Make a Gram stain from a colony of each yeast. Make drawings of the microscopic appearance of each. Be sure to include a few budding cells. Also include a short, written description comparing the microscopic morphology of *S. cerevisiae* to *R. rubra.* Submit the drawings and descriptions of the cultural and microscopic characteristics of these yeasts to your file.

Activity 2: Physiological Differences of Yeast Fungi When Grown in Grape Juice

Carefully pipet 10 ml of grape juice into two clean screw-cap test tubes. Inoculate one tube of grape juice with a loopful of *S. cerevisiae* broth. Inoculate the other tube with a similar amount of *R. rubra* broth. Be sure to put the caps on tightly and incubate at 30°C for 48 hours (no longer). After incubation, look for bubbles of carbon dioxide gas before opening the tube. Remove the cap next, look for gas again, and smell each tube for the characteristic odor of alcohol. Tabulate the results for carbon dioxide and alcohol production for both yeasts. Submit a table similar to Table 28-1 to your file.

TABLE 28-1   Results of Grape Juice Fermentation

Yeast	CO_2*	Alcohol odor†
Rhodotorula rubra		
Saccharomyces cerevisiae		

*To fill in the CO_2 column, use the terms "Bubbles" or "No Bubbles."
†To fill in the Alcohol odor column, use the terms "Present" or "Absent."

Activity 3: Effect of Pasteurization on the By-Products of Metabolism

You will be putting two to three grapes into each of three clean screw-cap test tubes. Wash your hands with soap and water before handling the grapes. Now rinse off the grapes with cold tap water. *Do not* wash off the bloom (grayish film on skins). You may need to cut each grape into pieces in order to get them into the test tube. Cleanliness, but not asepsis, is important here since in the past, people successfully mashed grapes for wine with their feet. You, however, will mash grapes in all three tubes with a glass-stirring rod. Now that you have released the fermentable sugars in the grape pulp, label the three tubes 1, 2, and 3.

Loosen the caps, and place Tubes 1 and 2 in a boiling water bath, as shown in Figure 28-4. Immediately turn off the heat source, and leave the tubes in the hot water for 5 minutes to kill the mixed flora in the bloom. Cool the tubes rapidly in cold water. Inoculate Tube 1 with one loop of *S. cerevisiae* broth culture. Do nothing more to Tube 2 after heating. Tube 3 has not been heated in order to determine the effect of the still active bloom.

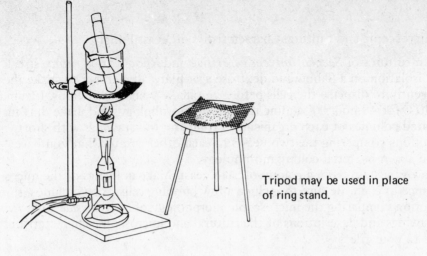

FIGURE 28-4
Pasteurization setup.

Tripod may be used in place of ring stand.

Now tighten the caps on all three tubes, and incubate them at 30°C for 48 hours. Make a table similar to Table 28-2, and fill in the 48 hour results. Submit the completed table to your file.

TABLE 28-2   Results of Pasteurization of Grape Juice

By-products of metabolism	Tube 1 Pasteurized, inoculated with *Saccharomyces*	Tube 2 Pasteurized, not inoculated	Tube 3 Not pasteurized, bloom alive
CO_2 bubbles			
Alcohol odor			

Gram stain the liquid from each of the three tubes. Make drawings of the organism, if present, in each Gram stain. Did you find different organisms in Tube 3 than in Tube 1? If you did, give a short explanation of this, and submit it along with your Gram stain drawings to your file.

Now that you have completed the activities, take the post test so that you can determine if you have learned what this module was designed to teach.

POST TEST

The post test is a self-evaluation. It is not used for a grade. It is designed only to let you decide if you have successfully completed this module.

Part I: True or False

_____   1. Conjugation of yeast cells of different mating types is an example of asexual reproduction.

_____   2. The budding of yeast cells is identical to the fission of bacteria.

_____   3. Isolated yeast colonies look more like bacterial colonies than molds.

_____ 4. The fermentation of sugars by all genera of yeast results in the production of alcohol and carbon dioxide.

_____ 5. The bloom on grapes refers to the blossoms that form on the grape vines.

_____ 6. "Sick wine" means unaged wine that is more likely to give you a hangover.

_____ 7. The ascus is formed in asexual reproduction of yeasts.

_____ 8. The daughter bud begins as a protrusion of the mother yeast cell.

_____ 9. In essence, pasteurization in wine-making is the heating of the substrate to kill only non-alcohol-producing yeasts.

_____ 10. When alcohol and carbon dioxide are produced by yeast, it is an anaerobic process.

_____ 11. *Rhodotorula rubra* produces acid end products of sugar fermentation.

_____ 12. Bubbles in grape sugar fermentation indicate that acid is being produced.

_____ 13. *Saccharomyces cerevisiae* colonies have a red pigmentation.

_____ 14. Leavened bread is a result of the by-products of metabolism of *Rhodotorula rubra*.

_____ 15. Asepsis is essential in the preparation of wine mash.

Part II

List two pathogenic yeasts (give the genus and species name) and the diseases that they cause.

Yeast name

1. _____

2. _____

Diseases produced

1. _____

2. a. _____

   b. _____

Part III

Give a characteristic microscopic feature for each of the two pathogenic yeasts that you have just listed in Part II.

1. _____

2. _____

MODULE 29

Filamentous Fungi

PREREQUISITE SKILL

Successful completion of Module 8, "Aseptic Transfer of Microbes," and Module 10, "Pour Plates."

MATERIALS

7 day Sabouraud slant culture of *Rhizopus* sp. (any species of the genus *Rhizopus)*

Penicillium sp. (any species of the genus *Penicillium)*

sterile petri dishes containing bent glass rod and microscope slide (2)*

sterile water (25 ml) in screw-cap bottle or cotton-stoppered flask*

glass cover slips (2)

small beaker of absolute alcohol (isopropyl)

slide forceps

inoculating equipment

sterile Sabouraud dextrose agar plates (2)*

sterile, empty petri dish

sterile Sabouraud dextrose agar (50ml)*

lactophenol cotton blue, dropper bottle*†

*To be prepared by the student if the instructor so indicates.
†Formula is at end of this module.

OVERALL OBJECTIVE

Recognize colony and structural differences of two common, contaminating molds and learn the scientific names of several pathogenic molds, the diseases that they cause, and the specimens used for their identification.

Specific Objectives

1. Draw or describe a microculture preparation.
2. Recognize and describe different specialized microscopic structures of the molds used in this module.
3. Describe different gross morphology of two selected molds.

4. Define the terms, *hypha, mycelium, sporangiophore, sporangiospore, sporangium, diphasic, dermatophyte, mycosis, conidiophore, conidiospore, aseptate, coenocytic, fruiting heads,* and *rhizoids.*

5. List the four classes of fungi and a genus representative of each class.

6. List the genus and species names of five filamentous fungi that cause systemic mycoses and list the common name of the disease caused by each.

7. Give the common name of a superficial mycosis of hair and skin caused by *Microsporum* sp.

8. Give the common name of an interdigital skin infection caused by *Trichophyton* sp.

9. List the specimens usually collected for the identification of systemic mycoses and superficial mycoses.

10. List three members of genus *Penicillium* that are useful to humans and describe how they are used.

Definitions

Aerial	Growing, forming, or existing in the air
Ascomycetes	A large class of higher fungi distinguished by septate hyphae and by their sexual spores formed in asci or spore sacs
Ascospore	A spore formed as a result of sexual reproduction developed in a sac-like cell, known as an ascus
Ascus	A sac-like structure containing (usually eight) ascospores developed during sexual reproduction in the Ascomycetes
Aseptate	Lacking cross walls
Coenocytic	Hollow, tubelike, nonseptate hyphae containing numerous nuclei in free-flowing protoplasm
Conidiophore	A specialized, aerial hypha-bearing conidia
Conidiospore	An asexual spore produced by a pinching off of the conidiophore or sterigma; not enclosed in a sac; also called conidia
Dermatophytes	Fungi that cause superficial mycoses
Diphasic (dimorphic)	The ability of some fungi to grow in the yeast or filamentous stage, depending on conditions of growth
Fruiting heads	The specialized aerial hyphae (conidiophore or sporangiophore) and the asexual spores they bear
Fungi Imperfecti	A large class of fungi with septate hyphae in which the asexual state of reproduction is known, but not the sexual state. Also called Deuteromycetes.
Hypha	A branching, tubular, or threadlike structure of the fungi
Mycelium	A mass of hyphae forming the vegetative portion of the fungus
Mycosis	A disease caused by a fungus
Phycomycetes	A class of fungi forming a coenocytic mycelium with stiff sporangiophores that bear sporangiospores contained in a sporangium
Rhizoids	Rootlike structures
Septate	Divided by cross walls
Sporangiophore	A special aerial hypha or stalk bearing a sporangium
Sporangiospore	An asexual spore produced at the end of a sporangiophore and contained in a sporangium
Sporangium	A sac or cell containing spores produced asexually
Sterigma	A specialized structure that arises from a conidiophore and supports conidiospores
Zygospore	A thick-walled spore formed during sexual reproduction in the Phycomycetes

DISCUSSION

Molds are among the largest organisms studied in microbiology. Often no special staining or biochemical studies are necessary to identify the genus since their structural differences are often readily visible microscopically *if* the mold colony is not disturbed. You will be growing microcultures to look at undisturbed colonies. These slide cultures can be placed directly on your microscope stage for examination of the structural differences in different mold types. The structural differences that are most often helpful in the identification of filamentous fungi are the different types of spores, spore arrangement, and hyphae.

The molds, also called filamentous fungi in order to differentiate them from yeast fungi, usually have a large, light, fluffy, colonial morphology. The light, fluffy appearance of the colony is due to its hyphae, aerial mycelia, and fruiting heads. So the molds differ from yeasts and most bacteria in the following ways:

1. They have different colonial morphology
2. They are multicellular
3. They have specialized structures that have specialized functions
4. Their asexual reproduction is a result of the pinching off or fragmentation of one of these specialized structures.

There are four classes of fungi:

1. Phycomycetes—have coenocytic hyphae, produce asexual spores in a sporangium
2. Ascomycetes—have septate hyphae, produce unenclosed asexual spores, produce sexual spores in a sac-like structure known as an ascus
3. Basidiomycetes—mushrooms, toadstools, smuts, and rusts
4. Deuteromycetes *(Fungi Imperfecti)*—same as Ascomycetes, except no sexual spores formed.

Following the activities in this module, you will find a much more informative taxonomic key.

In this module, you will be examining the cultural and structural differences of a representative genus *(Rhizopus)* of the class Phycomycetes and the genus *Penicillium* in the class Ascomycetes. *Rhizopus* and *Penicillium* are among the many ubiquitous contaminants that are considered nonpathogenic. You often see them as food spoilage organisms growing on bread, fruit, and many other foods. You will recall the discussion of the important role of fungi in the balance of nature from Module 28 on yeasts. Therefore, as food spoilage organisms, they are merely returning protoplasm to inorganic molecules so that they can be used again.

Other molds are used in the making of some cheeses. For example, the characteristic flavor and aroma of Roquefort cheese are caused by the growth of *Penicillium roqueforti* in curdled milk. The greenish-blue strands that run through the cheese are the actual mold growth itself. The mold, of course, is dead if the cheese has been pasteurized. The same is true of *Penicillium camemberti,* which is used to make Camembert cheese.

Another close relative of the cheese-making mold and the most famous of all molds is *Penicillium notatum.* Sir Alexander Fleming in 1929 showed that if this mold were grown in the appropriate substrate, the end product of metabolism was the wonder drug penicillin. This began the era of the discovery of antibiotics.

Although most molds are harmless or useful to humans, a few are pathogenic. Most of the pathogenic fungi are in the class *Fungi Imperfecti.* For examples of the pathogenic fungi, consult the taxonomic key.

In general, the specimens observed and cultured for laboratory identification of the systemic mycoses are sputums, lesion tissue, and lesion aspirations, whereas the superficial mycoses (dermatophytes) are usually cultured and identified from skin scrapings, nail scrapings, and broken-off hair shafts.

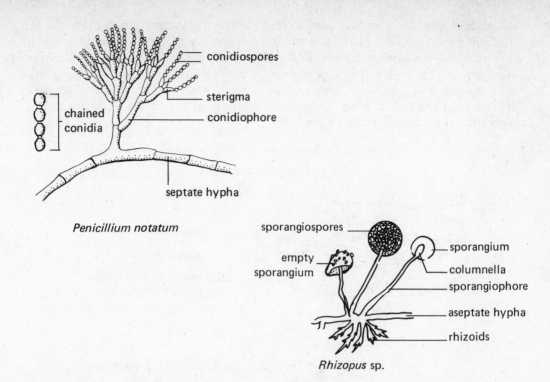

FIGURE 29-1
Two genera of nonpathogenic fungi.

Before you begin the activities, reread the Definitions and inspect Figure 29-1. This will allow you to distinguish between the two classes and to identify the genus of the two nonpathogenic molds that you will be using in this module.

ACTIVITIES

Activity 1: Procedure for Preparing a Microculture Using *Rhizopus* sp.

Pour the Sabouraud dextrose agar into the empty, sterile petri dish, and allow it to solidify. You will cut your microculture agar blocks from this plate so use enough medium to pour a thick plate. You will need only two small agar blocks from this plate so you may wish to share it with your partner.

Check your sterile microculture plate. Make sure that the slide is balanced horizontally across the bent glass rod as shown in Figure 29-2. Now cut a block of Sabouraud dextrose agar *1 cm square,* no larger, with your flamed inoculating loop. Lift the agar block onto the slide using your inoculating loop and/or flamed scalpel. After you have centrally located the agar block on the microscope slide, inoculate all four upper edges of the block with *Rhizopus* sp. Refer to Figure 29-2 for inoculating sites. Sterilize a cover slip in the following manner. While holding the cover slip with a slide forceps, dip the cover slip into a small beaker containing absolute isopropyl alcohol. Let most of the alcohol drain off, and then quickly pass the cover slip through the flame of the Bunsen burner once. This will burn off the remaining alcohol and complete the sterilization process. Place the sterile cover slip on top of the inoculated agar block. Some of the hyphae will grow away from the agar block and cling to the underside of the cover slip so that they will be easily seen through your microscope.

Next pour about 5 to 10 ml of sterile water into the bottom of the microculture plate so that the agar block will not dehydrate upon prolonged incubation. Take care not to get any water on the agar block, slide, or the cover slip. Replace the petri dish lid; your microculture is now complete and ready to incubate. Incubate it *right side up* at room temperature for one week. If the sterile water in the bottom of the plate evaporates, you may have to replace it. After incubation, remove the slide from the petri dish, and place it on the microscope (agar block, cover slip, and all). Be sure to wipe any moisture off the bottom of the slide. Using both the low-power objective and the high-power objective, look carefully at the hyphae, fruiting structures, spores, and rhizoids, if present. Examples of what you are likely to see are depicted in Figure 29-1. Make microscope drawings of the specialized structures, label them, and submit them to your file.

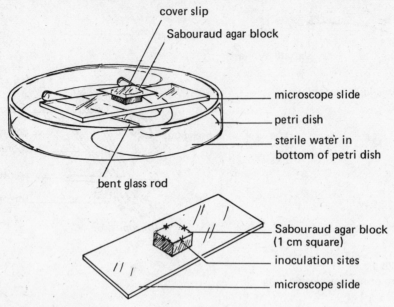

FIGURE 29-2
A microculture preparation and inoculating sites.

Activity 2: Microculture of *Penicillium* sp.

Repeat Activity 1 exactly, except use *Penicillium* sp. for the inoculation of this microculture. Make labeled drawings, and submit them to your file. Refer to Figure 29-1.

Activity 3: Microculture Cover Slip Wet Mount Using Lactophenol Cotton Blue

Follow this procedure to prepare your lactophenol cotton blue wet mount:

1.   Place two drops of lactophenol cotton blue on a clean microscope slide.
2.   With your fingers, remove the cover slip from the microculture agar block. Handle the cover slip by the edges.
3.   Place the cover slip with fungus clinging to it on the pool of lactophenol cotton blue on the slide. *Precaution:* To avoid trapping large air bubbles under the cover slip, place one edge of the cover slip into the lactophenol cotton blue first, and then lower the cover slip gently until it is flat on the slide.
4.   Remember that the fungus is clinging to the cover slip. Examine your wet mount under low power and under high power.

Repeat this procedure with both organisms. Make drawings of all the specialized structures of each fungus that you can find. Do this activity thoroughly because it is the visual support that results in memorized knowledge for you. Label the structures appropriately, and submit the drawings to your file.

Activity 4: Gross Colonial Morphology

Make a single dot inoculation of both molds in the middle of separate plates of Sabouraud dextrose agar, as shown in Figure 29-3.

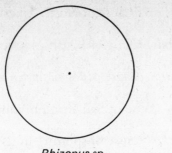

Rhizopus sp.   *Penicillium* sp.

FIGURE 29-3
Single dot inoculation on Sabouraud dextrose
agar.

Incubate the plates right side up at room temperature for several days. Make drawings and give a written description of both plates. Submit them to your file.

Take the post test now.

SIMPLIFIED TAXONOMIC KEY OF THE MORE IMPORTANT FUNGI

Phylum: Thallophyta—Entire plant is somatic, that is, no roots, stems, or leaves.

Subphylum: A. Algae—Possess chlorophyll.
B. Fungi—Yeasts, molds, mushrooms, toadstools, and such. All are devoid of chlorophyll.

Class: 1. Phycomycetes—Usually possess filamentous, nonseptate, multinucleate, tubular hyphae; also called coenocytic hyphae. Asexual spores (sporangiospores) form within a sporangium by a pinching off of the sporangiophore. Sexual reproduction takes place in a thick-walled structure called a zygospore, which forms between two hyphae of different mating types.
Genera: 1. *Rhizopus* sp.
2. *Mucor* sp.
3. Others.

Class: 2. Ascomycetes—Have septate (cross-walled) hyphae; exception is yeasts. All members (molds and yeasts) produce sexual spores (ascospores) within a sac (ascus). Asexual reproduction varies from budding to the formation of conidiospores.
Genera: 1. *Saccharomyces* sp.
2. *Penicillium* sp.
3. *Aspergillus* sp.
4. Many others.

Class: 3. Deuteromycetes *(Fungi Imperfecti)*—Resemble the Ascomycetes, except that no sexual stage has been observed. Most of the fungi pathogenic for humans are in this class. The following are notable pathogens:
a. Yeast—Reproduce by budding; bud may elongate to form a pseudomycelium.

Genus   1. *Cryptococcus neoformans*—Budding yeast cells sur-
and   rounded by a large capsule; cause of a fatal meningitis.
Species: 2. *Candida albicans*—Bud elongates to form a pseudo-
               hypha; cause of vulvovaginitis and thrush; attacks skin
               and nails; may also become a pulmonary infection.
b. Diphasic fungi—Have a yeast stage *in vivo* or *in vitro* at 37°C and a filamentous stage when grown *in vitro* at 25°C. Most systemic (deep) mycoses are in this group.

Genus   1. *Blastomyces dermatitidis*—Cause of North American
and   blastomycosis.
Species: 2. *Blastomyces brasiliensis*—Cause of South American
             blastomycosis.
         3. *Histoplasma capsulatum*—Cause of Mississippi Valley
             fever.
         4. *Coccidioides immitis*—Cause of San Joaquin Valley
             fever.
         5. *Sporotrichum schenckii*—Cause of sporotrichosis of
             lymph nodes.

c. Filamentous fungi—No budding yeast stage. Most dermatophytes
   (superficial mycoses) are in this group.
Genera: 1. *Geotrichum* sp.—Can cause oral, intestinal, or pul-
             monary geotrichosis.
         2. *Microsporum* sp.—Causes ringworm of scalp and skin.
         3. *Trichophyton* sp.—Attacks hair, skin, and nails; causative
             organism of athlete's foot found in this genus.
         4. *Epidermophyton* sp.—Attacks skin and nails; hair
             never affected.

Class: 4. Basidiomycetes—Have septate branched hyphae. The typical
            reproductive structure is a basidium (a club-shaped structure)
            that gives rise to basidiospores. No pathogens for humans are in
            this class. Some produce fatal toxic substances; some are edible.
            Mushrooms, toadstools, smuts and rusts are in this group.
Genera: 1. *Amanita* sp.—Many poisonous toadstools.
         2. *Agaricus* sp.—Edible mushrooms.
         3. Others.

POST TEST

The post test is a self-evaluation. It is not used for a grade. It is designed only to let you decide if you have successfully completed this module.

Part I

List the four classes of fungi, and give a representative genus for each.

1. class _____ genus _____
2. class _____ genus _____
3. class _____ genus _____
4. class _____ genus _____

Part II

1. From the following choices, pick out and list below the specialized structures
   of the mold *Rhizopus*.

   conidiospore   conidiophore
   sporangiospore   rhizoid
   sterigma   septate hyphae
   columnella   coenocytic hyphae
   sporangium   ascus
   sporangiophore

   a. _____   d. _____
   b. _____   e. _____
   c. _____   f. _____

2. From the same list, pick out the specialized structures of the mold *Penicillium*.

   a. _____   d. _____
   b. _____   e. _____
   c. _____

Part III

Put the following list in the order in which it would be placed or admitted into a petri dish microculture:

sterile water   bent glass rod
microscope slide   cover slip
inoculum   Sabouraud agar block

1. _____

2. _____

3. _____

4. _____

5. _____

6. _____

Part IV

List the genus and species names of five diphasic fungi and the common name of the disease each one causes.

Genus and species   Disease

1. _____   _____

2. _____   _____

3. _____   _____

4. _____   _____

5. _____   _____

Part V

Give the common name of a disease caused by:

1. *Microsporum* sp. _____

2. *Trichophyton* sp. _____

Part VI

List the specimens most often used to identify:

Systemic Mycoses   Superficial mycoses

1. _____   _____

2. _____   _____

3. _____   _____

Part VII

List three species of the genus *Penicillium* that are advantageous to humans, and state how they are helpful.

1. *Penicillium* _____

    used for: _____

2. *Penicillium* _____

    used for: _____

3. *Penicillium* _____

    used for: _____

FORMULA FOR REAGENT

LACTOPHENOL COTTON BLUE

phenol crystals (C.P.)	20.0 gm
lactic acid (C.P.)	20.0 ml
glycerin (C.P.)	
(glycerol)	40.0 ml
distilled water	20.0 ml
cotton blue	
(aniline blue)	0.5 gm

Combine ingredients in the order listed, and mix thoroughly. Store in a brown screw-cap bottle.

MODULE 30

Protozoans: Sarcodines, Flagellates, and Ciliates

PREREQUISITE SKILL

Successful completion of Module 4, "Compound Microscope for the Study of Microbes," and Module 6, "Preparing a Wet Mount."

MATERIALS

living cultures* of
   Amoeba proteus
   Euglena viridis
   Paramecium caudatum
stained smears* of
   Entamoeba histolytica cysts
   Trichomonas hominis trophozoites
   Giardia lamblia cysts or trophozoites
   Balantidium coli cysts
blood smear* of *Trypanosoma gambiense*
   or another *Trypanosoma* sp.

microscope slides
cover slips
Vaseline
toothpicks
Protoslo* or methyl cellulose†

*Purchased from a biological supply company.
†Formula is at the end of this module.

OVERALL OBJECTIVE

Describe nonpathogenic, representative protozoans and identify some parasitic representatives of the same classes and the diseases that they cause.

Specific Objectives

1. List the three groups of organisms studied in a parasitology course.
2. Describe how protozoans are classified.
3. List the four classes of protozoans and describe how they move.
4. Give the genus and species name of a nonparasitic representative in each class.
5. List the genus and species names of all the parasitic representatives of the three classes presented in this module; also name and describe the disease each causes and the specimen of choice for laboratory identification of each.

6. Draw and label the most identifying structural characteristics of the nonparasitic representatives of each class described in this module. List the following number of identifying structural characteristics:

    Amoeba proteus: two characteristics
    Euglena viridis: three characteristics
    Paramecium caudatum: three characteristics

7. Describe the difference between the trophozoite stage and the cyst stage and tell when each stage occurs in the parasitic protozoans.

8. Draw and label the most identifying cyst structural characteristics of all the parasitic protozoans described in this module.

9. Define the terms *protozoan, metazoan, free-living,* and *helminth.*

10. Give the genus name of two parasitic protozoans introduced into man via the bite of an insect.

DISCUSSION

The study of parasites that are larger than bacteria and are in the animal kingdom involves a two-semester course at some schools. In this course you will be introduced to only the most common representatives of this vast specialty.

Parasites may be one-celled animals (protozoans) or multicellular animals (metazoans). Parasitology courses include the study of protozoans, helminths (parasitic worms), and parasitic arthropods such as lice, mites, and ticks. Helminths and arthropods are metazoans. Classical parasitology courses are slanted toward the zoological aspect, which emphasizes the parasite rather than the effect of the parasite on the host. This module will be a medically oriented presentation of a few protozoan-caused diseases.

Protozoans are classified according to the type of motility they exhibit or their mode of reproduction. The following is a simplified classification:

    Kingdom: Animal
      Phylum: Protozoa
        Class 1:   Sarcodina—Locomotion by means of pseudopods.
        Class 2:   Mastigophora—Locomotion by means of flagella.
        Class 3:   Ciliata—Locomotion by means of cilia.
        Class 4:   Sporozoa—Multiply by spore formation; no locomotor organelles; complex life cycle.

Table 30-1 shows the free-living (nonparasitic) as well as the parasitic representatives that you will study in each class. This table, along with Figure 30-1, summarizes the clinical and laboratory significant characteristics of each parasite. From Table 30-1, you can see why the specimen of choice to study parasitic protozoans is not readily available for classroom use. It would be most difficult to have a fresh stool specimen containing *Entamoeba histolytica* or *Balantidium coli* and a vaginal discharge of *Trichomonas vaginalis* available when you need it. Therefore, it will be necessary for you to study the parasitic, disease-producing protozoans from stained smears. For the same reason, you will be using the nonparasitic representatives for the study of living forms of each class.

The parasitic sarcodines, flagellates, and ciliates are motile in a freshly obtained specimen that is still warm from body heat. This motile form of the protozoans is called the trophozoite. As the specimen cools after leaving the host, the trophozoite becomes inactive, rounds up, forms thick walls, and is then called a cyst. It is the trophozoite that causes the pathological condition in the host. Most of the stained smears that you will be examining of the parasitic protozoans will be of the cyst stage, with the exception of *Trichomonas hominis (T. vaginalis)* and *Giardia lamblia.* *T. hominis* has no cyst stage. The inactive cysts are often more useful than the trophozoites for laboratory identification of the parasite.

TABLE 30-1 Summary of Significant Characteristics of Protozoans Presented in Modules 30 and 31

Class	Classification characteristics	Free-living nonparasitic representative	Parasitic representative	Portal of entry or mode of entry	Parasitic condition in humans	Specimen of choice for identification of parasite
Sarcodina	Locomotion — pseudopods	Amoeba proteus	Entamoeba histolytica	Ingestion of cysts	Amoebic dysentery	Fresh stool
Mastigophora	Locomotion — flagella	Euglena viridis	Trichomonas hominis (T. vaginalis)	Fecal contamina-tion of the vagina or sexual intercourse	Vulvo-vaginitis	Vaginal discharge Urethral discharge
			Giardia lamblia	Ingestion of cysts	Enteritis and diarrhea	Fresh stool
			Trypanosoma gambiense or any trypanosomal species	Bite of insect vector	African sleeping sickness	Blood smear
Ciliata	Locomotion — cilia	Paramecium caudatum	Balantidium coli	Ingestion of cysts	Recurrent diarrhea alternating with consti-pation	Fresh stool
Sporozoa*	No locomotor organelles; multiply by forming spores; complex life cycle	None	Plasmodium sp.	Bite of insect vector	Malaria	Blood smear

*Sporozoa are presented in a separate module because of the complexity of their life cycles.

In the following activities, you will not see all the structures that are labeled in Figure 30-1. Since this is not a zoology course, just search for enough structures to fulfill Specific Objective 6.

ACTIVITIES

Activity 1: Class: Sarcodina (Rhizopoda)

Make a wet mount of the trophozoite stage of the nonparasitic, living *Amoeba proteus*. Refer to Module 6, "Preparing a Wet Mount," for review if you feel it necessary. First search the wet mount preparation with your low-power objective to locate an amoeba. Under low power, an amoeba looks like a mass of granular material. Sometimes it is necessary to examine more than one wet mount before finding an amoeba. Once you have located an amoeba, move the organism to the absolute center of the microscope field before turning to your high-power objective. Look for forming pseudopods, streaming protoplasm, nucleus, contractile vacuole, and inclusions, as shown in Figure 30-1.

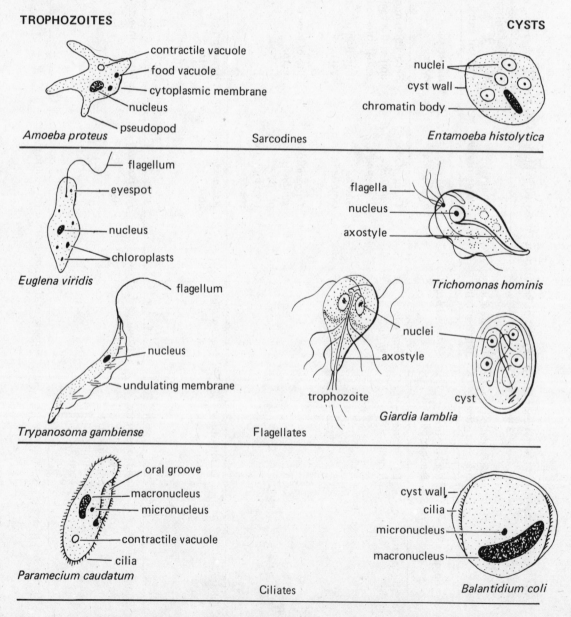

TROPHOZOITES

Amoeba proteus
- contractile vacuole
- food vacuole
- cytoplasmic membrane
- nucleus
- pseudopod

Sarcodines

CYSTS

Entamoeba histolytica
- nuclei
- cyst wall
- chromatin body

Euglena viridis
- flagellum
- eyespot
- nucleus
- chloroplasts

Trichomonas hominis
- flagella
- nucleus
- axostyle

Trypanosoma gambiense
- flagellum
- nucleus
- undulating membrane

Flagellates

Giardia lamblia
- nuclei
- axostyle
- trophozoite
- cyst

Paramecium caudatum
- oral groove
- macronucleus
- micronucleus
- contractile vacuole
- cilia

Ciliates

Balantidium coli
- cyst wall
- cilia
- micronucleus
- macronucleus

FIGURE 30-1
Identifying structures of representative protozoans.

Make a drawing of the trophozoite, using arrows to show the direction of the movement of protoplasm. Label the internal structures that you were able to see, and submit the drawing to your file.

Next examine a stained smear of the parasitic representative of this class, *Entamoeba histolytica.* Draw as much detail of the cysts as possible using your oil-immersion objective. Refer to Figure 30-1 for identifying structures. Remember that it is the trophozoite of *E. histolytica* that invades the mucosa of the large intestine, producing ulcers, which are accompanied by dysentery. Trophozoites and red blood cells can be found in the freshly caught fluid feces. Amoebic liver abscesses occur on rare occasions. Submit your drawing of the parasitic cysts to your file with a short, written description of the disease that they cause.

Activity 2: Class: Mastigophora

Make a wet mount of the trophozoite stage of the living *Euglena* by adding one drop of culture to one drop of Protoslo. Euglena move very rapidly by means of their whip-like flagella. The Protoslo slows down the organisms so you will be able to keep them in your microscope field. Except for the addition of the drop of Protoslo, make the wet mount the same as you did in Activity 1. Refer to Figure 30-1 for the structures you are likely to see. Look especially for the flagellum and chloroplasts. *Euglena* is thought to be a transitional organism linking the plant and animal kingdoms since it has characteristics of each. *Euglena* are small so draw what you see, and then submit this to your file.

Review Table 30-1 again for the most common or most famous parasitic flagellates. Using your oil-immersion objective, examine the stained slides of the flagellates that you have available.

Note that *Trichomonas hominis* trophozoites are depicted in Figure 30-1. Review Table 30-1 again to associate this organism with the area that it parasitizes. *T. hominis* infestations in females remain limited to the vulva, vagina, and cervix. The major symptoms in females are a profuse, frothy, yellowish vaginal discharge with tenderness, pruritis, and burning of the vulva. In males the prostate, seminal vesicles, and urethra are affected. From Table 30-1 you can see that the genital discharge is the specimen sent to the laboratory for identification of the causative organism. Immediately after arrival in the laboratory, a wet mount is made of the vaginal or urethral discharge in a drop of saline. The wet mount is then inspected microscopically for the actively motile trichomonad flagellates. It is important to distinguish *Trichomonas hominis* infestations from other organisms that cause profuse discharge and irritation of the genitalia, such as *Candida albicans* and *Neisseria gonorrheae.*

Make drawings of the stained *Trichomonas hominis* trophozoites, along with a written description of the condition that they cause, and then submit them to your file.

If you have stained smears of *Giardia lamblia* and *Trypanosoma gambiense,* examine them with your oil-immersion lens also. As you look at the cysts and trophozoites of *Giardia lamblia,* refer to Figure 30-1 for identifying structures and to Table 30-1 for the pathological conditions that they cause. Note the eyelike appearance of the nuclei. Giardiasis is found more often in children than in adults. Make drawings of the Giardia cyst and trophozoite, give a written description of the disease, and submit both to your file.

Next inspect a blood smear of *Trypanosoma gambiense* or another *Trypanosoma* species with your oil-immersion objective. You will see many of these flagellates among the erythrocytes of a patient who has had African sleeping sickness. These trypanosomal flagellates are injected into the host via the bite of a tsetse fly. The parasites multiply at the site of inoculation, where a primary lesion appears. Regional lymphadenitis occurs, followed by dissemination of the flagellates via the bloodstream. The flagellates release a toxin, and the CNS (central nervous system) is seriously affected. The progressive symptoms are malaise, drowsiness, coma, and finally death.

Make a drawing of the trypanosomes among the blood cells, accompany the drawing with a written description of the disease, and submit this to your file.

Activity 3: The Ciliata

Make a wet mount of the living *Paramecium* culture using Protoslo as you did in Activity 2. Examine the wet mount preparation for the characteristic structures and movements of this nonpathogenic representative of the ciliated protozoans. See Figure 30-1 for characteristic structures. Make a drawing of the organism, and label the structures that you see. If you have stained smears of paramecia, you may wish to look at them with your oil-immersion objective also.

Next, study the stained smears of the cysts of *Balantidium coli*, the only parasitic ciliate of humans. It is the largest of parasitic protozoans. The cysts of *Balantidium coli* are ingested by the host, the cyst wall dissolves, and the trophozoite evolves to invade the intestinal lining of the large bowel, causing abscesses and ulcerations. The most common clinical manifestation of balantidial dysentery is chronic, recurrent diarrhea alternating with constipation. Intermittent attacks of severe dysentery with bloody, mucoid stools accompanied by abdominal pain can also occur. Make a drawing of a *Balantidium coli* cyst, along with a written description of the disease, and submit both to your file.

Remember that the parasitic protozoa, class Sporozoa, will be presented in Module 31.

When you have completed the drawings and the written descriptions of activities in this module, take the post test. Repeat this module if necessary in order to complete the post test to your satisfaction.

POST TEST

The post test is a self-evaluation. It is not used for a grade. It is designed only to let you decide if you have successfully completed this module.

Part I: True or False

_____ 1. All protozoans are classified according to their modes of locomotion.

_____ 2. *Giardia lamblia* is a ciliate.

_____ 3. All of the following cause a form of enteritis: *Entamoeba histolytica, Balantidium coli, Giardia lamblia,* and *Trypanosoma gambiense.*

_____ 4. Laboratory identification of the parasitic protozoans is usually made from microscopic examination of the trophozoite stage.

_____ 5. Specimens containing the parasitic trophozoite stage are more readily available for classroom study than those containing the cyst stage.

_____ 6. The cyst stage causes the pathological condition in the host.

_____ 7. *Paramecium caudatum* is thought to link the plant kingdom to the animal kingdom.

_____ 8. An identifying characteristic of *Trichomonas vaginalis* in the cyst stage is that the two nuclei appear as eyes.

_____ 9. The identifying symptoms of giardiasis are recurrent diarrhea alternating with constipation.

_____ 10. African sleeping sickness is contracted by ingestion of cysts.

_____ 11. The cyst stage is the active, motile stage of protozoans.

_____ 12. All metazoans are multicellular and therefore are not parasitic.

_____ 13. Helminths are protozoans.

_____ 14. Free-living is a synonym for parasitic.

Part II

Match the disease or symptoms of the disease with the causative organism.

_____ 1. *P. vivax*   a.   vulvovaginitis

_____ 2. *B. coli*   b.   drowsiness, coma, and death

_____ 3. *E. histolytica*   c.   giardiasis

_____ 4. *T. gambiense*   d.   recurrent diarrhea alternating with

_____ 5. *G. lamblia*   constipation

_____ 6. *T. vaginalis*   e.   malaria

                                        f.   dysentery with fluid feces

Part III: Completion

1. Give the genus and species names of three parasitic protozoans that enter the host via the oral cavity.

   a. _____

   b. _____

   c. _____

2. Give the genus and species names of two parasitic protozoans in different genera that are transmitted to humans via an insect vector.

   a. _____

   b. _____

3. Give the genus and species name of a parasitic protozoan that can be transmitted by sexual intercourse.

   _____

4. List three groups of organisms studied in parasitology.

   a. _____

   b. _____

   c. _____

FORMULA FOR REAGENT

10% METHYL CELLULOSE

      methyl cellulose power   10 gm

      tap water   100 ml

      Heat tap water to 85°C. Add methyl cellulose powder. Cool the mixture in an ice bath to approximately 5°C as you stir rapidly and constantly. This solution is stable at room temperature. Store in a tightly closed, screw cap bottle.

      Dilute this stock solution 1:5 for use in protozoan wet mounts. Add water slowly and stir constantly as you make this dilution to prevent the formation of lumps.

MODULE 31

Protozoans: Sporozoa

PREREQUISITE SKILL

Successful completion of Module 4, "Compound Microscope for the Study of Microbes."

MATERIALS

stained blood smears of any species of
   Plasmodium

OVERALL OBJECTIVE

Learn the life cycle of a sporozoan and recognize the parasite in a stained blood smear.

Specific Objective

1. Describe and sketch the life cycle of *Plasmodium* sp.
2. Give the genus and species names of three malaria-causing organisms.
3. Define the terms *merozoite, sporozoite, amoeboid, schizogony, sporogony,* and *Anopheles*.
4. Describe the symptoms of malaria and the cause of the symptoms.
5. Name four ways that malaria, in theory, could be eradicated.
6. Name, draw, and label the most common stages of the life cycle of the parasite in a stained blood smear.

DISCUSSION

The organisms in this class (Sporozoa) are all obligate parasites; therefore, there are no free-living representatives for you to examine. Plasmodia, the malaria-causing protozoans, are the representative parasites that you will study. The plasmodia have a complex life cycle requiring two very different hosts. They have a sexual cycle, in which spores are formed, and an asexual cycle. The sexual cycle takes place in the gut and abdominal wall of the female of some species of mosquito in the genus *Anopheles*. The asexual cycle takes place in the liver and erythrocytes of humans and causes the symptoms of the disease. There are several species (at least five) of the genus *Plasmodium* that cause malaria. Some examples are *Plasmodium vivax, Plasmodium malariae,* and *Plasmodium falciparum.* In order for malaria to remain endemic in the human population, both the asexual cycle and the sexual cycle of this parasitic protozoan must be completed.

Humans contract malaria from the bite of a plasmodial-infected, female anopheline mosquito. As the mosquito inserts its proboscis into a human to take its blood meal, it injects the plasmodial sporozoite at the same time via its saliva. The sporozoites begin the asexual cycle by the pre-erythrocytic development of merozoites in the parenchymal cells of the liver. The merozoites can repeat the pre-erythrocytic cycle in liver cells, or they can enter the erythrocytic cycle. Once the merozoites penetrate the erythrocytes, the parasite undergoes several morphological changes, as shown in Figure 31-1. First a ring form develops, which enlarges to become a mature amoeboid trophozoite filling most of the parasitized red blood cell. Next, asexual multiplication takes place by the splitting of nuclear material and cytoplasm of the amoeboid-appearing parasite to form more merozoites. Depending on the species, this multiple fission (schizogony) results in 6 to 36 new merozoites per parasitized erythrocyte. As the erythrocyte ruptures, the merozoites are freed into the blood plasma to infect many other erythrocytes. This cycle of erythrocyte infection and new merozoite production is repeated several times before the human shows symptoms, usually about 10 days after being bitten by the infected mosquito.

During the erythrocytic cycle, some merozoites differentiate as male and female gametocytes. Figure 31-2 is a simplified schematic presentation of the asexual cycle. For the sexual cycle to evolve, the gametocytes of both sexes must be ingested in the blood meal of another female *Anopheles* mosquito, as shown in Figure 31-3. In the gut of the mosquito, the male gametocyte forms spermatozoa, and the female forms an ovum. Fertilization of the ovum takes place, and the resulting zygote changes

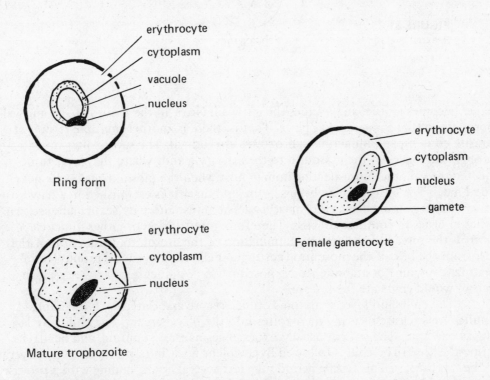

Ring form

Mature trophozoite

Female gametocyte

FIGURE 31-1
Stages of the plasmodium life cycle most commonly seen in blood smears.

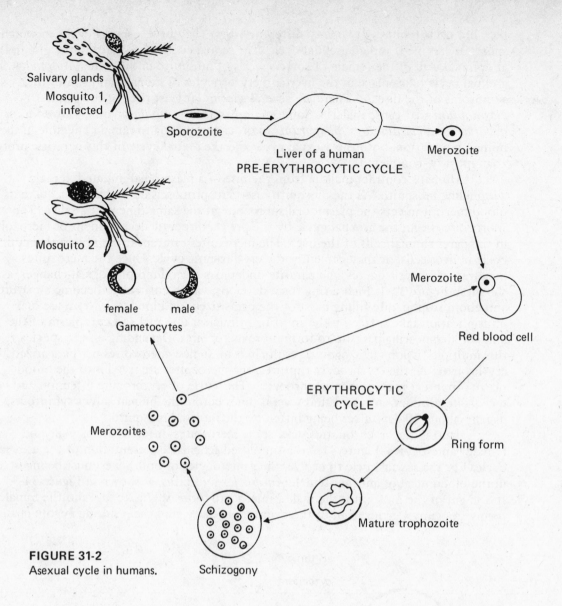

FIGURE 31-2
Asexual cycle in humans.

Labels within figure: Salivary glands, Mosquito 1, infected; Sporozoite; Liver of a human PRE-ERYTHROCYTIC CYCLE; Merozoite; Mosquito 2; female male Gametocytes; Merozoite; Red blood cell; ERYTHROCYTIC CYCLE; Ring form; Merozoites; Mature trophozoite; Schizogony

shape, becomes motile, and invades the gut wall. Next, in the tissues of the gut wall, sporogony of the parasite takes place. That is, there is another multiple fission of parasite content, and numerous sporozoites are formed. The sporozoites migrate through the tissues of the mosquito to the salivary glands where they wait to be injected into another unsuspecting human host when the mosquito takes its next blood meal. The asexual cycle begins again, and malaria is established in a new host.

An organism such as the malaria parasite must alternate sexual and asexual cycles in order to continue to exist. Therefore, interruption of either life cycle controls the spread of the disease. Elimination of the mosquito, protection of the host from the bite of the mosquito (netting, repellants, and such), prophylactic treatment (quinine or atabrine) for exposed persons, and cure of active cases, in theory, would eradicate the disease.

The chills and fever symptoms of malaria are associated with the almost simultaneous release of many merozoites into the bloodstream. The chill may last as long as one hour. The patient usually experiences nausea, vomiting, and headache during this time. The chill is followed by a violent fever in response to toxins liberated by the merozoites. This febrile period may last several hours, ending with a profuse sweating stage. The patient is exhausted and falls asleep but awakens feeling relatively well. In time, these attacks recur at regular intervals of 48 to 72 hours, depending upon the species of *Plasmodium* causing the disease.

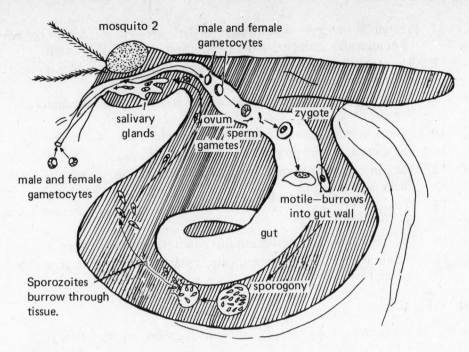

FIGURE 31-3
Sexual cycle in gut of mosquito.

ACTIVITY

Activity 1: Examination of the Malaria Parasite in Stained Blood Smears

Inspect a stained blood smear for the different stages of the *Plasmodium* parasite as shown in Figure 31-1. Use your oil-immersion objective since you must look inside the erythrocytes. Make drawings of as many stages as you find. The ring form is the most common stage. You will have to be a persistent microscopist to find other stages of the plasmodial parasite. Submit your drawings to your file, and take the post test when you feel ready.

Related Experience

Draw and label the life cycle of a plasmodial parasite. If you can do this from memory, you have memorized two-thirds of this module.

POST TEST

The post test is a self-evaluation. It is not used for a grade. It is designed only to let you decide if you have successfully completed this module.

Part I: True or False

_____ 1. Merozoites form in the liver and erythrocytes of humans.
_____ 2. The multiple fission of the mature trophozoite results in the production of sporozoites.
_____ 3. The sexual cycle begins with the ingestion of the male and female gametes by the mosquito.
_____ 4. Sporogony takes place in the lumen of the mosquito's gut.
_____ 5. Fertilization takes place in the lumen of the mosquito's gut.
_____ 6. Schizogony and sporogony are both multiple fission of parasite cell content.

_____ 7. The chills and fever symptoms of malaria are associated with the release of numerous merozoites from many rupturing red blood cells.

_____ 8. The most common stage of the parasite seen in stained blood smears is the mature trophozoite.

_____ 9. A few organisms in the class Sporozoa are free-living and therefore nonparasitic.

_____ 10. Sporozoites are formed in the asexual cycle.

_____ 11. The asexual cycle takes place in the mosquito.

_____ 12. The gametes of both sexes must be ingested before the asexual cycle in the mosquito begins.

_____ 13. The pre-erythrocytic cycle results in the production of merozoites in the liver of humans.

_____ 14. The saliva of an infected mosquito contains the sporozoites.

_____ 15. The amoeboid stage of the parasite in the erythrocytic stage is called the mature trophozoite.

Part II: Completion

1. Give the genus and species names of three malaria-causing parasites.

   a. _____

   b. _____

   c. _____

2. List four major symptoms that always occur during a malaria attack.

   a. _____

   b. _____

   c. _____

   d. _____

3. Name the three structures of the ring form of the parasite that you saw in Activity 1.

   a. _____

   b. _____

   c. _____

4. List four means by which malaria may theoretically be eradicated.

   a. _____

   b. _____

   c. _____

   d. _____

MODULE 32

Some Platyhelminthic Infestations of Humans

PREREQUISITE SKILL

Successful completion of Module 4, "Compound Microscope for the Study of Microbes."

MATERIALS

prepared slides* of:
- *Clonorchis sinensis* adult
- *Clonorchis sinensis* ova
- *Taenia saginata* composite adult
- *Taenia saginata* ova
- *Taenia solium* composite adult
- *Taenia solium* ova
- *Taenia solium* cysticercus
- *Echinococcus granulosus* adult
- *Echinococcus granulosus* ova
- Hydatid sand

preserved specimens of adult helminths listed

*Available from a biological supply company.

OVERALL OBJECTIVE

Recognize some parasitic flatworms and their ova, as well as the diseases they cause and the major symptoms of each infestation.

Specific Objectives

1. Explain why helminths are studied in microbiology.
2. Define the terms *helminth, monoecious, dioecious, trematode, cestode, scolex, proglottid, cysticercus, distal,* and *proximal.*
3. Discuss Chinese liver fluke infestations and list the diagnostic stage.
4. Discuss beef tapeworm infestations and give the name for the stage that is infective for humans.

5. Describe clonorchiasis, including the tissues parasitized in humans, major symptoms of the disease, and the method by which it is contracted.
6. Compare beef tapeworm infestation with pork tapeworm infestation.
7. Describe hydatid disease, including symptoms and treatment.
8. Draw and describe the adult parasitic worms and their characteristic ova.

DISCUSSION

Two phyla of worms are of medical significance because they contain genera that are parasitic in humans. Members of the phylum Platyhelminthes (flatworms) cause fluke and tapeworm infestation. Roundworms, which cause a variety of infestations, are placed in the class Nematoda of the phylum Aschelminthes. The collective term for all these parasitic worms is "helminths," and the science that studies them is helminthology. Helminthology is a vast specialty in itself, and, once again, we will deal primarily with the medically significant organisms most frequently encountered in the United States. We will study the Chinese liver fluke *(Clonorchis sinensis),* however, because it is probably the best known fluke infestation of humans. Most fluke infestations seen in the United States occur in recent immigrants from areas of the world where they are more common, for example, the Orient or South America.

This module is limited to the flatworms, and the next module will describe the roundworms.

Helminths, which are metazoans (sometimes quite large), are studied in microbiology because diagnosis of helminthic infestations in the clinical laboratory is usually by microscopic examination of stool specimens, body fluids, or tissues (biopsy) for ova or larvae of the parasite (see Table 32-1).

TABLE 32-1 A Summary of the Parasitic Platyhelminthes

Parasite	Disease	Clinical symptoms	Diagnostic stage	Infective stage for humans
Clonorchis sinensis liver fluke	Chinese liver fluke infestation	Blocking of bile ducts, jaundice, cirrhosis	Ova in feces	Metacercaria in raw, fresh-water fish
Schistosoma mansoni *Schistosoma japonicum* blood flukes	Schistosomiasis	Spleen and liver enlargement, cirrhosis, schistosomal dysentery	Ova in feces	Free-swimming cercaria in fresh water penetrate skin and enter circulatory system
Taenia saginata beef tapeworm	Beef tapeworm infestation	Diarrhea, increased appetite, intestinal obstruction	Ova or proglottids in feces	Cysticercus
Taenia solium pork tapeworm	Pork tapeworm infestation	Persistent diarrhea, serious complications with bladderworm encystment	Ova or proglottids in feces, surgical detection of bladderworm	Cysticercus or ova
Echinococcus granulosus	Hydatid disease or echinococcosis	Symptoms vary depending on location of cysts.	Precipitin, skin tests	Ova

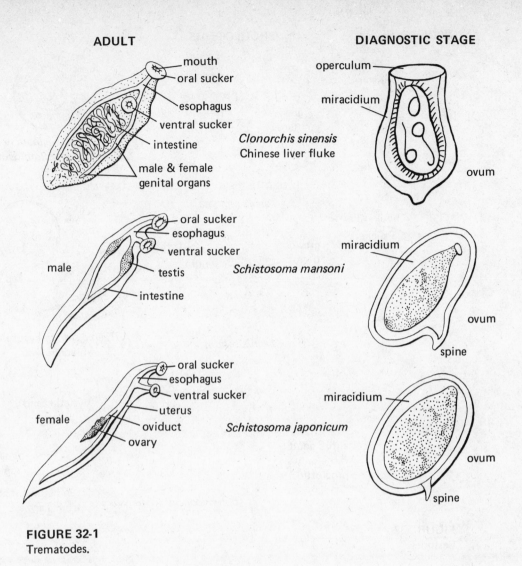

ADULT

mouth
oral sucker
esophagus
ventral sucker
intestine
male & female genital organs

Clonorchis sinensis
Chinese liver fluke

oral sucker
esophagus
ventral sucker
testis
intestine

male

Schistosoma mansoni

oral sucker
esophagus
ventral sucker
uterus
oviduct
ovary

female

Schistosoma japonicum

DIAGNOSTIC STAGE

operculum
miracidium
ovum

miracidium
ovum
spine

miracidium
ovum
spine

FIGURE 32-1
Trematodes.

The parasitic flatworms may be classified by the following simplified scheme:

Kingdom: Animal
  Phylum: Platyhelminthes—Body flattened dorsoventrally, thin and soft.
    Class 1: Trematoda—Flukes, all parasitic (liver flukes, blood flukes, lung flukes).
    Class 2: Cestoda—Tapeworms, all parasitic (beef tapeworm, pork tapeworm, fish tapeworm, hydatid worm).

The leaf-shaped trematodes are mostly *monoecious* (both sexes in one animal), but the schistosomes (blood flukes) are *dioecious,* that is, males and females are separate animals. Most trematodes have two suckers at the anterior end of the body for attachment to the host. The major portion of the body is occupied by the extensive reproductive system, as shown in Figure 32-1.

Most fluke infestations are diagnosed from the characteristic ova in the feces of the parasitized host. One or two intermediate hosts may be required in the complex life cycles of the flukes. We will make no attempt to study life cycles in this module but rather will stress the effect of the parasite on the host. You may check any standard zoology textbook for details of the life cycle of any of the specific helminths if you are sufficiently interested.

Cestode or tapeworm infestations are common in humans. Adult tapeworms live in the small intestine where they attach themselves by suckers and, in some species, hooks on the *scolex* or head of the worm. The neck or growing region is directly behind the scolex and is, in turn, followed by immature, mature, and then gravid

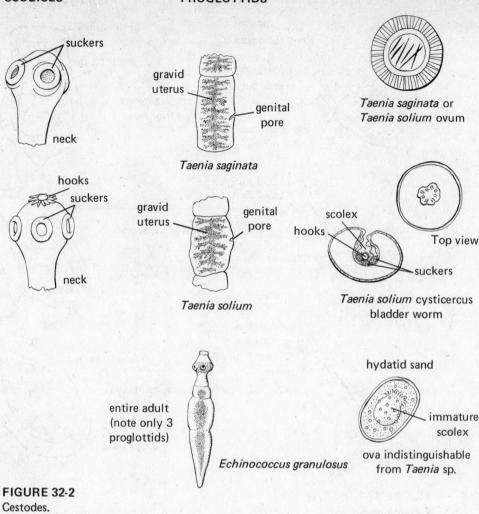

FIGURE 32-2
Cestodes.

proglottids (segments), in that order. That is, the immature proglottid is located **proximally** and the gravid proglottid is located distally to the budding scolex. The uterine pattern in gravid proglottids is frequently species specific. The scolex and sometimes proglottids are characteristic and hence diagnostic for a given species, as shown in Figure 32-2.

    The complex life cycles of the tapeworms include one or more intermediate hosts that harbor the larval stage. This larval stage is infectious for humans and is usually contracted by consuming the flesh of the intermediate host, such as cattle or swine. The cysticercus or larval stage that occurs in the muscle tissue of the intermediate host animal is a small bladder containing an inverted scolex, often called a bladderworm. When a human ingests the cysticercus in rare meat, it passes into the intestine where the cyst everts, allowing the scolex to attach to the intestinal mucosa and develop into a mature tapeworm.

ACTIVITIES

Activity 1: *Clonorchis sinensis*

The Chinese liver fluke, *Clonorchis sinensis (Opisthorchis sinensis),* inhabits the bile ducts, gall bladder, and pancreatic ducts of humans where it causes biliary cirrhosis and jaundice. The adults are about 20 mm long by 4 mm wide. Human infestation begins when the larva, the metacercaria, are ingested with the raw or undercooked flesh of freshwater fish. Once ingested, the metacercaria move to the bile ducts where they become adult flukes. *Clonorchis* is monoecious so a single fluke can produce fertilized ova, each containing a viable miracidium, an early larval form. The ova leave the human body in feces and enter fresh water. A single genus of snail serves as the first intermediate host in the life cycle of *Clonorchis* when it ingests the miracidium that hatches from the ovum in the water environment. The miracidium undergoes further asexual multiplication in the snail and finally emerges as numerous cercaria. These cercaria are free-swimming until they finally encyst in the muscle tissue of freshwater fish and become metacercaria, and the cycle begins again. This infestation is not uncommon in the Orient where human feces are used for fertilizer (commonly called "night soil"). This use allows the fluke eggs to be deposited in rice paddies so that the life cycle of the fluke can begin. As you know, raw fish is a basic food in the Orient. If the raw fish harboring the larval stage is eaten by humans, Chinese liver fluke disease is established.

Obtain a prepared slide of the adult *Clonorchis*. Examine it carefully under your scanning lens or a dissecting microscope if available. Compare your prepared slide with Figure 32-1. It is not necessary to locate all the structures that are labeled in the figure. It is more important that, in the future, you will be able to rcognize and identify the parasite. Make a sketch of what you see, and label it as fully as possible.

Obtain a slide of *Clonorchis* ova, and examine it under your high-power objective. Note the characteristic lidlike dome at one end of the ovum and the small knob at the other end. Distinguish the miracidium within the ovum if present. Sketch what you see, and label it appropriately. Submit both drawings to your file, accompanied by a brief written description of the disease caused by *Clonorchis* infestation.

Activity 2: *Taenia saginata,* the Beef Tapeworm

Humans contract the beef tapeworm by eating rare meat of infested cattle. The cysticerci in the beef enter the intestine of a human, and the scolex everts and attaches to the intestinal mucosa where the worm begins to grow. Normally only one worm develops to maturity in the intestine, even though dozens have been ingested. Ova are deposited into the environment in human feces where they can be ingested by grazing cattle. Larvae hatch from the ova in the intestine of the cattle. The larvae then penetrate the intestinal wall and circulate throughout the body in the bloodstream and finally encyst in the striated muscles as cysticerci. These bladderworms are easily visible to the naked eye and can be detected by meat inspectors in slaughter houses. Infested animals are called "measly beef" because of the appearance of the numerous cysticerci in the beef. Such cattle are discarded, thus preventing human infestations. The ovum of *Taenia saginata* is not infective for humans. Proper sanitation and disposal of human excreta can control the spread of beef tapeworm infestations by preventing the deposition of the ova in grazing areas where they can be ingested by cattle.

Human infestation with the beef tapeworm is accompanied by diarrhea, vomiting, and epigastric pain. Weight loss and increased appetite are common in prolonged infestations. Successful treatment must eliminate the scolex, or the worm will regenerate. The post-therapy stool specimen must be systematically searched for the scolex to ensure that treatment is complete.

Obtain a slide of composite sections of *Taenia saginata,* and examine it with a scanning lens or a dissecting microscope if available. Pay particular attention to the scolex and gravid proglottids. Refer to Figure 32-2. Sketch what you see, and label it appropriately.

Examine preserved specimens of *Taenia saginata* if they are available. Obtain a slide of *Taenia saginata* ova, and examine it under your high-power objective. Make a sketch of the ova, and be able to recognize and identify them in the future. Remember that tapeworm infestations are most often diagnosed by identification of the ova. Write a short description of the disease caused by *Taenia saginata,* and submit it to your file along with your sketches.

Activity 3: *Taenia solium,* the Pork Tapeworm

The life cycle of the pork tapeworm is essentially identical to that of the beef tapeworm. That is, when the ova are ingested by swine, larvae hatch from them in the intestine of the swine. The larvae penetrate tissue to the bloodstream where they circulate throughout the body, finally reaching striated muscle tissue where they encyst as bladderworms. If a human ingests these cysticerci in rare pork, they develop into adult worms, and pork tapeworm infestation is established. The principal difference between the life cycles of *Taenia solium* and *Taenia saginata* is that the ova of *T. solium* are infectious for humans, but the ova of *T. saginata* are not. Because it is not possible to distinguish ova of one species of *Taenia* from another, extraordinary care must be exercised by nurses and laboratory personnel when handling or disposing of feces from humans with *Taenia* infestations.

Human cysticercosis results from ingestion of the ova of *Taenia solium.* Once the ova hatch and the larvae penetrate the intestinal wall to the bloodstream, they encyst in striated muscle tissue, subcutaneous sites, and sometimes in vital organs. Cysticercosis in humans or in swine is similar to the bladderworm condition in cattle (measly beef). However, the cysticerci of *T. solium* are much smaller than those of *T. saginata* and are not readily visible macroscopically. In human cysticercosis the life cycle of *Taenia solium* is arrested at the bladderworm stage; that is, a human as the intermediate host is also the final host unless he is cannibalized. The symptoms of cysticercosis depend upon the location of the cysts. They can occur in the eye, brain, muscles, and visceral organs, but they have been reported most frequently in subcutaneous tissue.

Obtain a slide of composite sections of *Taenia solium,* and examine it carefully with a scanning lens or a dissecting microscope if available. Pay especial attention to the scolex and gravid proglottids. Refer to Figure 32-2. Sketch what you see, and label the structures that you are able to recognize. Examine preserved specimens of *Taenia solium* and cysticerci if they are available. Obtain a cysticercus slide, and examine it with low power. Sketch what you see and label it.

Inspect a slide of *Taenia solium* ova with your high-power objective. You will note that they are indistinguishable from *Taenia saginata* ova, as shown in Figure 32-2. Write a short description of the diseases caused by *Taenia solium* infestations, and submit it to your file along with your sketches.

Activity 4: *Echinococcus granulosus*—Hydatid Disease

The adult worm occurs in dogs, usually in great numbers, and is quite small—normally only three proglottids. The dog is usually infested by feeding on the viscera of another animal that is infested with the larvae of *Echinococcus granulosus.* Dog feces contain the ova that, when ingested, are the infective stage for humans. Sheep, cattle, and occasionally humans serve as intermediate hosts for the larval stage. This larval stage is called the hydatid cyst. This cyst is the cause of hydatid disease, which is usually quite serious. Symptoms, however, depend upon the size and location of the hydatid cysts.

The hydatid cyst, often the size of a football, is a fluid filled sac. Inside the cyst, buds may form and grow into brood capsules where many immature scolices develop but are unable to mature. These immature scolices are called "hydatid sand." Thus in *E. granulosus,* multiplication occurs in both the adult and the larval stages. As the cyst enlarges, pressure and structural damage to surrounding tissues result. If the cyst ruptures, the contents extend to adjacent tissue, and new cysts form. In humans, the liver is the most common site of hydatid cysts, and the lung is the next most common.

Surgical excision of those cysts that are in operable sites and have not extended too widely into surrounding tissues is the only relief for hydatid disease. Great care must be exercised to avoid puncturing the cysts and spilling the infectious contents into the surgical field.

Examine slides and preserved specimens as available, and make appropriately labeled sketches. Submit them to your file with a written description of hydatid disease.

Take the post test now if you feel ready.

POST TEST

The post test is a self-evaluation. It is not used for a grade. It is designed only to let you decide if you have successfully completed this module.

Part I: True or False

_____ 1. A helminth is a parasitic worm.

_____ 2. Helminths are studied in microbiology because they are microscopic in size.

_____ 3. A dioecious animal contains both sexes within itself.

_____ 4. The diagnostic stage for fluke infestations is ova in feces.

_____ 5. Both ova and proglottids in patient feces are diagnostic for tapeworm infestations.

_____ 6. *Clonorchis sinensis* parasitizes the small intestine of humans.

_____ 7. The ova of *Taenia saginata* are infective for humans.

_____ 8. Cysticercosis results when a patient ingests a tapeworm cysticercus.

_____ 9. Ingested ova of *Echinococcus granulosus* cause hydatid disease in sheep, cattle, and humans.

_____ 10. Tapeworm infestations result from ingestion of cysticerci in the improperly cooked flesh of an intermediate host animal.

_____ 11. A hydatid cyst is approximately the same size as a tapeworm cysticercus.

_____ 12. Surgical excision can be a successful treatment of hydatid disease.

_____ 13. The proglottids proximal to the scolex are more mature than the distally located ones.

_____ 14. The ova of *Taenia saginata, Taenia solium,* and *Echinococcus granulosus* are indistinguishable microscopically.

_____ 15. Human feces containing the ova of *Clonorchis* must be ingested by the intermediate host to establish the life cycle. Therefore, elimination of the use of night soil would eradicate this fluke infestation.

Part II: Completion

1. List two clinical symptoms of Chinese liver fluke infestation.

   a. _____

   b. _____

2. List four symptoms of tapeworm infestations.

   a. _____

   b. _____

   c. _____

   d. _____

Part III

Define the following terms.

1. monoecious _____

   _____

2. dioecious _____

   _____

3. scolex _____

_____

4. proglottid _____

_____

5. cysticercus _____

_____

6. distal _____

7. proximal _____

MODULE 33

Some Nematode Infestations of Humans

PREREQUISITE SKILL

Successful completion of Module 4, "Compound Microscope for the Study of Microbes."

MATERIALS

prepared slides* of
  Enterobius vermicularis adults
  Enterobius vermicularis ova
  Ascaris lumbricoides ova
  Necator americanus adults
  Necator americanus ova
  Trichuris trichiura adults
  Trichuris trichiura ova
  Trichinella spiralis adults
  Trichinella spiralis
    encysted in skeletal muscle

preserved specimens* of
  Ascaris lumbricoides adults
  Necator americanus adults
  Trichuris trichiura adults

*Available from a biological supply company.

OVERALL OBJECTIVE

Recognize some representative parasitic roundworms and their ova, as well as the diseases and major symptoms caused by these infestations.

Specific Objectives

1. Discuss *Enterobius vermicularis* infestations, including modes of infection, major symptoms, diagnosis, treatment, and control measures.
2. Discuss *Ascaris lumbricoides* infestations, including mode of infection, major symptoms, diagnosis, treatment, and control measures.

3. Discuss hookworm infestations, including the genus and species name of the causative agent and all information listed under Specific Objectives 1 and 2.
4. Describe trichuriasis, including information listed under Specific Objectives 1 and 2.
5. Describe trichinosis, including all details requested in the preceding objectives.
6. Recognize and name the characteristic ova presented in this module and the disease each ovum causes.

DISCUSSION

The parasitic roundworms of humans are all in the class Nematoda of phylum Aschelminthes. These animals are cylindrical in shape, have unsegmented bodies, and are tapered at each end. The body is covered by a tough layer of cuticle that protects the parasite from the gastric juices and enzymes of digestion in the intestinal tract of the host. Nematodes are dioecious, with the male usually smaller and more slender than the female. Males of most species are sharply curved at the posterior end of the body. Females produce large quantities of ova daily, a fact that is of great help to the clinician attempting to diagnose the infestation. The ova of the various genera that commonly parasitize humans are distinctly different and hence have much diagnostic significance. Study Table 33-1 and Figures 33-1 and 33-2 regarding parasitic nematodes. In fact, all infestations discussed in this module are diagnosed in the clinical laboratory from ova present in the patient's feces or on the perianal region, with the exception of trichinosis. *Trichinella spiralis* is a parasite of the tissues rather than of the intestinal tract and so must be detected by muscle biopsy and serologic tests.

There are numerous parasitic nematodes, but we will confine our study to the five representatives included in Table 33-1 because these infestations are encountered in the United States with some regularity.

Enterobius vermicularis causes pinworm or seatworm infestations that occur in persons of all economic levels. It is widely distributed in families and institutions

TABLE 33-1 Some Parasitic Nematodes

Parasites	Disease	Clinical symptoms	Diagnostic stage	Source of infection
Enterobius vermicularis	Pinworm infestation	Pruritis ani, diarrhea, or asymptomatic	Ova from perianal region by Graham scotch tape method	Ingestion of ova on hands and fomites or linens
Ascaris lumbricoides	Ascariasis	Allergic symptoms, abdominal pain or discomfort, intestinal blockage, vomiting, diarrhea, pneumonitis, fever	Ova in feces	Ingestion of embryonated ova in soil, often in contaminated water or food
Necator americanus	Hookworm disease	Pulmonary or intestinal pain, anemia or asymptomatic	Ova in feces, rarely—larvae in feces	Larvae in soil burrow into skin of bare feet
Trichuris trichiura	Trichuriasis or whipworm disease	Allergic symptoms or asymptomatic	Ova in feces	Same as *Ascaris*
Trichinella spiralis	Trichinosis	Mild gastrointestinal symptoms, painful respiration, heart muscle damage, muscle pain	Early infection-- adults in feces, later—muscle biopsy and serologic tests	Ingestion of larvae in raw or under- cooked pork or bear meat

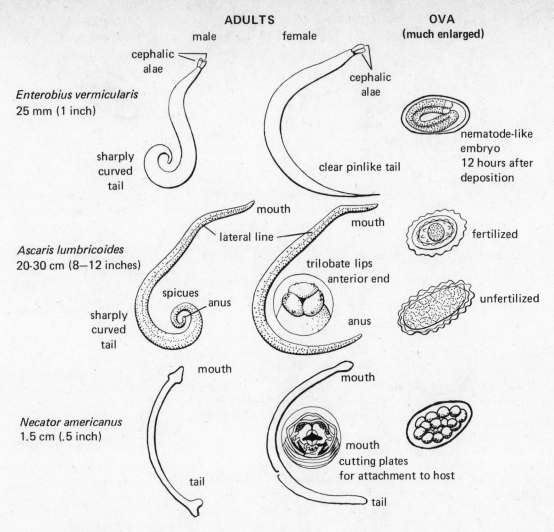

ADULTS

male   female

OVA
(much enlarged)

cephalic
alae

Enterobius vermicularis
25 mm (1 inch)

cephalic
alae

sharply
curved
tail

clear pinlike tail

nematode-like
embryo
12 hours after
deposition

mouth

lateral line

mouth

Ascaris lumbricoides
20-30 cm (8—12 inches)

fertilized

trilobate lips
anterior end

spicues

anus

sharply
curved
tail

anus

unfertilized

mouth

mouth

Necator americanus
1.5 cm (.5 inch)

mouth
cutting plates
for attachment to host

tail

tail

FIGURE 33-1
Some parasitic nematodes.

where neglect of individual hygiene rather than sanitation *per se* is a primary factor in its dissemination. The adults inhabit the cecum of the host from which the gravid female migrates to the anus to deposit large numbers of ova, each containing a living embryo, on the perianal folds. These ova are immediately infective and, if swallowed, will reinfect the host or establish an infestation in a new host. Pinworm infestations are common in children because hand cleanliness is so often neglected and because young children have the overwhelming tendency to scratch when and where they itch.

The Graham scotch tape method is commonly used to recover pinworm specimens from the perianal region of patients for laboratory examination and diagnosis since the ova and adults are rarely seen in feces. The Graham preparation is inexpensive, and the technique of specimen collection is simple enough that mothers can collect specimens from their children with a minimum of instruction. See Figure 33-3 for Graham's scotch tape method. Mothers are instructed to take the specimens on three successive days as the child awakens, that is, before the child is bathed or uses the bathroom.

Ascaris lumbricoides is an intestinal parasite, but it does not actually parasitize the tissues of the intestine. Rather, it resides in the intestinal tract of the host and absorbs the predigested nutrients from the host's food supply.

Necator americanus, the hookworm, attaches to the intestinal lining of the host by means of cutting plates, and feeds on the blood, lymph, and mucous membrane. The hookworm sucks its food by means of a pumping pharynx facilitated by a secretion that prevents coagulation of the host's blood. Anemia often accompanies hookworm infestations.

Some Nematode Infestations of Humans   213

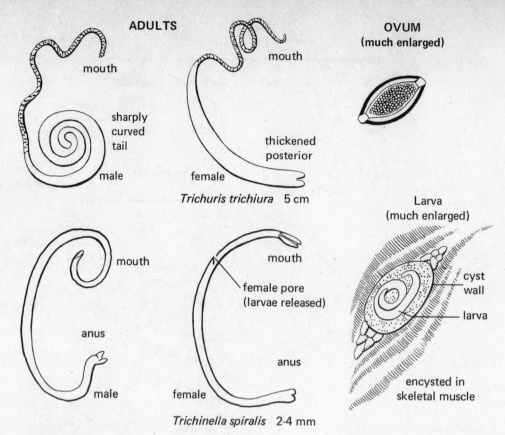

ADULTS

mouth

sharply
curved
tail

male

mouth

thickened
posterior

female

Trichuris trichiura 5 cm

OVUM
(much enlarged)

mouth

anus

male

mouth

female pore
(larvae released)

anus

female

Trichinella spiralis 2-4 mm

Larva
(much enlarged)

cyst
wall

larva

encysted in
skeletal muscle

FIGURE 33-2
Some parasitic nematodes.

Trichuris trichiura is commonly called whipworm because it is shaped like a whip; the anterior is relatively long and thin, and the posterior is thickened and rounded like the handle of a whip. Adult whipworms attach to the lining of the cecum.

As previously mentioned, *Trichinella spiralis* is a tissue parasite, as opposed to the other nematodes we will be studying here. The female trichina worm differs from the other parasitic roundworms in that she releases fully developed, viable larvae instead of ova. She migrates from the intestine by burrowing through the mucosa of the villi and deposits these larvae in the lymph nodes. The larvae reach the circulatory system and are distributed throughout the body, eventually reaching striated muscle tissue where they encyst. The cysts are slowly calcified, but the larvae remain viable for many months although they cannot develop into adults unless they are ingested by another host.

ACTIVITIES

Activity 1: *Enterobius vermicularis,* the Pinworm

Enterobius vermicularis is commonly called the pinworm because the female is shaped very much like a small, straight pin; the tail is long, narrow, and sharply pointed, as shown in Figure 33-1. It is also commonly called the seatworm for obvious reasons! Adults are about 25 mm long.

Refer to Table 33-1 to review the source of infection and the clinical diagnosis. The major symptom of pinworm infestation is an intense, local pruritis caused by the female's migration to the perianal region to deposit ova. Because the female migrates at night the pruritis is most aggravated then, often causing restlessness, insomnia, and incessant scratching with consequent reinfection.

Children are often infected by other children and can transfer the infestation to an entire household. If several members of a family are infested, they should be treated simultaneously to prevent new adult worms from regenerating by reinfection.

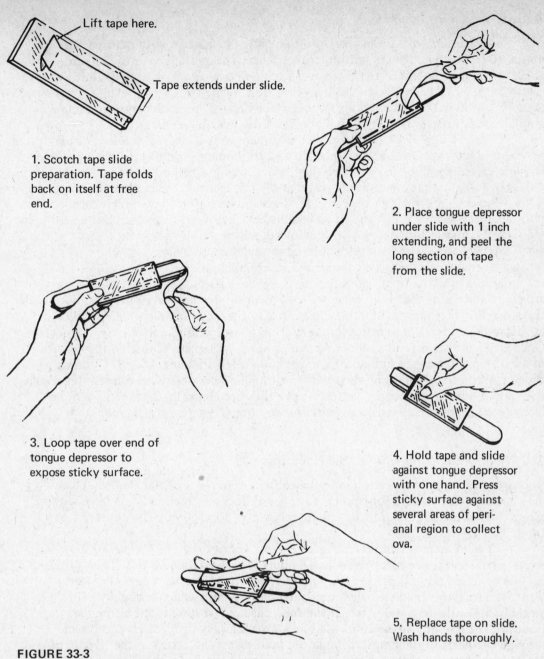

Lift tape here.

Tape extends under slide.

1. Scotch tape slide preparation. Tape folds back on itself at free end.

2. Place tongue depressor under slide with 1 inch extending, and peel the long section of tape from the slide.

3. Loop tape over end of tongue depressor to expose sticky surface.

4. Hold tape and slide against tongue depressor with one hand. Press sticky surface against several areas of perianal region to collect ova.

5. Replace tape on slide. Wash hands thoroughly.

FIGURE 33-3
Graham's scotch tape method for collection of pinworm specimens.

Simultaneous chemotherapy should be accompanied by vigorous household disinfection of bed linens and bedroom dust, as well as bathrooms and towels. Careful personal hygiene is necessary, especially with regard to cleanliness of hands and fingernails.

Obtain a prepared slide of *Enterobius vermicularis* adults, and examine it carefully under your scanning lens or dissecting microscope. Refer to Figure 33-1 for comparison, and sketch what you see on the slide. Label the distinguishing features that you recognize.

Now examine a slide of *Enterobius vermicularis* ova with your high-power objective. Sketch what you see, and label it with the aid of Figure 33-1. Since the ova are diagnostic, it is more important to recognize and identify them than the adults.

Write a brief description of pinworm infestation, and submit it and your two sketches to your file.

Activity 2: *Ascaris lumbricoides*

Ascaris lumbricoides is the largest intestinal nematode of humans; adults are about 20 to 30 cm in length and sometimes more. The adults live in the lumen of the small intestine, unattached to the host. Obtain and examine a preserved adult ascarid worm. Most preserved ascarids are recovered from swine and are somewhat larger than those found in human infestations. Notice the trilobate lips at the anterior end of the animal. Sketch the adult worm, and label its distinguishing features.

The ova of *Ascaris* are infective for humans only after two weeks or a longer period of maturation in the soil. Once the ova containing the embryos are ingested, larvae hatch in the duodenum where they penetrate the small intestine and enter the portal circulation that carries them through the right side of the heart to the lungs. *Ascaris* larvae may remain in the lungs for several days before they penetrate the pulmonary capillaries, reach the alveoli, and work their way up the trachea where they are coughed up through the glottis, swallowed, and travel down the esophagus to the small intestine, where they develop into adult males and females.

The larvae in the lungs often cause severe pneumonitis, occasionally with secondary bacterial complications. Fever, coughing spasms, and asthmatic breathing are characteristic of the pulmonary phase of ascarid infestations. Hypersensitivity to the ascarids is often manifested by allergic reactions, including rashes. Control of ascariasis depends on preventing contamination of soil with infective human feces. Once again, sanitation and proper disposal of human feces are critical in the control of ascarid infestations of humans. Examine and sketch an adult *Ascaris*. Obtain a slide of *Ascaris* ova, and examine it with your high-power objective. Sketch representative ova, and be able to recognize them since they are clinically diagnostic. Write a brief description of ascariasis, and submit it to your file with your sketches.

Activity 3: *Necator americanus,* the Hookworm

Hookworm disease is still common in the southeastern parts of the United States where the climate is moist, from subtropical to tropical. It is confined to rural areas without proper human sanitation and where many people go barefooted. The adult *Necator* is about 1 to 1.5 cm long.

Fertile ova pass from the infested person's body with the feces, and in warm, moist, shady soil larvae hatch in 24 to 48 hours. The larvae feed on the feces or other organic debris in the soil and become infective for humans after a period of development. The infective larvae usually enter the body by burrowing through the soft skin on the sides of the feet and between the toes, causing "ground itch" of the feet. Once in the body, the larvae enter the bloodstream and lymph vessels and travel through the heart to the lungs, through the pulmonary capillaries to the alveoli, up the bronchial tree to the trachea, through the glottis and down the esophagus to the intestine. Once in the intestine the larvae mature to adult males and females where they mate and produce more fertile ova (several thousand daily).

The adult worm attaches to the intestinal mucosa by its cutting plates and feeds on the blood and lymph of the host. If the worms pump more blood than they can digest, the wounds will bleed excessively because of the anticoagulant secretion of the parasite. This condition can result in anemia of the host.

Heavily infested children (100 or more adult worms) are retarded mentally and physically. Any person infested with many worms will become anemic, lethargic, and more susceptible to other diseases. Chemotherapy can cure the infestation, and proper sanitary disposal of human feces plus the routine wearing of shoes will control and prevent reinfestation.

Obtain a prepared slide of a *Necator americanus* adult, and examine it with your low-power objective. Sketch an adult, and label only the readily distinguishable features. Also examine preserved specimens of *Necator* adults if available. Now examine a slide of *Necator americanus* ova with your high-power objective, then sketch an ovum. Be able to recognize it again because of its diagnostic value. Write a brief description of hookworm infestation, and submit it to your file with your sketches.

Activity 4: *Trichuris trichiura,* the Whipworm

The whipworm is about twice as long as the pinworm, measuring up to 5 cm in length. Adults live partially embedded in the mucosa of the large intestine. Each female produces about 5,000 fertile ova daily, which are discharged in the feces of the infested person.

Whipworm ova are immature when passed and require a period of maturation (10 to 14 days) in the soil before they become infective. The mature, embryonated ova must be ingested to establish an infestation. The outer layers of the ova are then digested, and larvae emerge and develop into adults in the small intestine. They pass on to the cecum or large intestine, where they attach to the mucosa.

Light whipworm infestations or infestations of adults are usually asymptomatic. Massive infestations of children may produce inflammation of the mucosa, mucous or bloody diarrhea, systemic toxicity, allergic reactions, and anemia. Symptoms often resemble those of hookworm infestation.

Trichuriasis prevails in rural areas where sewage disposal facilities are poor and social conditions uncontrolled. Unsupervised small children may contaminate the ground frequently and later infect themselves with soiled hands or playthings. Once again, control depends upon preventing contamination of the soil with infective human feces.

Obtain a prepared slide of *Trichuris trichiura* adults, and examine it under your scanning lens or dissecting microscope. Sketch an adult of either sex. Examine preserved specimens of *Trichuris* adults if available. Select a prepared slide of *Trichuris trichiura* ova, and study it with your high-power objective. Sketch a characteristic ovum. Be able to recognize these ova since they are diagnostically important. Write a brief description of whipworm infestation, and submit it to your file along with your sketches.

Activity 5: *Trichinella spiralis*

Trichinella spiralis is about half the size of the hookworm, or approximately 2 to 4 mm long, just barely visible to the naked eye in fecal material. During the first few weeks of infestation, the adults can be observed microscopically in the stools of patients complaining of diarrhea.

Humans are infected by eating insufficiently cooked meat from an infected animal, most commonly pork or occasionally bear meat. The flesh of these mammals contains encysted, viable larvae that are freed from their cysts by gastric digestion of the pork that harbored them. Larvae pass to the duodenum where they develop into mature male and female adults within four or five days. After mating in the intestine of a human, the female burrows into the mucosa of the intestinal villi where she enters the lymphatics and deposits fully developed, viable larvae in the lymph nodes to be carried throughout the body by the bloodstream until they ultimately encyst in striated muscle tissue. Development is arrested at this point since the larvae cannot develop into adults unless they are ingested by an alternate host, which is unlikely in the case of human infestations.

The severity of trichinosis symptoms depends on the number of viable larvae ingested. There may be gastrointestinal discomfort as the intestinal mucosa is invaded. Larval migration is accompanied by muscle pain, chills, weakness, and occasionally prostration. Respiratory distress and myocardial involvement result if the larvae encyst in the diaphragm or heart muscle and can cause death. After three to eight weeks, the larval migration ceases, and recovery begins. There may be a residual handicap of muscle function, depending on the number and location of encysted larvae. Muscles most frequently involved are the diaphragm, thoracic and abdominal walls, the tongue, biceps, and deltoid.

Precipitating antibodies form in the patient's serum after two or three weeks and compose the basis for serologic diagnosis of trichinosis. There is no specific treatment for the disease. Control of trichinosis depends on eliminating sources of infection for hogs by providing sanitary feed, destroying contaminated animals, and, finally, by thorough cooking of pork. One pink pork chop can contain 10,000 larvae!

Obtain a prepared slide of *Trichinella spiralis* adults, and examine them with your low-power objective. Sketch an adult worm of either sex if available. Examine a prepared slide of *Trichinella spiralis* larvae encysted in striated muscle tissue under your high-power objective. Sketch what you see, and label the sketch as thoroughly as possible. Write a brief description of trichinosis, and submit it to your file with your sketches.

This module contains much information, so take the post test and then repeat any parts of the module necessary to enable you to achieve 100% on the post test.

POST TEST

The post test is a self-evaluation. It is not used for a grade. It is designed only to let you decide if you have successfully completed this module.

Part I: True or False

_____ 1. Parasitic roundworms of humans are all nematodes.

_____ 2. Trichinosis is the only nematode infestation discussed in this module that is not diagnosed in the laboratory by the ova of the parasite.

_____ 3. Specimens for diagnosis of whipworm infestations are taken by the Graham scotch tape method.

_____ 4. *Enterobius vermicularis* ova are fully embryonated and infective when they are deposited.

_____ 5. *Necator americanus* ova are fully embryonated and infective when they are deposited.

_____ 6. *Trichinella spiralis* females bear fully developed, viable larvae instead of depositing ova.

_____ 7. Ascarids, hookworms, and trichina worms are all carried throughout the host's body in the bloodstream at some stage of the infestation.

_____ 8. Ascarids, hookworms, and trichina worms all produce ova that require a period of maturation in the soil before they are infectious for humans.

_____ 9. *Trichuris trichiura* ova require a 10 to 14 day period of maturation before they are infective.

_____ 10. Trichinosis is diagnosed by clinical symptoms, serologic tests, and muscle biopsy.

Part II: Completion

1. List two major symptoms of pinworm infestation.

    a. _____

    b. _____

2. List three major symptoms of hookworm infestation.

    a. _____

    b. _____

    c. _____

3. List four major symptoms of ascariasis.

    a. _____

    b. _____

    c. _____

    d. _____

4. List three major symptoms of trichinosis.

   a. _____

   b. _____

   c. _____

Part III: Completion

1. What is the principal means of control of ascariasis, hookworm, and whipworm infestations?

   _____

2. List the three major means of control of trichinosis.

   a. _____

   b. _____

   c. _____

3. What is the main means of control of pinworm infestations?

   _____

PART SIX

Selected Physiological Reactions of Bacteria

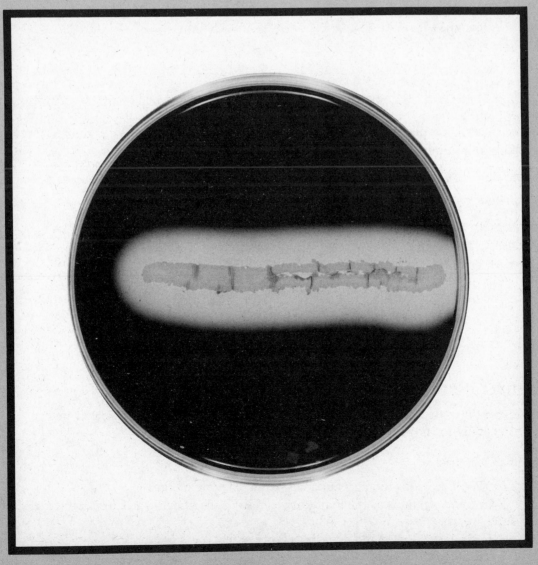

Effect of a bacterial exoenzyme on starch agar.

MODULE 34

Exoenzymes and the Major Food Groups

PREREQUISITE SKILL

Successful completion of Module 8, "Aseptic Transfer of Microbes."

MATERIALS

For Activity 1:
    sterile starch agar plate*
    Gram's iodine, diluted 1:1 (10 ml)*
    nutrient agar slant cultures of
        Bacillus subtilis
        Escherichia coli

For Activity 2:
    sterile skim milk agar plate*
    nutrient agar slant cultures of
        Bacillus subtilis
        Corynebacterium xerosis

For Activity 3:
    sterile nutrient gelatin (7 ml) stabs (2)*
    nutrient agar slant cultures of
        Pseudomonas aeruginosa
        Escherichia coli

For Activity 4:
    sterile tributyrin agar plate
    nutrient agar slant culture of
        Escherichia coli
        Pseudomonas aeruginosa

*To be prepared by the student if the instructor so indicates. Formulae are at the end of this module.

OVERALL OBJECTIVE

Understand the effect of exoenzymes on the major food groups and be able to detect experimentally the presence of these exoenzymes.

Specific Objectives

1. List the three major food groups and give an example of each.
2. List the end products of the hydrolytic action of the exoenzymes for all three food groups.
3. Name the reagent used and describe a positive test for starch.

4. Describe two tests used to demonstrate protein hydrolysis and include a positive result for each test.

5. Describe a positive test for lipolysis.

6. Define the terms *hydrolysis, biosynthetic, disaccharide, monosaccharide, lipase, carbohydrase, amylase, catabolic, anabolic, lipolysis, endoenzyme, exoenzyme, proteinase,* and *proteolytic.*

DISCUSSION

Bacteria, like all living things, have individuality. Some examples of their individual differences that you have already see are colonial morphology; pigment production; staining properties, such as Gram stain and acid-fast stain; and structural differences, such as spores and capsules. Bacteria also demonstrate their individuality in the nutrients they are able to use and the end products that result from the metabolism of these nutrients, all of which is dependent on the type of enzymes they are able to produce. The enzymes produced by different bacteria are a result of individual genetic differences.

All bacteria have endoenzymes (enzymes that function within the cell). Their differing endoenzyme systems result in different end products of metabolism. Some bacteria produce exoenzymes also. The exoenzymes are secreted out of the cell and diffuse into the surrounding medium. Most, but not all, exoenzymes are hydrolytic, which means that they degrade large molecules into smaller ones. These small molecules are then available to the bacterial cell and can cross through the cell membrane to be acted upon by endoenzymes. Not all endoenzyme systems are catabolic (hydrolytic, that is, energy producing). Some are anabolic (biosynthetic) and are used by the cell to make protoplasm and other cell parts. So both exoenzyme and endoenzyme systems can be either catabolic or anabolic.

In this module you will be studying hydrolytic exoenzymes. Keep in mind that the exoenzyme individuality demonstrated by the bacteria used in the following activities can be used to aid in the identification of many unknown bacteria.

The three major food groups used by some bacteria and other living organisms, including yourself, for energy and growth are complex carbohydrates, such as starch; proteins; and fats. In this module you will be studying representative bacteria able to produce the necessary exoenzymes that function to reduce the molecular size of the three major food groups to a size that can be used by the bacterium.

ACTIVITIES

Activity 1: Starch Hydrolysis

This activity is a demonstration of the disappearance of a complex carbohydrate due to the action of the hydrolytic exoenzyme, amylase. Amylase is an enzyme that cleaves the starch molecule into disaccharides (double sugars) and some monosaccharides such as glucose. Disaccharides and monosaccharides are small enough in molecular size to cross through the semipermeable membrane into the cytoplasm of the bacterial cell and there be used by the endoenzymes.

Make a single streak inoculation of *Escherichia coli* and *Bacillus subtilis* on one starch agar plate, as shown in Figure 34-1.

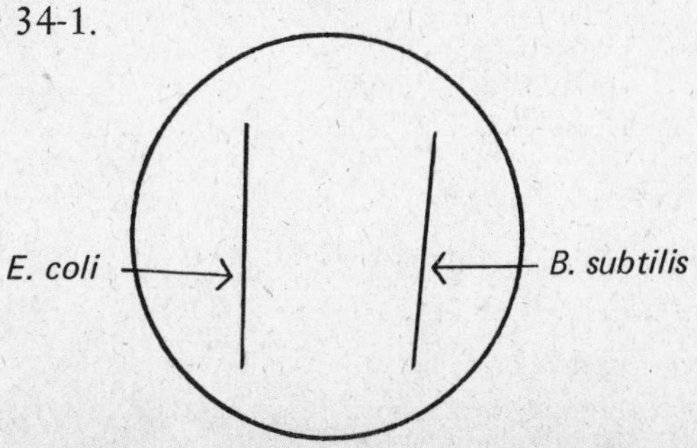

E. coli →   ← B. subtilis

FIGURE 34-1
Single streak inoculations.

Incubate the inoculated plate for 48 hours at 37°C. After incubation, the exoenzyme, if produced by the organism, has diffused into the medium surrounding the bacterial growth. Next, flood the plate with a thin layer of iodine, and look for a color change of the medium. Iodine gives a blue color if the starch remains unaltered.

Make a drawing of the iodine-flooded plate, and accompany the drawing with a short, written description. In the description, conclude why the area around the bacterial growth does or does not turn blue. Submit this to your file.

Activity 2: Hydrolysis of Milk Protein

The action of proteinase can also be measured by the disappearance of the substrate. This activity will demonstrate the hydrolysis of casein, milk protein. Skim milk agar is opaque due to the molecular size of casein molecules. If caseinase is produced by the bacterium, the medium surrounding the colony will become clear since the exoenzyme has cleaved the large protein molecules into smaller molecules (peptides and amino acids), making them invisible.

Make a single streak inoculation of *Corynebacterium xerosis* and *Bacillus subtilis* on one skim milk agar plate, just as you did in Activity 1. Incubate the plate at 37°C for 48 hours. After incubation, look for clear areas denoting protein hydrolysis around the growth of both organisms. Make a drawing of the plate, and accompany it with a descriptive conclusion about the action of this enzyme. Submit this to your file.

Activity 3: Hydrolysis of Gelatin Protein

Gelatinase degrades the protein, gelatin, by liquefying it. In this instance, the hydrolytic activity of the exoenzyme is manifested by the liquefaction of the nutrient, not by its disappearance.

Make separate stab inoculations of two nutrient gelatin tubes using slant cultures of *Pseudomonas aeruginosa* and *Escherichia coli*. A stab inoculation is done by taking the organism from the slant with your inoculating *needle* and then inserting the needle into the center of the gelatin to the bottom of the tube and withdrawing the needle along the line of the stab.

Make the two stab inoculations now. Incubate the gelatin stabs at 37°C for 48 hours or longer until one stab shows positive results, which may take a week or longer. The elevated temperatue of incubation or even room temperature will liquefy gelatin. Therefore, to determine whether gelatinase has been produced, cool the tubes in the refrigerator for 10 minutes. Return the tubes to room temperature, and inspect for liquefaction immediately. If gelatinase has been produced by the organisms, the hydrolyzed gelatin will remain fluid. Liquefaction of tubed gelatin usually proceeds from the surface down.

Precaution: Take care not to agitate the tubes as you move them from the incubator to the refrigerator and back to your work table. Agitation can mix the hydrolyzed gelatin with the unhydrolyzed gelatin to produce a combination that solidifies.

After examination of the gelatin surface of both tubes, make a table similar to Table 34-1, record your results, and submit them to your file. Include a written description of the proteolytic action that you observed in this activity.

TABLE 34-1 Results of Gelatinase Production

	P. aeruginosa	E. coli
Gelatinase produced*		

*Use the terms positive or negative.

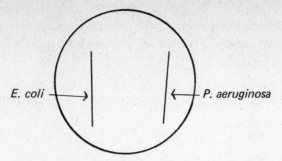

FIGURE 34-2
Single streak inoculations.

Activity 4: Fat Hydrolysis

Some microorganisms can decompose fats by hydrolyzing the fat molecule to glycerol and fatty acids. Tributyrin agar plates will show the disappearance of the fat if the organism is capable of producing the exoenzyme, lipase. Tributyrin agar is an opaque, fat-agar emulsion and, once again, a clear zone around the bacterial growth denotes lipolysis, that is, fat hydrolysis.

Make a single streak inoculation of the two organisms used in Activity 3 on a single tributyrin agar plate, as shown in Figure 34-2. Incubate the inoculated plate at 37°C for 48 hours, and look for the lipolytic activity. Make a table similar to Table 34-2, record your results, and submit them to your file along with a short, written description of the meaning of lipolysis.

Now that you have completed the activities, you should reflect upon the individuality of bacteria. Remember that you have observed the action of only a few exoenzymes. In order for unknown bacteria to be identified, many other enzymatically controlled, physiological reactions must be studied. You will encounter other physiological individualities in subsequent modules.

Now take the post test, and evaluate yourself.

TABLE 34-2 Results of Lipase Production

	E. coli	P. aeruginosa
Fat hydrolysis*		

*Use the terms positive or negative.

POST TEST

The post test is a self-evaluation. It is not used for a grade. It is designed only to let you decide if you have successfully completed this module.

Part I: True or False

_____ 1. Lipolysis is a result of lipase activity on complex carbohydrates.

_____ 2. Starch is a complex carbohydrate.

_____ 3. Amylase is a carbohydrase.

_____ 4. Some endoenzymes are active outside of the cell.

_____ 5. Anabolic enzyme reactions result in biosynthesis.

_____ 6. Catabolic enzyme reactions result in cleavage of the substrate molecule into smaller molecules.

_____ 7. Catabolic enzyme reactions are energy producing reactions.

_____ 8. Peptides and amino acids are end products of protein hydrolysis.

_____ 9. Iodine is a positive test for protein.

_____ 10. Tributyrin is a fat.

_____ 11. Disaccharides are too large in molecular size to pass through the cell membrane.

_____ 12. The enzymatic activity of caseinase is manifested by liquefaction of the skim milk agar.

_____ 13. Lipase degrades the fat molecule to glycerol and fatty acids.

_____ 14. Gelatinase activity results in disappearance of the protein substrate.

_____ 15. Gelatin is liquid at room temperature.

Part II

List the three major food groups studied in this module, the exoenzymes that hydrolyze them, and the subunits that result from the hydrolytic activity of the enzymes.

Food group	Exoenzyme	Subunits of hydrolysis
1. _____	_____	a. _____
		b. _____
2. _____	_____	a. _____
		b. _____
3. _____	_____	a. _____
		b. _____

FORMULAE FOR MEDIA

1. **STARCH AGAR**
    1. Add nutrient agar to water according to directions on the media bottle label.
    2. Mix with *no heat.* The starch must not be added to hot or boiling liquid.
    3. Add 3 gm of soluble starch per liter of the nonheated NA-water mixture (i.e., 0.3 gm/100 ml).
    4. Turn on the heat.
    5. Heat the starch-NA mixture to boiling, dispense, and autoclave at 15 psi for 15 minutes at sea level.

2. **SKIM MILK AGAR**
    1. Rehydrate nonfat, dry milk according to the directions on the package. Dispense 2 ml in a test tube, cap, and autoclave at 10 psi for 15 minutes. (If you are working in groups of 4, dispense 10 ml milk in a test tube, cap and autoclave as above.)
    2. Prepare nutrient agar according to the directions on the bottle label. Dispense 15 ml in a large screw cap test tube. Autoclave at 15 psi for 15 minutes. (For groups of 4 prepare 100 ml NA and dispense in an 8 oz. bottle. Autoclave as above.)
    3. Aseptically pour the sterile skim milk into the sterile, liquid nutrient agar when you are ready to pour your plate. Replace the screw cap tightly on the agar tube (or bottle) and shake it vigorously up and down to combine the milk with the agar. Pour the plate immediately, and pass the flame from your burner over the surface of the liquid milk agar to "pop" the bubbles.

3. **NUTRIENT GELATIN**
    Prepare the nutrient gelatin as directed on the label of the bottle. Dispense and autoclave at 15 psi for 15 minutes.

4. **TRIBUTYRIN AGAR**
    1. Prepare nutrient agar as directed on the label of the bottle. Autoclave at 15 psi for 15 minutes.
    2. In a separate bottle, autoclave 10 ml of tributyrin per liter of nutrient agar at 15 psi for 15 minutes.
    3. In a sterile blender, emulsify the tributyrin and liquid agar. Pour the plates at once.

MODULE 35

Carbohydrate Fermentation

PREREQUISITE SKILL

Successful completion of Module 2, "Preparing and Dispensing Media," and Module 8, "Aseptic Transfer of Microbes."

MATERIALS

sterile fermentation tubes*† of
   glucose (dextrose) broth (5)
   lactose broth (5)
   sucrose broth (5)
slant cultures of
   Escherichia coli
   Proteus vulgaris
   Staphylococcus aureus
   Pseudomonas aeruginosa

For related experiences:
1. tablets for rapid carbohydrate fermentation tests
2. carbohydrate fermentation disks for rapid test
   phenol red agar plate (4)*
   tubes containing 1 ml sterile water and swab (4)*

*To be prepared by the student if the instructor so indicates.
†Directions are at the end of this module.

OVERALL OBJECTIVE

Demonstrate that different bacteria produce enzymes capable of fermenting specific carbohydrates.

Specific Objectives

1. Describe why different bacteria produce different enzymes.
2. Correlate Specific Objective 1 with the individuality of bacteria.
3. Describe how the individuality of bacteria is used in the identification of closely related species.
4. List the ingredients of a fermentation tube.

5. Give all possible information that can be obtained by growing bacteria in fermentation broths.

6. List seven fermentable carbohydrates.

7. Define the terms *substrate, end product, endoenzyme, Durham tube, pH indicator, catabolic,* and *altered substrate.*

8. Describe a rapid tablet test for carbohydrate fermentation.

DISCUSSION

The most important criteria in the identification of bacteria are their physiological differences. Physiological differences in bacteria are manifested in the nutrients they are able to utilize and the end products that result from their metabolism of these nutrients, all of which depends on the enzymes the bacteria are able to manufacture. The type and number of enzymes produced depend on the genetic code of the bacterial cell.

Thus not all bacteria produce the same enzymes; therefore, different substrates (and/or nutrients) are degraded, resulting in different end products. For example, certain microbes produce gelatinase, while others can produce lipase but not gelatinase. It is by careful study of the substrates utilized and the resulting end products of the metabolism of these substrates that a pattern of metabolic activities can be established. All this can be successfully employed in the identification of bacteria of closely related species.

Most bacteria are very much like your tissue cells in that they use various carbohydrates as their main source of energy. The carbohydrates you will be studying in this module are small enough in molecular size to diffuse from the surrounding environment to the interior of the cell and there be used for energy. The catabolic endoenzymes within the cell remove the energy from these carbohydrates, which results in an alteration of the substrate molecule. When this happens, the substrate molecule usually becomes smaller, and the altered substrate could be an end product of metabolism.

$$\text{Substrate} \xrightarrow{\text{enzyme}} \text{Altered substrate} + \text{energy}$$
$$\text{(e.g., carbohydrate)} \qquad\qquad \text{(end product)}$$

Actually, in the fermentation of a nutrient, a series of enzymatic reactions (enzyme systems) are involved in the production of the end products.

The activities in this module will demonstrate clearly that different bacteria can ferment the same or different carbohydrates, depending on the enzymes they are able to produce. The activities will also show that even though different bacteria ferment the same carbohydrate, the end products of metabolism can be different for each bacterial type. Thus, it is the intent of this module to exemplify the individuality of bacteria by studying the nutrients they use and the end products they produce.

To study the physiological differences of bacteria you will be using fermentation tubes. A fermentation tube is a culture tube that contains the following:

1. *Durham tube*—A small tube that is placed upside down in the culture tube. The purpose of the inverted Durham tube is to show if gas has been produced as an end product of metabolism. This does not allow you to determine which gas is being evolved. The most common gases produced by bacteria are hydrogen, carbon dioxide, and methane.

2. *Phenol red broth base*—A medium that contains the ingredients of nutrient broth, plus a pH indicator. The nutrient-broth-like ingredients support the growth of most bacteria. The pH indicator is phenol red, which is red in color at a pH near neutral and turns yellow if organic acids are produced.

3. *A specific carbohydrate*—Such as glucose, lactose, maltose, sucrose, or mannitol. If a single carbohydrate such as glucose is fermented, an acid

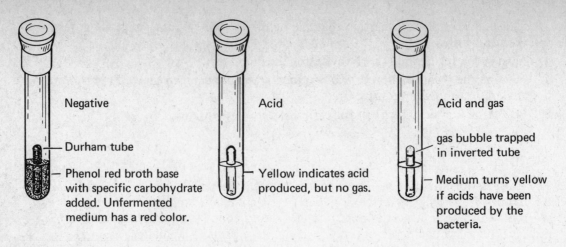

FIGURE 35-1
Possible reactions in fermentation tubes.

environment results, and the broth turns yellow. At the end of this module, a description is given of exactly how to prepare the different fermentation tubes for Activity 1.

Figure 35-1 shows the most typical results for Activity 1.

ACTIVITY

Activity 1: Inoculation of the Different Fermentation Tubes

Using the organisms listed both in the Materials section of this module and in Table 35-1, inoculate the three different sugar fermentation broths with each bacterium. Keep one uninoculated tube of each fermentation broth as a comparative control. Incubate all tubes at 37°C for 48 hours. Make a table similar to Table 35-1, and use the symbols beneath the table for recording the results. After incubation, compare the inoculated sugar tubes to the control tubes. Only a decided yellow color change occurring within 48 hours is considered positive.

If necessary, refer again to Figure 35-1 for a schematic representation of these different results. Consult *Bergey's Manual of Determinative Bacteriology* * (or

TABLE 35-1   Collection of Data

	Glucose	Lactose	Sucrose
E. coli			
P. vulgaris			
S. aureus			
P. aeruginosa			
Control			

A = acid only, that is, the broth has turned yellow.
AG = acid and gas, that is, the broth is yellow and the gas is a bubble trapped in the inverted tube.
V = variable (can be A, AG, or —).
— = no change.

*Robert S. Breed, et al., *Bergey's Manual of Determinative Bacteriology,* 7th ed. (Baltimore: Williams & Wilkins, 1957).

Table 53-1 in Module 53.) to determine if your results are correct. Submit the completed table to your file. Accompany this with a short summary of how physiological differences can be used to identify bacteria.

Take the post test after you have performed or read thoroughly the related experiences.

Related Experiences

1. Rapid tablet test for carbohydrate fermentations*

The tablets incorporate specific substrates, a gelling substance, and the pH indicator, phenol red.

Procedure:

1. Place the sugar or sugar alcohol tablet (dextrose, lactose, mannitol, or other fermentable substrate) in a clean test tube. Sterilization of materials is not necessary.
2. Add 1 ml of distilled water.
3. Inoculate with a *loopful* (a massive inoculum) of pure culture.
4. Mix with an inoculating loop or a stirring rod until the tablet dissolves.
5. Incubate at the optimum temperature for the organism used.

As the tablet dissolves, a red, semisolid compound is formed. After incubation, the phenol red turns yellow if acid is formed, and gas can be detected as bubbles near the bottom of the tube. Fermentation may be apparent within 30 minutes and usually is complete in 6 hours, but tubes should be held for 24 hours before discarding to detect possible reversion of reactions.

If fermentation test tablets are available to you, it would be prudent to perform this related experience if the identification of unknown bacteria is part of your lab course.

2. Rapid disk test for carbohydrate fermentation†

Small filter paper disks are impregnated with different sugars or sugar alcohols. They may be used in a variety of carbohydrate-free nutrient broths, semisolid media, or solid media.

Procedure for solid medium:

1. Inoculate a phenol red agar plate for confluent growth. Organisms from solid media should be dispersed in a small amount of sterile water to obtain an even seeding of the agar surface. Use tubed sterile water and a sterile swab to inoculate for confluent growth.
2. Place the desired carbohydrate disks on the surface of the inoculated plate with your flamed forceps. Press the disks gently to ensure contact with the agar surface.
3. Incubate at the optimum temperature for 4 hours.
4. Make observations after 4 to 8 hours and again after 18 hours of incubation to detect possible reversion of reactions.
5. Repeat this procedure for the remaining three organisms used in Activity 1.

A color change surrounding the disk (yellow) indicates acid production. Often this is enough information to differentiate sugar-fermenting bacteria from nonfermenters.

---

*Purchased from Key Scientific Products Company, P.O. Box 66307, Los Angeles, CA 90066.

†Purchased from Difco Laboratories, Detroit, MI 48201.

POST TEST

The post test is a self-evaluation. It is not used for a grade. It is designed only to let you decide if you have successfully completed this module.

Part I: True or False

_____ 1. The nutrients used by bacteria are determined by their genetic code.

_____ 2. An altered substrate can be an end product of metabolism.

_____ 3. A nutrient fermented to end products is usually due to a series of enzymatic reactions.

_____ 4. Carbohydrates such as glucose or lactose cannot diffuse through the semipermeable membrane of the bacterial cell because their molecular size is too large.

_____ 5. A Durham tube is the same size as a culture tube.

_____ 6. Individual bacterial types have differing characteristics similar to the varying characteristics of the different races or cultures of human beings.

_____ 7. Catabolic reactions are usually energy producing.

_____ 8. Catabolic reactions usually result in an increase in molecular size of the substrate.

_____ 9. The endoenzymes of fermentation are secreted into the environment.

_____ 10. The term "substrate" can be defined as the molecule that an enzyme is able to alter.

Part II: Completion

A. List the three significant ingredients of a fermentation tube.

1. _____

2. _____

3. _____

B. List the two significant ingredients of phenol red broth base.

1. _____

2. _____

C. Describe the meaning of the following symbols.

1. − _____

2. AG _____

3. A _____

4. V _____

D. List seven fermentable carbohydrates.

1. _____

2. _____

3. _____

4. _____

5. _____

6. _____

7. _____

Part III

List two major advantages in the tablet fermentation test as compared to the tube fermentation test in Activity 1.

1. _____

2. _____

PREPARING CARBOHYDRATE FERMENTATION BROTH TUBES

1. According to the directions on the label of the medium bottle, rehydrate 300 ml of phenol red broth base.

2. Divide your phenol red broth into three 100 ml aliquots.

3. To each 100 ml aliquot add 0.7 gm of one of the sugars (carbohydrates) listed in Table 35-1 (glucose, lactose, or sucrose). Mix to dissolve thoroughly.

4. Place an inverted Durham tube in each of 15 culture tubes.

5. Dispense about 7 ml of glucose broth into five tubes. Label the glucose broth tubes immediately.

6. Dispense and label five tubes of lactose broth and five tubes of sucrose broth into the other culture tubes with inverted gas tubes. Label the tubes appropriately. Care is needed since all tubes will look alike although the substrates differ.

7. Autoclave your carbohydrate fermentation tubes at *10 psi* for 20 minutes. The pressure of the autoclave will force the air out of the Durham tube. Note the reduced autoclaving pressure for these three sugars. Other carbohydrates such as fructose and xylose are decomposed by heat and must be sterilized by using a bacteriological filter.

MODULE 36

Nitrate Reduction

PREREQUISITE SKILL

Successful completion of Module 8, "Aseptic Transfer of Microbes."

MATERIALS

slant cultures of
  Escherichia coli
  Pseudomonas aeruginosa
  Gaffkya tetragena
sterile nitrate broth tubes (5 ml) (4)*
sulfanilic acid*†
a-naphthylamine reagent*†

powdered zinc
flat toothpick to dispense zinc powder
1 ml pipets (2)

*To be prepared by the student if the instructor so indicates.
†Formulae are at the end of this module.

OVERALL OBJECTIVE

Ascertain the capability of microbes to reduce nitrates.

Specific Objectives

1. Define biological reduction of a molecule or a compound.
2. Name two enzymes produced by microbes that are active in the alteration of potassium nitrate to ammonia or free nitrogen.
3. Describe the ingredients of nitrate broth and give the purpose of each.
4. Describe the procedure used to perform a nitrate test and describe the appearance of a positive nitrate test.
5. Describe the procedure used to perform a nitrite test and describe the appearance of a positive nitrite test.
6. Describe the procedure used to determine if ammonia or free nitrogen is produced by microbes from potassium nitrate and describe the appearance of a positive test for both.

7. Discuss the purpose of the control tube.
8. Discuss the importance of the nitrate reduction determination to the clinical microbiologist.

DISCUSSION

A physiological characteristic of many microbes is their ability to reduce certain molecules or compounds. In this module, you will be studying the capability of different bacteria to produce the enzymes nitrate reductase and/or nitrite reductase. Certain microbes can only reduce nitrate to nitrite. Other microbes cannot reduce nitrate at all. Still other microbes not only reduce nitrate to nitrite, but they are also able to reduce the nitrite to ammonia or gaseous nitrogen. The detection of nitrate reduction by bacteria is another valuable diagnostic tool used in the identification and differentiation of bacterial types. Reduction is the addition of hydrogen and electrons to the molecule or the removal of oxygen from the molecule.

Reduction of nitrate can be detected by testing for the presence of nitrite after bacteria have grown in a broth to which potassium nitrate (KNO_3) has been added. Commercially prepared nitrate broth is similar to nutrient broth that contains a small amount of potassium nitrate.

If a microbe is growing in nitrate broth, one of the following results may occur, depending upon which enzymes the microbe is capable of producing, if any.

1. Nitrate remains unaltered.
2. Nitrate is reduced to nitrite.
3. Nitrate has been rapidly reduced to nitrite, then further reduced to ammonia or free nitrogen.

It is expedient to test first for Result Number 2, the reduction of nitrate (NO_3) to nitrite (NO_2), because if a positive nitrite test is obtained, no further testing is necessary. If the test is negative, you must then determine if Number 1 or 3 has taken place. (See Table 36-1.)

The following activities are designed to clarify the enzyme behavior of different bacteria in the presence of nitrate.

ACTIVITIES

Activity 1: Testing for the Reduction of Nitrate to Nitrite

Aseptically inoculate three tubes of nitrate broth, one with *Escherichia coli,* another with *Pseudomonas aeruginosa,* and the third with *Gaffkya tetragena.* Use the fourth, uninoculated tube as a comparative control. Incubate all tubes at 37°C for 48 hours.

After incubation, test for the presence of nitrite by adding approximately 1 ml of sulfanilic acid and approximately 1 ml of a-naphthylamine reagent to each of the nitrate broth cultures and the control. *Do not shake* the cultures since introduction of oxygen interferes with reduction. If the test is positive for nitrite, a distinct red color (which may turn brown rapidly) will appear, indicating that NO_3 has been reduced to NO_2. See Table 36-1 for clarification.

Make a table similar to Table 36-2, and record the positive nitrite test in the appropriate place. If no color change occurs, perform Activity 2.

Activity 2: Determining if Nitrite Has Been Further Reduced or Has Remained Unaltered

To the negative nitrite tests from Activity 1, add a small amount of powdered zinc. Zinc reduces nitrate to nitrite so if a red color appears now, it means that the nitrate remained unaltered by the bacterial growth. The red color still indicates a positive test for nitrite, but the reduction of NO_3 to NO_2 was chemically induced by the zinc. This would not be possible if the nitrate were not present in the broth in its original form, KNO_3. Therefore, the nitrate remained unaltered by the bacteria. If no color change occurs upon the addition of zinc, this means that the nitrate has been reduced by the bacteria beyond the nitrite stage to NH_3 or N_2.

TABLE 36-1 Summary of Expected Results

Reagent or reagents	Red color change	No color change
Activity 1: sulfanilic acid and alpha-naphtylamine	Bacteria produce only nitrate reductase, and NO_3 becomes NO_2.	If no color change, test for further reduction of NO_2 by doing Activity 2.
Activity 2: powdered zinc	KNO_3 in broth remains unaltered by bacterial growth. The red color change is a chemical reduction of $NO_3 \rightarrow NO_2$ by the zinc.	$NO_3 \rightarrow NO_2 \rightarrow NH_3$ or N_2 Therefore, the bacterial cells produce both nitrate reductase and nitrite reductase.

Remember that all tests must always be controlled by comparison with an uninoculated tube of nitrate broth. This means that you must treat the control tube in the same way that you treat the pure culture broth tubes.

On a duplicate of Table 36-2, record your results in the appropriate places, and submit the completed table to your file.

Now take the post test.

TABLE 36-2 Collection of Data on Nitrate Reduction

	Activity 1 Addition of reagents*	Activity 2 Subsequent addition of zinc*
E. coli		
P. aeruginosa		
G. tetragena		
Control		

*Fill in the appropriate blocks of Table 36-2 with one of the following results: NO_3 reduced to NO_2, NO_3 reduced to NO_2 and on to NH_3 or N_2, or NO_3 not reduced. Check your results with *Bergey's Manual of Determinative Bacteriology* only after you have recorded your test data.

POST TEST

The post test is a self-evaluation. It is not used for a grade. It is designed only to let you decide if you have successfully completed this module.

Part I: True or False

_____ 1. In testing for nitrate reduction, the zinc is usually added first.

_____ 2. Nitrates can only be reduced to nitrites by microbial enzymes.

_____ 3. Nitrite reductase alters nitrate to nitrite.

_____ 4. The ingredients of nitrate broth are such that if a bacterium does not utilize potassium nitrate, it will be unable to grow in the broth.

_____ 5. A red color indicates a positive test for nitrite when the correct reagents are added.

_____ 6. A red color indicates a positive test for nitrate when the correct reagents and zinc are added.

_____ 7. A red color indicates a positive test for ammonia or free nitrogen when the correct reagents and zinc are added.

_____ 8. A complete definition of reduction is the addition of hydrogen and electrons to a molecule or compound.

_____ 9. In microbiology a control is usually an uninoculated tube of medium that is treated in the same way as the inoculated tubes of the same medium.

_____ 10. The addition of zinc to nitrate broth can make changes that some bacteria cannot make.

Part II: Completion

A. List the two reagents used to test for nitrite.

1. _____

2. _____

B. List the three reagents used to test for nitrate.

1. _____

2. _____

3. _____

Part III

Give the color changes, if any, of the following reactions for a *positive test* for reduction of potassium nitrate.

Reaction—Chemical or enzymatic   Color change

1. $NO_3 \longrightarrow NO_2$ _____

2. NO_3 _____

3. $NO_3 \longrightarrow NH_4$ or N_2 _____

REAGENTS FOR ACTIVITY 1

1. SULFANILIC ACID
   Dissolve 8 gm of sulfanilic acid in 1 liter of 5N acetic acid.

2. ALPHA—NAPHTHYLAMINE REAGENT
   Dissolve 5 gm of a-naphthylamine in 1 liter of 5N acetic acid.

3. 5N ACETIC ACID
   Add 294 ml of glacial acetic acid to 706 ml of distilled water.

MODULE 37

Urea Hydrolysis

PREREQUISITE SKILL

Successful completion of Module 8, "Aseptic Transfer of Microbes."

MATERIALS

For Activity 1:
urease test tablets*
nutrient broth cultures of
   Proteus vulgaris
   Escherichia coli
clean capped test tubes (3)
clean 1 ml pipet (3)
sterile nutrient broth tube†
inoculating equipment

*Available from Key Scientific Products Company, P.O. Box 66307, Los Angeles, CA 90066.
†To be prepared by the student if the instructor so indicates.

For related experience:
urea broth†
sterile capped test tubes (3)†
nutrient broth cultures (same as for
   Activity 1)
sterile 10 ml pipet†
apparatus for bacteriological filtration
   (see Figure 37-1); pieces of equipment
   that must be sterile:
      bacteriological filter (membrane type)
      cotton-plugged vacuum flask

OVERALL OBJECTIVE

Demonstrate that the production of urease by certain microbes can be used to differentiate them from other bacteria, especially enteric pathogens.

Specific Objectives

1. Describe how *Proteus* sp. can be confused with enteric pathogens.
2. Name two genera of enteric pathogens.
3. Explain the chemical action of urease on the ingredients of urea broth.
4. List the significant ingredients of urea broth.

5. Describe the color change of a positive test for urea hydrolysis.
6. Describe the color of a negative test for urea hydrolysis.
7. Discuss why no further testing is needed if a positive test for urea hydrolysis is obtained.
8. Describe the apparatus involved in filter sterilization.
9. List the ingredients added to a test tube to do a rapid tablet test for urea hydrolysis.
10. Write the formula for the hydrolysis of urea by the enzyme urease.
11. Name the molecule that must accumulate to give a positive test for urea hydrolysis.
12. Give the pH of uninoculated urea broth.
13. Give the pH change necessary to produce a positive test for urea hydrolysis.
14. List three diseases caused by the "true" enteric pathogens, which are G- rods.
15. List four pathological conditions that can arise if *Proteus* gets out of its normal habitat, the intestinal tract.

DISCUSSION

The detection of urea hydrolysis is especially helpful in the differentiation of the genus *Proteus* from the other Gram negative bacilli that inhabit the intestinal tract, especially those that cause disease. Members of the genus *Proteus* mimic most of the enteric pathogens in the property of not being able to ferment lactose. Media used in the identification of enteric pathogens incorporate lactose for the purpose of separating the intestinal pathogens from the nonpathogenic bacteria that make up the normal flora of the intestinal tract. Most nonpathogenic normal flora ferment lactose. *Proteus* is one of the few nonpathogenic intestinal bacteria that are lactose nonfermenters and therefore can be confused with the pathogens. As far as *Proteus* being non-pathogenic is concerned, let us note that *Proteus* bacilli ordinarily do no harm in the intestines. However, if the *Proteus* organisms get out of the intestinal tract and into the urinary tract, they can cause severe cystitis and other forms of urinary tract infections. They have also been isolated from other pathological conditions such as gangrenous wounds, peritonitis, and otitis media. *Proteus* differs from the intestinal pathogens clinically in that it does not produce devastating diseases such as typhoid fever *(Salmonella typhosa)*, severe gastroenteritis (numerous *Salmonella* species), and bacillary dysentery *(Shigella* species).

Thus, the true enteric pathogens, which are Gram negative rods, are found in the genera *Salmonella* and *Shigella*. All species of *Salmonella* and most species of *Shigella* are unable to ferment lactose. Most *Proteus* sp. can be easily mistaken for an intestinal pathogen because of this common characteristic, therefore it is necessary to distinguish one from the other. Fortunately, there is a *single* test that separates *Proteus* from all other lactose nonfermenting bacteria. This test is urea hydrolysis. *Proteus* produces the enzyme urease which splits urea to form ammonia and carbon dioxide

$$O=C \begin{matrix} NH_2 \\ \\ NH_2 \end{matrix} \quad + \quad H_2O \quad \xrightarrow{\textit{Proteus} \text{ urease}} \quad 2NH_3 \quad + \quad CO_2$$

$$\underset{\text{Urea}}{} \qquad\qquad\qquad\qquad\qquad \underset{\text{Ammonia}}{}$$

The conventional method of testing for urea hydrolysis is to grow bacteria in rehydrated, commercially prepared urea broth. Urea broth is a highly buffered medium that contains yeast extract, which supports bacterial growth; urea, which is the enzyme substrate; and phenol red, which is a pH indicator.

When *Proteus* is grown in this broth, proteal urease is produced, which hydrolyzes the urea. This results in the accumulation of sufficient ammonia to make the environment quite alkaline. As the pH of the broth becomes more alkaline (that is, the pH is 8.1 or higher), the phenol red (which is salmon colored at pH 6.8 before inoculation) turns a cerise color. The appearance of the cerise color is a positive test for urea hydrolysis. When a lactose nonfermenting, Gram negative bacillus gives a positive test for urea hydrolysis, no further testing is necessary to eliminate it from the prominent enteric pathogens, and its generic name, *Proteus,* has been determined.

The related experience will help you become familiar with filter sterilization. You should perform the related experience if your instructor so indicates. In any case, read the related experience, and be responsible for understanding the principles involved since they will be included in the post test.

ACTIVITY

Activity 1: Determination of Urea Hydrolysis Using Urease Test Tablets

Aseptically pipet 1 ml of each broth culture into separate test tubes and 1 ml of sterile broth into the third tube. Deposit one urease test tablet in each of the three test tubes. The third tube is an uninoculated control tube. Incubate all three tubes at 37°C for 2 to 4 hours.

It is not necessary to use sterile test tubes if you are able to examine your tests in 2 to 4 hours since the heavy inoculum will be the predominating organism. If it is not possible for you to observe your results within this short incubation time, it will be better to use sterile test tubes.

In this rapid test for urea hydrolysis, a positive test is again the appearance of a cerise color. A negative test is indicated by a persisting yellow to salmon color. Record your findings on a duplicate of Table 37-1, and submit this to your file.

Precaution: If you use an inoculum other than a broth culture, the incubation time required for urease production is prolonged considerably.

TABLE 37-1   Collection of Data on Urea Hydrolysis Using Urease Test Tablets

	Urea hydrolysis*	Color change
Proteus vulgaris		
Escherichia coli		
Control		

*Use the terms positive or negative to fill in this table.

Related Experience*

Sterilize a cotton-plugged vacuum flask, a bacteriological filter inserted through a rubber stopper, and three capped or plugged test tubes. Wrap the flask and bacteriological filter (brown paper will do) before autoclaving.

Rehydrate urea broth, modifying the amounts in the directions on the medium bottle for the amount you wish to prepare. Unwrap the cotton-plugged filter flask and the bacteriological filter. Remove the plug from the flask, and place it on the sterile surface of the wrap. Aseptically insert the stopper attached to the bacteriological filter into the mouth of the flask. Twist the stopper to ensure a tight seal. Fold the sterile wrap around the cotton plug from the flask. Connect the vacuum flask to the vacuum apparatus, which is often a faucet adapter.

Figure 37-1 shows a common arrangement used to create a vacuum for filter sterilization. When a vacuum has been created by turning the faucet on full force,

*This experience is optional unless otherwise indicated by your instructor.

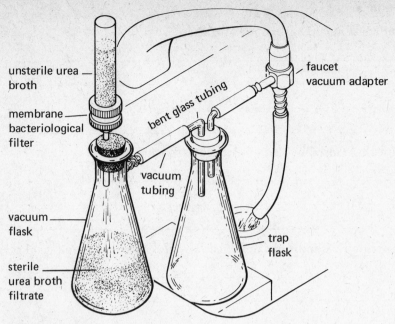

FIGURE 37-1
Apparatus for bacteriological filtration. All
tubing and stoppers must be tight fitting.

pour the urea broth into the funnel of the bacteriological filter. After all the broth has
been filter sterilized, remove the filter, and close the flask with the original sterile
cotton plug. Aseptically pipet 3 ml of the sterile urea broth into each of the three
sterile test tubes. Aseptically inoculate one tube of urea broth with *Escherichia coli.*
Inoculate another tube with *Proteus vulgaris,* and use the third tube for an uninoculated control. Incubate at 37°C for 48 hours.

Design a table to record your results, and submit it to your file. Then take the
post test.

POST TEST

The post test is a self-evaluation. It is not used for a grade. It is designed only to let you
decide if you have successfully completed this module.

Part I: True or False

_____ 1. *Proteus* is considered a member of the normal flora of the intestinal tract.

_____ 2. *Proteus* can be considered a pathogen in certain instances.

_____ 3. A negative test for urea hydrolysis is the appearance of a cerise color.

_____ 4. Urea broth is a highly buffered medium.

_____ 5. Urease splits urea into ammonia and water.

_____ 6. *Proteus* bacilli can be confused with *Salmonella* and *Shigella* because all
three genera ferment lactose.

_____ 7. If a lactose nonfermenting G- bacillus produces a positive test for urea
hydrolysis, no further testing is necessary to establish the genus of the
bacterium as *Proteus.*

_____ 8. When a cerise color appears as a result of urea hydrolysis, the pH of the
environment is 6.8.

_____ 9. The molecule that accumulates to give a positive test for urea hydrolysis
is ammonia.

_____ 10. Urea broth can be sterilized in the autoclave by reducing the temperature
and pressure of the autoclave considerably.

Part II: Completion

A. List the significant ingredients of urea broth.

1. _____

2. _____

3. _____

4. _____

B. List three diseases caused by "true" enteric pathogens.

1. _____

2. _____

3. _____

C. List four pathological conditions in which *Proteus* can be the etiological agent.

1. _____

2. _____

3. _____

4. _____

D. List the significant apparatus necessary to sterilize urea broth.

1. _____

2. _____

3. _____

Part III

Write the chemical formula for urea hydrolysis by urease.

MODULE 38

Litmus Milk Reactions

PREREQUISITE SKILL

Successful completion of Module 8, "Aseptic Transfer of Microbes."

MATERIALS

trypticase soy agar slant cultures of
  Streptococcus lactis
  Escherichia coli
  Proteus vulgaris
  Pseudomonas aeruginosa
10 ml of acetic acid—1:10 dilution of
  glacial acetic (10%)

medicine dropper
tubes of litmus milk (10 ml/tube) (6)
  autoclave at 7 psi for 20 minutes*

*To be prepared by the student if the instructor
so indicates.

OVERALL OBJECTIVE

Demonstrate the different kinds of bacteria-induced reactions in litmus milk and show
how these reactions can aid in the identification of bacteria.

Specific Objectives

1. Describe the reactions of litmus dye as a pH indicator.
2. Describe the reaction of litmus as a reducible molecule.
3. Name the two types of curds formed in milk by bacterial action.
4. Describe gas production in milk.
5. Describe peptonization of milk by bacteria.
6. Describe the effect of organic acids on litmus milk.
7. Define the terms *proteolysis, stormy fermentation,* and *comparative control.*

DISCUSSION

Litmus milk is an excellent liquid medium that does much to aid in the differentiation of bacteria because several different reactions can result when bacteria are grown in it. Milk itself supports the growth of most bacteria; it contains the milk protein, casein, and the milk sugar, lactose, as well as vitamins and minerals.

Litmus milk medium consists of 10% powdered skim milk and the dye molecule, litmus. Litmus, when added to rehydrated skim milk, turns the colloidal milk suspension from white to lavender. Litmus is a pH indicator, as well as a reducible dye molecule. Therefore, the two roles of litmus dye in litmus milk are the following:

1. As a pH indicator, litmus turns from lavender to red in an acid environment and from lavender to purplish-blue in alkaline conditions.

2. As a reducible molecule, the lavender litmus dye turns colorless when the dye molecule becomes reduced by the addition of hydrogen to the molecule. When reduction takes place, the lavender colored milk returns to its normal white color again. When litmus is in the reduced state, it can no longer function as a pH indicator. The word equation for reduction of the litmus molecule is as follows:

$$\text{Litmus} \xrightarrow{\text{H}_2} \text{Litmus H}_2$$
$$\text{(lavender)} \qquad\qquad \text{(colorless)}$$

In addition to these litmus dye reactions, many other reactions can be determined by the manner in which different bacteria utilize the ingredients of milk itself. Some bacteria ferment the lactose, some hydrolyze the casein, others produce a rennin-like enzyme, while still others simultaneously ferment sugar and hydrolyze the protein. The following is a list of possible bacterial reactions, depending upon which milk ingredients are utilized by the bacteria. Once again, this is dependent upon the type of enzymes that the bacteria are able to produce.

1. Curd formation
    a. Acid curd: If lactose is fermented to organic acids, a very firm, acid curd is formed. As the acidity increases, the curd becomes so solid that there is a squeezing out of a *clear liquid* (whey) from the curd. The formation of the curd is preceded by the litmus turning from lavender to red as the pH becomes lower.

    b. Rennet curd: Some bacteria produce a rennin-like enzyme. Rennin is a casein clotting enzyme. A rennet curd is soft and occurs at a neutral pH. Upon prolonged incubation, peptonization of the casein takes place with the consequent alkaline reaction. (See the description of peptonization listed under 2.)

2. Peptonization
    Some bacteria cannot ferment lactose and do not produce rennin but do possess proteolytic enzymes capable of hydrolyzing casein (peptonization). Peptonization results in the release of large amounts of ammonia; an alkaline condition then ensues, and the litmus turns a purplish-blue. With further incubation, an opaque clearing of the milk occurs as the casein is hydrolyzed to peptides and amino acids. The opaque (turbid) liquid supernatant often turns brown.

3. Gas production
    Gas occurs if an organism is capable of fermenting lactose to acid and gas. Gas production is usually detected by bubbles in the acid curd. Proteolytic bacteria, such as *Clostridia,* produce so much gas from degradation of the protein that the curd is blown into shreds. This blowing apart of the curd by the large amounts of gas produced is known as "stormy fermentation of milk."

A bacterium may cause more than one of the described reactions. Usually the combination of reactions occurs in a sequence. The sequence of reactions develops as the incubation time is prolonged.

A correlation of litmus milk reactions is useful in the identification of bacteria. A given species of bacteria will alter the litmus, milk sugar, or casein in a characteristic way, once again demonstrating the individuality of bacterial types.

The changes in litmus milk, caused by microbes growing in it, may occur very rapidly or may take a considerable length of time. For instance, the microbes that ferment the small amount of lactose in milk do so in a few hours, while complete peptonization of milk protein may take a week or longer. Therefore, you should look at your litmus milk cultures daily if possible. It is especially important that you observe them near the end of the first 24 hours of incubation and again after 48 hours of incubation.

ACTIVITIES

Activity 1: Microbial Reactions in Litmus Milk

Inoculate a tube of litmus milk with each of the bacterial types listed in the Materials section of this module. Incubate the four inoculated tubes at 37°C for one week or longer. Make frequent observations of the incubating tubes, as indicated in Table 38-1. Refrigerate one uninoculated tube of litmus milk for a comparative control. Make a table similar to Table 38-1, and fill it in at the times specified if possible, using the terms defined below. The table is designed as a guide. The 24-hour reading and the 48-hour reading are most important; the others can be adjusted to your convenience.

The following is a summary of microbial reactions in litmus milk.

acid pH	=	pink to red color.
alkaline pH	=	purplish blue color.
reduction	=	white.
acid curd	=	hard curd with clear supernatant (whey).
rennet curd	=	soft curd followed by peptonization.
peptonization	=	an opaque clearing of the milk because the casein is being hydrolyzed. It occurs at an alkaline pH. Supernatant is turbid and often turns brown eventually.
gas production	=	bubbles in coagulated milk, or "stormy fermentation."
no change	=	color and consistency remain the same.

Often you may need to use combinations of these terms: acid curd with gas production, alkaline pH followed by peptonization, or other combinations.

TABLE 38-1 Litmus Milk Reactions

	24 hours	48 hours	96 hours	7-14 day summary
S. lactis				
E. coli				
P. vulgaris				
P. aeruginosa				
Control				

Consult *Bergey's Manual of Determinative Bacteriology* to check the correctness of the results with which you filled in Table 38-1. Submit a completed, duplicate copy of Table 38-1 to your file.

Activity 2: The Effect of an Organic Acid on Litmus

Streptococcus lactis is one of many organisms that produce lactic acid as an end product of metabolism. Certain other bacteria, such as those found in the genus *Acetobacter,* produce acetic acid as an end product of their metabolism. Lactic acid and acetic acid are both organic acids. To demonstrate the effect of a known organic acid on litmus and on milk add one medicine dropperful of a 1:10 dilution of acetic acid to a tube of litmus milk. Allow the tube to set for 30 minutes or more without agitation. Submit a short, written description of your observation to your file.

Take the post test when you feel ready.

POST TEST

The post test is a self-evaluation. It is not used for a grade. It is designed only to let you decide if you have successfully completed this module.

True or False

_____ 1. Acetic acid turns litmus a purplish blue color.

_____ 2. When litmus is not reduced, it is a lavender color.

_____ 3. When litmus is reduced, hydrogen is added to the molecule, and the medium turns red.

_____ 4. An acid curd is firmer than a rennet curd.

_____ 5. Rennet is an enzyme that coagulates milk.

_____ 6. In milk, proteolysis means digestion of milk protein.

_____ 7. In milk, peptonization is a synonym for proteolysis.

_____ 8. Stormy fermentation means an extremely rapid acidification of milk lactose.

_____ 9. An uninoculated tube of medium used to compare against the inoculated tubes of the same medium is called a control.

_____ 10. *Bergey's Manual of Determinative Bacteriology* lists the litmus milk reactions for prolonged incubation.

MODULE 39

Hydrogen Sulfide Production

PREREQUISITE SKILL

Successful completion of Module 8, "Aseptic Transfer of Microbes."

MATERIALS

slant cultures of
  Escherichia coli
  Proteus vulgaris
  Salmonella typhimurium
7 ml sterile peptone iron agar stab (4)*

inoculating needle and other inoculating equipment

*To be prepared by the student if the instructor so indicates.

OVERALL OBJECTIVE

Determine the capability of various microorganisms to produce hydrogen sulfide.

Specific Objectives

1. Give a word equation yielding a positive test for H_2S production in peptone iron agar.
2. List four media designed to show hydrogen sulfide production.
3. Name the type of compound that microbes attack in order to produce hydrogen sulfide.
4. Describe the visible results of hydrogen sulfide production in peptone iron agar.
5. Discuss why the determination of hydrogen sulfide production is important to the clinical microbiologist.
6. Give the genus and species name of two different bacteria that manufacture the necessary enzymes to liberate hydrogen sulfide.
7. Give the general name of the molecule that makes up the black precipitate of a positive test for H_2S production when peptone iron agar is used.

DISCUSSION

You will be using peptone iron agar (PIA) to demonstrate the production of hydrogen sulfide (H_2S). Many other media have been designed for this purpose. Lead acetate agar is so employed but has been shown to be less sensitive than peptone iron agar. Kligler iron agar (KIA) and triple sugar iron agar (TSI agar) will also show H_2S production. The latter two media, however, are multipurpose media and are used to demonstrate the fermentation of certain sugars, as well as gas production other than hydrogen sulfide gas. The use of TSI agar will be presented in Module 42 on the identification of enteric pathogens, since it is most often used in this capacity.

In this module, using PIA, you will be able to determine *only* H_2S production by microorganisms. You are probably very familiar with the revolting odor of H_2S gas; do you remember its rotten egg odor?

Many organisms manufacture the enzymes necessary to liberate H_2S as a gas from sulfur-containing, organic compounds. When the H_2S is liberated by the enzyme, the sulfur of the H_2S combines with the iron in the PIA medium to form an iron sulfide. The iron sulfide is manifested as a black precipitate in the medium.

The production of H_2S by certain bacteria gives you another clue that you can utilize in the separation and identification of bacteria.

Caution: *Salmonella typhimurium* is a potential pathogen causing a mild gastroenteritis; therefore, use your best aseptic technique as you perform the following activity.

ACTIVITY

Activity 1: Demonstration of Hydrogen Sulfide Production

Using your inoculating needle, make aseptic stab inoculations of three peptone iron agar tubes. Stab one tube of medium with *Escherichia coli,* stab another with *Proteus vulgaris,* and stab the third tube with *Salmonella typhimurium.* Keep the fourth tube of medium as your uninoculated control medium. Incubate the three stab cultures and the control tube at 37°C for 48 hours. After incubation, make your observations, record your results on a duplicate of Table 39-1, and submit this to your file. When you have completed the table, take the self-evaluating post test.

TABLE 39-1 Data Collection for Hydrogen Sulfide

Stab culture	H_2S production*
E. coli	
P. vulgaris	
S. typhimurium	
Control	

*Use the terms positive or negative to fill in the blanks.

POST TEST

The post test is a self-evaluation. It is not used for a grade. It is designed only to let you decide if you have successfully completed this module.

Part I: True or False

_____ 1. H_2S is the chemical symbol for hydrogen sulfate.

_____ 2. *Escherichia coli* manufactures the enzymes necessary to produce H_2S.

_____ 3. Iron sulfide can be a by-product of metabolism of sulfur-containing organic compounds.

_____ 4. Determination of H_2S production is useful in the identification of bacteria.

_____ 5. A stab inoculation is executed by streaking the slanted area of the tubed medium.

_____ 6. When H_2S is released by bacteria, it is a gas.

_____ 7. An inoculating loop is used to make a stab inoculation.

_____ 8. A positive test for H_2S production in most media is manifested as a black precipitate.

_____ 9. TSI agar and KIA are used solely to detect H_2S production.

_____ 10. The odor of H_2S is comparable to the smell of a rose.

_____ 11. H_2S production is a universal characteristic for bacteria.

_____ 12. *Salmonella typhimurium* has the necessary enzymes to release H_2S from sulfur-containing organic compounds.

Part II

Write the word equation that describes a positive test for H_2S production in PIA.

MODULE 40

The IMViC Tests

PREREQUISITE SKILL

Successful completion of Module 8, "Aseptic Transfer of Microbes," and Module 9, "Aseptic Use of a Serological Pipet."

MATERIALS

nutrient broth cultures of
   Escherichia coli and
   Enterobacter cloacae or *aerogenes*

For Activity 1: Indole production
   sterile tryptone broth, 5 ml/tube (3)*
   Kovac's reagent for indole, dropper
    bottle*†

For Activity 2: Methyl red test
   sterile MR-VP medium, 5 ml/tube (3)*
   methyl red pH indicator, dropper
    bottle†
   clean, empty test tubes and capalls (2)†

For Activity 3: Voges-Proskauer test
   clean, empty test tubes and capalls (2)†
   5 or 10 ml pipets, clean and cotton
    plugged (2)†
   Barritt's reagents,
    dropper bottles (1 set)*†
    Solution A: 5% alpha naphthol,
     alcoholic
    Solution B: potassium hydroxide-
     creatine

For Activity 4: Citrate utilization
   sterile Simmons citrate agar slants (3)*

*Media and reagents to be prepared in advance by the student if the instructor so indicates.
†For use after 48 hour incubation of your subcultures. See the end of this module for formulae for all media and reagents.

OVERALL OBJECTIVE

Demonstrate the usefulness of the IMViC tests in the identification of Gram negative bacilli, particularly *Escherichia coli* and the *Enterobacter-Klebsiella* group.

Specific Objectives

1. After a suitable incubation period, add the appropriate reagents, gather the results of the tests, and tabulate the data as indicated in Activity 5.
2. Describe the IMViC series and explain what each test is designed to demonstrate.
3. Describe the visual difference between a positive result and a negative result for each test in the IMViC.
4. List the media inoculated for the IMViC series and name the significant ingredients of each.
5. List the reagents used for each test of the IMViC.
6. Define the terms *organic synthetic medium* and *Enterobacter-Klebsiella* group.

DISCUSSION

The IMViC test series consists of four different tests; they are *i*ndole production, the *m*ethyl red test, the *V*oges-Proskauer test, and the *c*itrate utilization test. The name IMViC stands for the first letter of the name of each test in the series, with the lower case "i" included for ease of pronunciation.

The IMViC tests are designed to determine specific physiological properties of microorganisms. They are especially useful in the differentiation of Gram negative intestinal bacilli, particularly *Escherichia coli* and the *Enterobacter-Klebsiella* group. Until most recently, the genus *Enterobacter* was called *Aerobacter*. *Enterobacter aerogenes, Enterobacter cloacae,* and *Klebsiella pneumoniae* have virtually identical physiological reactions. They are so similar that they cannot always be adequately differentiated and so are treated together as the *Enterobacter-Klebsiella* group. The IMViC series, plus gelatin liquefaction, and urease production tests are an invaluable aid to identification of organisms within the family *Enterobacteriaceae.* The gelatin and urease tests are dealt with in other modules.

Indole is produced in tryptone broth by the metabolism of certain organisms. Tryptone broth is rich in the amino acid tryptophan, which can be used by some bacteria as a source of carbon and nitrogen, as well as energy. As these organisms grow in tryptone broth, they attack the tryptophan and degrade it to indole, pyruvic acid, and ammonia. Not all bacteria—indeed not even all Gram negative intestinal organisms—are able to utilize tryptophan in this manner, which is the reason that the test for indole production aids in the identification of bacteria.

It is important to read the indole test after 48 hours of incubation because the indole itself may be attacked and further degraded in cases of prolonged incubation. If this occurs, the indole will eventually disappear, and you will then get a misleading negative test result.

The methyl red and Voges-Proskauer tests must be considered together since they are physiologically related and are inoculated into the same medium—MR-VP broth. You will, in fact, be making both tests from a single inoculated tube for each organism that you work with in this module. Opposite results are usually obtained for the methyl red and Voges-Proskauer tests, that is, MR+, VP- or MR-, VP+.

MR-VP medium contains peptone, dextrose (glucose), and dipotassium phosphate. The dextrose is a significant ingredient which is designed to determine what types of end products an organism forms from the degradation of glucose. Of course, the tests are not significant if an organism cannot utilize glucose, since dextrose is the only fermentable carbohydrate incorporated into the medium.

Typical strains of *Escherichia coli* ferment carbohydrates, including glucose, and produce a variety of acids as by-products or end products. After 48 hours of growth of *E. coli,* the pH of the culture reaches 4.5 or below (because of these acid end products), which changes the methyl red indicator to its acid color (red). The *Enterobacter-Klebsiella* organisms, utilizing the same quantity of glucose, do not produce a pH so low. Many of the end products of the *Enterobacter-Klebsiella* organisms are nonacidic, such as ethyl alcohol, acetoin, and acetyl methyl carbinol. Because the pH does not reach 4.5 the methyl red does not display its acid color (red).

The low pH produced by *Escherichia coli* is limiting for the continued growth of the organism. The higher pH produced by the *Enterobacter-Klebsiella* organism is not limiting, and when the glucose is exhausted, the organisms attack the peptone, causing the pH to rise above 6.2. The methyl red pH indicator then turns yellow, denoting a less acid condition in the medium. The range of methyl red as a pH indicator is 4.4 to 6.4 (red to yellow, respectively).

The Voges-Proskauer test confirms this difference. The reagents for the VP test react chemically with acetyl methyl carbinol to produce a pink or red color. *Escherichia coli* does not produce acetyl methyl carbinol, but the *Enterobacter-Klebsiella* group does. The VP color change is slow developing, and you should observe your tubes for at least 15 to 20 minutes before you can consider the test negative.

The MR-VP tests demonstrate the practical use of a good knowledge of microbial metabolism. The combination of the two tests helps confirm the separate identification of the *Enterobacter-Klebsiella* group and *Escherichia coli,* which appear virtually identical except for certain physiological differences demonstrated by the IMViC.

The citrate test is performed by inoculating the microorganisms into an organic synthetic medium in which sodium citrate is the only source of carbon and energy. An organic, synthetic medium is one that is chemically defined, containing known amounts of mineral salts and known amounts of simple organic compounds as the sole nutritional substrates. In sodium citrate broth (Koser's citrate medium), the presence of growth (turbidity) is a positive test result. When Simmons citrate medium is used, a color change from green to blue occurs, and this constitutes a positive test. Therefore, in Simmons citrate agar, the pH indicator bromthymol blue is also a significant ingredient. The pH range of bromthymol blue is 6.0 to 7.6, and it is yellow at the more acidic end of its range and blue at the more alkaline end. Uninoculated Simmons citrate agar is adjusted to a pH of 6.9, and so it displays an intermediate green color.

The classic results of the IMViC tests for *Escherichia coli* and the *Enterobacter-Klebsiella* group may be tabulated as shown in Table 40-1. Some strains of *Enterobacter aerogenes* are indole variable.

Now gather your media and stock cultures for the first four activities as indicated in the Materials section of this module. You will be doing *only* the inoculations for the first four activities at this time. The reagents will be added at the next lab session, at which time you will record your results and complete Activity 5. Use Figures 40-1 and 40-2 as guides for inoculations and results.

Remember: Do only the inoculations at this time.

TABLE 40-1 Classic Results for IMViC Tests

Organism	Indole	Methyl red	Voges-Proskauer	Citrate
Escherichia coli	+	+	–	–
Enterobacter-Klebsiella group	–	–	+	+

ACTIVITIES

Activity 1: Indole Production

Using good aseptic technique, inoculate one loopful of each organism into its own tube of tryptone broth; that is, of your three tubes of tryptone broth, inoculate one with *E. coli* and one with *E. cloacae*. The third tube is an uninoculated comparative control. Be sure to label your tubes carefully and include the organism used, the type

Indole: tryptone broth, 3 tubes

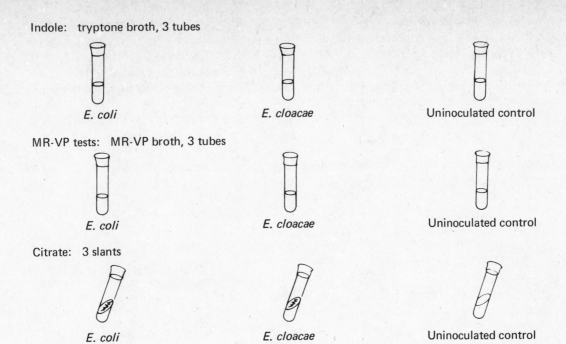

| E. coli | E. cloacae | Uninoculated control |

MR-VP tests: MR-VP broth, 3 tubes

| E. coli | E. cloacae | Uninoculated control |

Citrate: 3 slants

| E. coli | E. cloacae | Uninoculated control |

FIGURE 40-1
Inoculation scheme for IMViC tests to be made
during the first lab period.

of medium, the module and activity number, your name or initials, and the date.

Incubate your tryptone broth cultures at 37°C for 48 hours. In your next lab period, you will add the reagents and record your results.

After 48 hours of incubation, test each tube of tryptone broth for indole production as follows. Add about a dropperful of Kovac's reagent to each culture, and shake the tube gently from side to side. In the presence of indole, a deep red color will develop in the alcohol layer and rise to the surface of the medium. The development of color may take several minutes, and repeated shaking may be necessary. Do not consider the test negative until you have observed your tubes for 10 to 15 minutes. Add the reagents to the control tube and treat it the same as you did the broth cultures. Record your results carefully, and include them in a table similar to Table 40-2 in Activity 5.

Activity 2: Methyl Red Test

Inoculate one loopful of each organism into its own tube of MR-VP medium; that is, for your three tubes of MR-VP broth, inoculate one with *E. coli* and one with *E. cloacae.* The third tube is again a comparative control. Label your tubes carefully and completely as you did in Activity 1. These three tubes of MR-VP medium will be used for *both* the methyl red and the Voges-Proskauer test. The Voges-Proskauer test will be described in the next activity.

Incubate your freshly inoculated tubes of MR-VP medium at 37°C for 48 hours. You will divide the cultures and add reagents for the methyl red and Voges-Proskauer tests during your next lab period.

After 48 hours of incubation, aseptically pipet 2 ml of your MR-VP *E. coli* culture into a clean, empty test tube. Although the pipet is not sterile, asepsis should always be observed for your protection. Discard this pipet in the appropriate place. Set aside this tube containing 2 ml of *E. coli* culture to be used for the Voges-Proskauer test in Activity 3. Be sure to label these test tubes as you pipet the cultures into them. Now, aseptically pipet 2 ml of *E. cloacae* culture into a separate tube as you did for *E. coli.* When you have done this, you should have two tubes of each culture. See

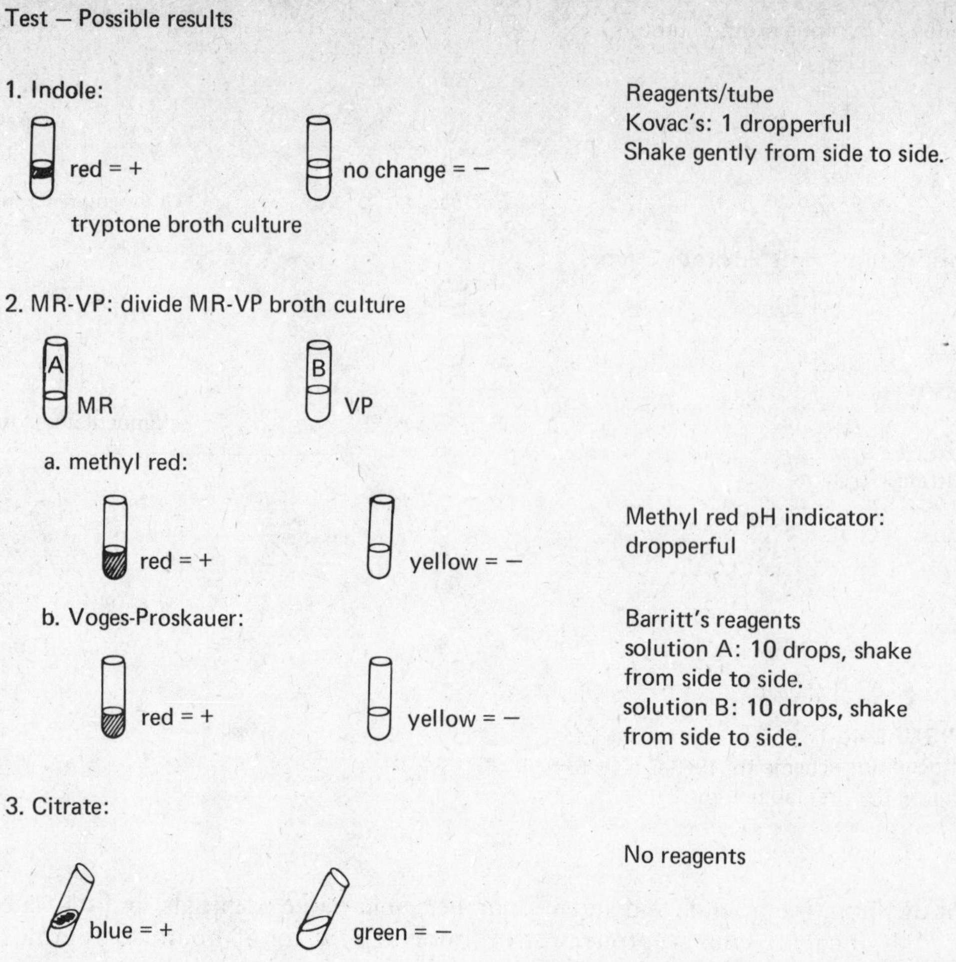

Test — Possible results

1. Indole:

red = +   no change = −

tryptone broth culture

Reagents/tube
Kovac's: 1 dropperful
Shake gently from side to side.

2. MR-VP: divide MR-VP broth culture

A   B
MR   VP

a. methyl red:

red = +   yellow = −

Methyl red pH indicator:
dropperful

b. Voges-Proskauer:

red = +   yellow = −

Barritt's reagents
solution A: 10 drops, shake
from side to side.
solution B: 10 drops, shake
from side to side.

3. Citrate:

blue = +   green = −

No reagents

FIGURE 40-2
Addition of reagents and results after 48 hours
incubation. Include a comparative control for
all of the above tests.

Figure 40-2 for clarification. Set aside the tube containing 2 ml of *E. cloacae*.
These 2 ml tubes of *E. coli* and *E. cloacae* culture will be used in Activity 3.

To perform the methyl red test, add a dropperful of methyl red indicator to
the remaining tube of *E. coli* culture in MR-VP medium. A color change should appear
immediately; a red color is a positive test, and a yellow to orange color is a negative
test. Remember that you are testing for acid end products of the metabolism of glucose
by the organism.

Repeat this test on *E. cloacae* culture. Record your results carefully, and
include them in Table 40-2 for Activity 5. Treat the control tube the same as the broth
cultures.

Activity 3: Voges-Proskauer Test

Reminder: You need not inoculate media for this test since the cultures inoculated in Activity 2 (the methyl red test) will also be used here.

After 48 hours of incubation, use the second set of tubes containing 2 ml of
each of the cultures in MR-VP medium that you reserved from the methyl red test
in Activity 2. Test each tube for acetyl methyl carbinol as follows:

1. Add about 10 drops of Barritt's solution A (alpha naphthol) and shake the tube from side to side.
2. Add an equal amount of Barritt's solution B (KOH-creatine) and shake vigorously from side to side. Reshake vigorously at 2 to 3 minute intervals.
3. Observe the tubes for an intense rose pink color, which indicates a positive test. It may take several minutes for the color to develop so you should observe the tubes for at least 15 to 20 minutes before you consider the test negative. *Repeated shaking* during the 20 minute waiting time aids the development of color.

Treat the control tube the same as the broth cultures. Record your results carefully, and include them in your duplicate of Table 40-2 for Activity 5.

Activity 4: Citrate Utilization

Inoculate each of the same two organisms onto its own Simmons citrate agar slant, that is, one slant with *E. coli* and one slant with *E. cloacae.* The third tube is an uninoculated comparative control. Be sure to label your slants carefully and completely as you did in Activities 1 and 2. Incubate your Simmons citrate agar slants at 37°C for 48 hours.

After 48 hours of incubation, examine your slants for growth, and compare them to your uninoculated control tube for any color change. A change from green to royal blue is a positive test for the utilization of citrate as a carbon source by the microbe. Record your results carefully, and include them in Table 40-2 for Activity 5.

Activity 5: Collection of Data

Record your data from Activities 1 to 4 in a table similar to Table 40-2, using + or − for a positive or negative test. Submit a copy of this table to your file. Then take the post test.

TABLE 40-2 Collection of Data

Organism	Indole	Methyl red	Voges-Proskauer	Citrate
E. coli				
E. cloacae				

POST TEST

The post test is a self-evaluation. It is not used for a grade. It is designed only to let you decide if you have successfully completed this module.

Part I

List the media used for the IMViC series and the significant ingredients of each.

Test	Media	Significant ingredients
a. indole	_____	_____
b. methyl red	_____	_____
c. Voges-Proskauer	_____	_____
		_____
d. citrate	_____	_____
		_____

Part II

Describe the visual difference between a positive result and a negative result for each test in the IMViC.

Test	Positive	Negative
a. indole	_____	_____
b. methyl red	_____	_____
c. Voges-Proskauer	_____	_____
d. citrate	_____	_____

Part III

Define the following terms.

1. *Enterobacter-Klebsiella* group _____

_____

_____

2. organic synthetic medium _____

_____

_____

Part IV: True or False

_____ 1. Methyl red is added to a culture to test for the presence of acetyl methyl carbinol.

_____ 2. Indole may be attacked and degraded to other products during prolonged incubation of a culture.

_____ 3. Kovac's reagent must be added to MR-VP medium to detect the results of the Voges-Proskauer test.

_____ 4. *Escherichia coli* can use sodium citrate as its sole source of carbon and energy.

_____ 5. A color change of inoculated Simmons citrate agar from green to blue constitutes a positive test for citrate utilization.

FORMULAE FOR ALL MEDIA AND REAGENTS

1. KOVAC'S REAGENT (INDOLE)

para-dimethylaminobenzaldehyde	5.0 gm
butyl alcohol	75.0 ml
hydrocholoric acid, concentrated	25.0 ml

Dissolve the aldehyde completely in the butyl alcohol. Slowly add the hydrochloric acid, stirring constantly. Alcohols resulting in reagents that become deep brown in color should not be used. The reagent should be light yellow in color. Store in refrigerator when not in use.

2. METHYL RED INDICATOR (4.4 to 6.4; red to yellow)

methyl red	0.1 gm
95% ethanol	300.0 ml
distilled water	200.0 ml

Dissolve the methyl red in the ethanol. Dilute to a final volume of 500 ml with distilled water.

3. BARRITT'S REAGENTS (V-P)

SOLUTION A

alpha naphthol	5.0 gm
ethanol, absolute	95.0 ml

Dissolve alpha naphthol in ethanol with constant stirring.

SOLUTION B

potassium hydroxide	40.0 gm
creatine	0.3 gm
distilled water	100.0 ml

Dissolve potassium hydroxide (KOH) in 75 ml of water. The solution will become quite warm, so allow the solution to cool to room temperature. Add creatine and stir to dissolve. Add remaining water. Store in refrigerator when not in use. This solution may be used for 4 to 6 weeks if kept refrigerated. It deteriorates rapidly at temperatures above 50°C.

4. TRYPTONE BROTH

tryptone	10.0 gm
distilled water	1.0 liter

Dissolve tryptone in distilled water. Dispense in tubes, and autoclave at 15 psi pressure for 15 minutes.

MR-VP medium and Simmons citrate agar are commercially available. Follow the directions on the labels of the media bottles. Autoclave at 15 psi for 15 minutes.

PART SEVEN

Medical Bacteriology

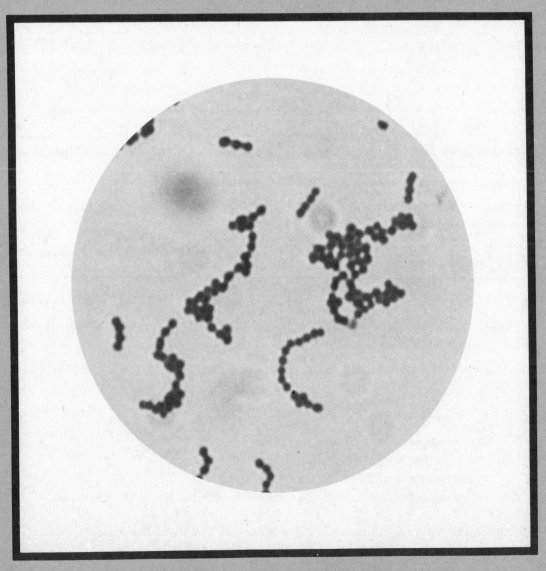

Long chained streptococci
(1000 x).

MODULE 41

Selective and Differential Media

PREREQUISITE SKILL

Successful completion of Module 8, "Aseptic Transfer of Microbes," and Module 14, "Streaking for Isolation."

MATERIALS

nutrient broth cultures of
   Salmonella typhimurium
   Enterobacter aerogenes
   Staphylococcus aureus
   Staphylococcus epidermidis
inoculating equipment
sterile blood agar plates* (BAP) (2)
sodium chloride agar plates (2)*†

eosin methylene blue agar plates (EMB) (3)*
   (Surface must be dry.)

*Formulae for media are at the end of this module.
†To be prepared by the student if the instructor so indicates.

OVERALL OBJECTIVE

Demonstrate a working knowledge of the principle and uses of selective and differential media.

Specific Objectives

1.   List three examples of enrichment ingredients.
2.   List two ways that selective agents function.
3.   List three selective ingredients.
4.   List the significant ingredients of the three media used in the activities of this module.
5.   Explain the function of the media ingredients that you listed in Specific Objective 4.

6. Define the terms *enrichment, selective medium, differential medium,* and *all-purpose* medium.

7. Discuss the value to the clinical microbiologist of the types of media used in this module.

DISCUSSION

Most of the plating media that you have worked with previously (for example, nutrient agar, TSA, and plate count agar) have been all-purpose media designed primarily to provide the nutrients necessary to support the growth of most microbes. These all-purpose media are sometimes enriched by the addition of ingredients that greatly enhance the growth of more fastidious organisms. The enrichment is often of such a nature that it cannot be sterilized with the basic medium but must be added aseptically after the medium has been autoclaved. Examples of such enrichments are skim milk, defibrinated sheep blood, and albumin.

Some enrichments also contain a selective agent that produces a medium that is specific for the isolation or cultivation of a particular group of microorganisms. A selective agent usually operates by inhibiting the growth of undesirable organisms, which allows more vigorous growth of the desired organism. When a selective agent is incorporated into a medium, the medium is then called a *selective medium.* Therefore, a selective medium is one that permits the growth of certain organisms, while preventing or inhibiting the growth of others. An example of a strictly selective medium is sodium chloride agar, which is selective for staphylococci. Azide agar is another highly selective medium. It contains sodium azide, which *inhibits* the growth of G- bacteria. You should remember this medium and its use if you are asked to perform Module 53, "Identifying Unknown Bacteria from a Mixed Culture." The inoculation of an appropriate selective medium with a mixture of several different kinds of microorganisms is a great aid in the isolation of a specific organism from the mixture because most of the unwanted organisms will not grow.

In this module you will be working with sodium chloride agar, which is nutrient agar with 7% NaCl added. This high concentration of salt inhibits most organisms but allows the growth of staphylococci, which are very salt tolerant.

Differential media incorporate ingredients designed to cause certain organisms to develop a different appearance from other microbes growing on the same medium. It is possible to tell at a glance whether a certain colony is causing a specific chemical change or not. For example, blood agar, which is an all-purpose enriched medium, is also a differential medium because it allows you to determine if a bacterial colony has produced the enzyme hemolysin and whether the resulting hemolysis is partial or complete (alpha (α) or beta (β)). See Figure 41-1 for types of hemolysis. If no change is apparent in the blood agar, this is called gamma (γ) hemolysis or no hemolysis. As a differential medium, blood agar differentiates between hemolytic and

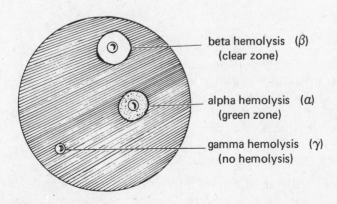

FIGURE 41-1
Types of hemolysis.

nonhemolytic organisms as well as between the types of hemolysis produced by hemolytic bacteria.

Other differential media often include a pH indicator to demonstrate that a specific nutrient has been attacked by the bacteria present, with a resultant change of pH. Carbohydrate fermentation broths, litmus milk, and triple sugar iron agar (TSI) are all differential media containing pH indicators. Differential media are especially useful when you are working with organisms with nearly identical morphology and other biological characteristics.

Most often, media are both selective and differential rather than one or the other. Eosin methylene blue agar (EMB), which is selective for G- organisms because of the bacteriostatic effect of the dye content on G+ bacteria, also contains lactose. Those G- bacteria that are capable of fermenting lactose take into the cell some of the dye, which causes the colonies to appear purple or dark centered with colorless edges. Colonies of the *Escherichia-Enterobacter* genera often have a metallic appearance, commonly called a green sheen. Colonies of organisms unable to ferment lactose appear colorless. Thus EMB is also a differential medium, differentiating between lactose-fermenting and lactose-non-fermenting bacteria.

MacConkey agar is another example of those media that are both selective and differential. MacConkey agar contains bile salts and a very small amount of crystal violet (1:100,000), both of which inhibit the growth of G+ organisms. It also contains lactose and neutral red pH indicator, which colors lactose-fermenting colonies red and may result in the formation of a red precipitate in the medium surrounding the growth. Colonies of bacteria unable to ferment lactose appear colorless. For isolation of *Proteus* sp., MacConkey agar instead of EMB is preferred because it inhibits the swarming effect of *Proteus*.

ACTIVITIES

Activity 1: Blood Agar as a Differential Medium

Gather together two sterile blood agar plates (BAP) and stock cultures of *Staphylococcus aureus* and *Staphylococcus epidermidis*. Label your BAP carefully, and inoculate a separate plate with each of the two bacteria. Streak for isolation, using as much of each plate as possible.

Caution: Staphylococcus aureus is a potential pathogen. Use your best aseptic technique as you make your inoculations.

Invert your plates and incubate at 37°C for 48 hours. After incubation, examine your BAP for hemolytic colonies, again noting Figure 41-1. A greening of the medium around the colonies is alpha hemolysis (partial), and a clear zone around the colonies is complete or beta hemolysis; no change is gamma hemolysis. In this module you will be observing only beta or gamma hemolysis.

Make a table similar to Table 41-1, and record your results in the appropriate spaces. When you have recorded the data from all three activities, submit the table to your file.

Activity 2: Sodium Chloride Agar—a Selective Medium

Obtain stock cultures of *Staphylococcus aureus* and *Enterobacter aerogenes*. Using your best aseptic technique, streak each organism for isolation on a separate NaCl agar plate. Label the plates carefully; invert and incubate them at 37°C for 48 hours. After incubation, examine the plates for growth. Compare and record a description of the growth in the appropriate spaces of your table.

Activity 3: EMB Agar—a Selective and Differential Medium

Obtain three sterile plates of EMB agar and stock cultures of *Salmonella typhimurium, Enterobacter aerogenes,* and *Staphylococcus aureus. Caution: Salmonella typhimurium* and *Staphylococcus aureus* are potential pathogens. Observe strict asepsis as you transfer them.

TABLE 41-1 Examples of Media for Selective and Differential Plating

	Differential medium blood agar	Selective medium NaCl agar		Selective-differential medium eosin methylene blue agar	
	Hemolysis	Growth	No growth	Growth	Lactose fermentation
Salmonella typhimurium					
Enterobacter aerogenes					
Staphylococcus aureus					
Staphylococcus epidermidis					

Carefully label an EMB agar plate for each organism, and streak for isolated colonies. Invert the plates, and incubate them at 37°C for 48 hours. After incubation, observe the plates for growth and any of the characteristic colors of colonies mentioned in the Discussion section of this module. Make a duplicate of Table 41-1, and record your observations in the appropriate places. Submit the completed table to your file.

Activity 4: Summary of Conclusions

Write a brief summary of your conclusions concerning selective media, differential media, and selective-differential media based on the activities that you have just completed. Submit the summary to your file also.

Related Experience

Repeat Activity 3 using MacConkey agar. Take the post test now.

POST TEST

The post test is a self-evaluation. It is not used for a grade. It is designed only to let you decide if you have successfully completed this module.

Part I: True or False

_____ 1. All-purpose media are formulated to support the growth of most bacteria.

_____ 2. Enriched all-purpose media will support the growth of more fastidious organisms than ordinary all-purpose media.

_____ 3. Enrichment ingredients are not selective.

_____ 4. Selective ingredients serve only to inhibit the growth of undesirable organisms.

_____ 5. Blood agar, EMB, NaCl agar, and MacConkey agar are examples of selective media.

_____ 6. Blood agar, EMB, and MacConkey agar are examples of selective-differential media.

_____ 7. Differential media often incorporate a pH indicator.

_____ 8. Media are rarely both selective and differential.

_____ 9. MacConkey agar incorporates bile salts and crystal violet as selective ingredients.

_____ 10. Lactose fermenting colonies appear red on EMB agar.

Part II

List the significant ingredients and their functions for the following media.

		Ingredient	Function
1.	Blood agar	_____	_____
2.	NaCl agar	_____	_____
3.	EMB agar	_____	_____
		_____	_____
		_____	_____

Part III

Define the following terms.

1. Selective medium _____

_____

2. Differential medium _____

_____

FORMULAE FOR MEDIA

1. BLOOD AGAR

Make the desired amount of nutrient agar with a 5% decrease in the amount of distilled water used. Add 0.8% salt (NaCl) to the NA. Autoclave the nutrient agar base in measured amounts in screw-cap bottles at 15 psi for 15 minutes.

Cool the agar to 60°C, and aseptically add 5% defibrinated sheep blood (5 ml/95 ml agar). Replace the cap on the bottle, and gently roll the bottle end over end to mix the blood thoroughly with the agar. Pour into sterile petri dishes. Flame the top of the agar to remove air bubbles.

2. SODIUM CHLORIDE AGAR (7% NaCl)

sodium chloride (NaCl)	7.0 gm
nutrient agar, rehydrated	100.0 ml

When making up the nutrient agar, add the 7.0 gm of NaCl. Dispense in screw-cap bottle, and autoclave at 15 psi for 15 minutes.

3. EOSIN METHYLENE BLUE AGAR (EMB)

This is commercially available. Follow the directions on the label.

MODULE 42

Intestinal Pathogens in the Family *Enterobacteriaceae*

PREREQUISITE SKILL

Successful completion of Module 8, "Aseptic Transfer of Microbes," and Module 14, "Streaking for Isolation."

MATERIALS

Use one of the mixed broth cultures:
1. *Pseudomonas aeruginosa*
   Escherichia coli
   Shigella dysenteriae
2. *Proteus vulgaris*
   Enterobacter cloacae
   Salmonella typhimurium

MacConkey agar plates (2)*
SS agar plates (2)*
selenite broth tubes (1 to 2)*†
TSI (triple sugar iron) agar slants (8)*

urease test tablet tests (5)*
motility test medium (5)*

For related experience:
   additional amounts of the above media
   must be calculated.

*To be prepared by the student if the instructor so indicates.
†May be made the day it is to be inoculated or not more than 48 hours in advance. Refrigerate.

OVERALL OBJECTIVE

Demonstrate an understanding of the principles of isolation and identification of intestinal pathogens by the use of significant selective-differential media and biochemical tests.

Specific Objectives

1. List four diseases caused by various species of intestinal pathogens.
2. Describe the cell morphology and Gram reaction of the intestinal pathogens.
3. Explain the use of selective and enrichment media in the identification of intestinal pathogens.

4. List four genera included in the normal flora of the human intestine.

5. Discuss specimen collection for culture and identification of suspected enteric pathogens.

6. Describe the biochemical changes that can be determined from triple sugar iron agar slants.

7. Describe the colony appearance of enteric pathogens on selective-differential media.

8. List three genera of intestinal flora easily confused with enteric pathogens and explain why they are easily confused.

9. Describe the characteristic reactions of enteric pathogens on TSI agar slants.

10. Describe a positive test on motility test medium.

11. Describe the positive test for urea hydrolysis used in this module and give the genus name of an organism that splits urea.

12. Write all possible conclusions that can be determined from all the significant media used in preliminary identification of enteric pathogens.

13. Define the terms *subplating, direct plating, streak plate,* and *picking a colony.*

DISCUSSION

The genera of enteric pathogens of medical significance found in the family *Enterobacteriaceae* are *Salmonella* and *Shigella. Salmonella typhi* causes typhoid fever, and *Salmonella paratyphi* causes less severe enteric fevers, while the many other species of *Salmonella* cause gastroenteritis of varying degrees of severity. *Shigella* species cause bacillary dysentery. All these enteric pathogens are Gram negative bacilli that may or may not be motile.

In order to identify the causative organism of enteric disease, the first accomplishment must be the separation of intestinal pathogens from the other Gram negative bacilli that comprise the normal flora of the intestinal tract (for example, *Escherichia, Proteus, Enterobacter,* and *Pseudomonas).* Selective and differential media are used for this purpose. For final identification of pathogens, several other significant media are used to study their biochemical activities. Final identification can be confirmed by serological typing (Module 49).

It is the purpose of this module to stress the biochemical tests, using a minimum of significant media, to separate the enteric pathogens from the normal fecal flora, as well as to identify to the genus and species the enteric pathogen in the mixed stock culture.

The stool specimen from patients suspected of having enteric disease should be collected early in the course of the disease, before antibiotic therapy has commenced. Stools that cannot be cultured shortly after collection should be placed in a transport medium such as buffered glycerol saline solution. Rectal swabs may also be used for specimen collection, especially from infants. The use of transport medium is even more critical for rectal swabs since they dry out so readily. It is generally agreed, however, that a freshly passed stool is the specimen of choice for bacteriological studies in enteric disease.

Upon delivery to the bacteriology laboratory, the stool specimen is streaked for isolation on at least two of many types of differential-selective media. At the same time, an enrichment broth is also inoculated with the fecal material. The differential-selective plating media that you will be using to streak for isolation in the following activities are MacConkey agar and SS agar *(Salmonella-Shigella* agar). You have only to look at any microbiology laboratory manual to see that many other media have been designed for this purpose.

Certain strains of *Shigella* do not grow on highly inhibitory media. Therefore, for isolation of *Shigella* strains, a mildly selective medium such as MacConkey (or EMB) agar is used. SS agar, the second plate of medium you will be using, is more highly selective, and certain fastidious species of *Shigella* grow very poorly on it. Hence it is advisable, if enteric disease is suspected, to use both types of media to avoid missing a possible *Shigella* infection.

Salmonellae grow well on both MacConkey and SS agar, and isolation can be made on direct plating of the stool specimen if large numbers of these organisms are present in the colon. However, especially in the case of carriers, the normal flora may greatly outnumber the *Salmonella,* and isolation of the pathogen may not occur on direct plating. It is for this reason that an enrichment broth such as selenite (or tetrathionate) is also inoculated with the patient's feces.

The enrichment broth is nutritive and highly selective. That is, these broths are designed to support the growth of enteric pathogens, while suppressing the growth of other intestinal bacteria. This results in the pathogens outgrowing the normal flora organisms. The enrichment broth is most useful in the isolation of *Salmonella,* while direct plating is indispensable for *Shigella* since *Shigella* can be inhibited by selenite. Fecally inoculated enrichment broths are incubated for 16 to 18 hours to allow the *Salmonella* to proliferate. The broth is then subcultured to plates of MacConkey agar and SS agar. A large number of *Salmonella* colonies will appear after incubation of the subcultured plates that were not present after the direct plating of the stool specimen.

Both media that you will be using to streak for isolation (MacConkey agar and SS agar) contain lactose and a pH indicator that will allow you to separate the lactose nonfermenting enteric pathogens from members of the intestinal flora that are lactose fermenters.

Lactose-fermenting colonies are red, while lactose-non-fermenting colonies are uncolored. (Read Module 41, "Selective and Differential Media.") Not all Gram negative bacilli of normal intestinal flora are lactose fermenters. Intestinal organisms such as *Proteus* and *Pseudomonas* are easily confused with enteric pathogens because of this common characteristic. The inability to ferment lactose makes the colony appearance of these nonpathogens similar to that of the pathogens; that is, all are uncolored colonies on the selective-differential media that you will be using to streak for isolation. However, colonies of H_2S producers, which may or may not be enteric pathogens, can have a dark center on SS agar because of the minute amount of iron in the medium.

Therefore, lactose-non-fermenting, colorless normal flora colonies must be differentiated from colorless *Salmonella* and *Shigella* colonies. This is accomplished by inoculating the uncolored colonies with or without dark centers on SS agar onto several different media to demonstrate biochemical differences. The media you will be using for this are triple sugar iron agar (TSI), motility medium, urease test tablets, and significant carbohydrate fermentation broths if necessary. The purpose of each of these media will be discussed in the activity designed for it.

Caution: You are now working with pathogens; hence you must observe all the rules of aseptic technique as you transfer these bacteria.

ACTIVITIES

Activity 1: Primary Isolation of Enteric Pathogens

This activity is designed to simulate the use of direct plating from fresh stool. It is a simulation because, for esthetic reasons, you will be using a mixed culture of enteric bacteria instead of a stool specimen.

Select one of the mixed broth cultures listed in the Materials section, and streak for isolation on a MacConkey agar plate and on an SS agar plate. Use a heavier original inoculum (two or three loops of broth culture) on the SS agar plate since it is more inhibitory than MacConkey agar (one loop is sufficient for original inoculum here). Streaking for isolation from the stool specimen is often called primary isolation. After you have streaked both plates for isolation, aseptically transfer three or four loops of the mixed broth culture to a selenite enrichment broth.

Incubate the streak plates at 37°C for 48 hours. Incubate the selenite broth at 37°C for 16 to 18 hours. The subplating from selenite broth culture will be described in Activity 2 and will require your attention before your streak plates from this activity have incubated 48 hours. After your streak plates have incubated 48 hours, select, if present, two different-appearing colony types that are uncolored pure colonies. You will be transferring these two pure colonies to TSI agar slants in Activity 3. The inoculation of motility medium and urease test tablets will be described in Activity 4.

Activity 2: Subplating from Enrichment Broth Culture

After the selenite enrichment broth culture has incubated 16 to 18 hours, aseptically transfer one loop of selenite broth culture to a MacConkey agar plate, and streak for isolation. Next aseptically transfer two or three loops of the enrichment broth culture to an SS agar plate, and streak for isolation. You may now discard the selenite broth culture. Incubate both plates at 37°C for 48 hours. From these subcultured streak plates, again select two colorless colonies, if two different types are present, for transfer to the medium described in Activity 3. If only one colony type is present, you will need to inoculate only one set of media described in Activities 3 and 4.

   Precaution: If you are unable to do this subplating from the enrichment broth after 16 to 18 hours of incubation, have your lab instructor refrigerate the selenite cultures until your next lab session, and streak them for isolation then. This works best if you do the subplating in the specified time.

Activity 3: TSI: a Significant Medium Used in Preliminary Identification of Enteric Pathogens

The two to five uncolored colonies that you selected in Activities 1 and 2 must now be transferred to separate TSI slants. With a wax glass-marking pencil, number the bottom of the plate under the colonies not fermenting lactose. Number only those different-appearing, uncolored colonies that you have selected for transfer. Label your TSI slants with numbers corresponding to those on the colonies that you have selected. Label completely as usual. Now use your inoculating needle to pick a few cells from the center of each colorless colony that you have selected. Be sure to touch just the center of the colony with the tip of your inoculating needle. (This is called picking a colony.) Inoculate the organisms clinging to the tip of your inoculating needle to a TSI agar slant by stabbing the butt and streaking the slant, as shown in Figure 42-1.

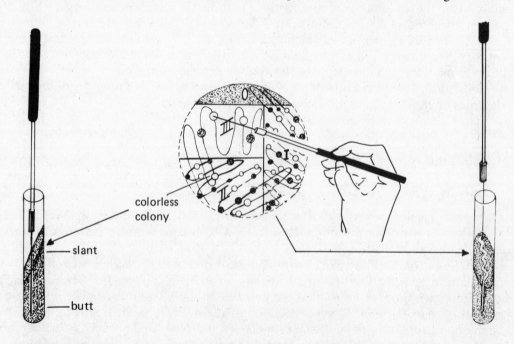

colorless colony

— slant

— butt

FIGURE 42-1
Pick cells from the center of the uncolored colony, and transfer them to the TSI slant by stabbing the butt and streaking the surface of the slant.

Do the same TSI slant inoculation with the remaining suspicious colonies that you have numbered. Incubate all the inoculated TSI slants at 37°C for 48 hours. Be sure the closures (capsuls or cotton plugs) are loose fitting so that the slant is surrounded by an aerobic environment.

The following list describes the data that you can determine from TSI slants. After careful study of these biochemical changes that can take place in TSI slants, compare them to your incubated slants, and record your results in tabular form. Make a table similar to Table 42-1 for collection of your data.

TABLE 42-1 Collection of Data*

| | TSI | | | | |
	Glucose	Lactose	H2S	Urea hydrolysis	Motility
Escherichia coli	AG	AG	–	–	V
Enterobacter cloacae	AG	AG	–	–	+
Pseudomonas aeruginosa	–	–	–	–	+
Proteus vulgaris	AG	–	+	+	+
Salmonella typhimurium	AG	–	+	–	+
Shigella dysenteriae	A	–	–	–	–

*A = acid production
 AG = acid and gas
 – = negative
 + = positive
 V = variable (may or may not be present)

The biochemical characteristics discernible from TSI slants are:

1. *Carbohydrate fermentation.* As the name of the medium implies, it contains three sugars: lactose, saccharose (sucrose), and a very small amount (0.1%) of glucose (dextrose). It also contains the pH indicator, phenol red. If an organism ferments any of the three sugars or any combination of them, the entire medium will turn yellow because of the acid pH caused by the end products of the fermentation. The enteric pathogens, however, are capable of fermenting only the dextrose. The small amount of dextrose present is usually utilized within the first 24 hours of incubation, and in the aerobic conditions of the slant, the reaction reverts and becomes alkaline (red). This is also true of bacteria such as *Proteus* that actively ferment sucrose and dextrose but not lactose. When the two sugars are used up, the pH also reverts to alkaline in an aerobic environment. The same organisms in the anaerobic butt are unable to cause a reversion, and so the butt remains acid (yellow). Thus, *Salmonella* and *Shigella* after 24 to 48 hours of incubation show a yellow butt and a red slant, indicating fermentation of *glucose only*. This is true for all enteric pathogens with the exception of a rare *Shigella* species (for example, *sonnei*). Because you are concerned only with detecting enteric pathogens you can discard all tubes displaying both yellow butt and slant after 48 hours of incubation unless an organism such as *Shigella sonnei* is suspected. No change in the medium indicates that none of the sugars has been fermented. Such tubes may also be discarded since the all alkaline reaction rules out the pathogens.

2. *Gas production.* If bubbles or a splitting appears in the butt, or if the entire slant has been pushed up from the bottom of the tube, then gas as well as acid has been produced by the fermentation.

3. *Hydrogen sulfide production.* The TSI medium also contains iron sulfate. If H_2S is produced by the organism, an iron sulfide results, which is manifested as a black precipitate. The enteric bacilli that produce H_2S also ferment glucose. The large amount of black precipitate can mask or obscure the yellow or acid condition in the butt. (Read Module 39 on H_2S production.)

Figure 42-2 is a summary of possible biochemical information that can be obtained from TSI agar slants.

Do not discard the TSI slants with reactions of possible pathogens. You will be using these culture tubes for the next activity.

Activity 4: Other Significant Media Used in Preliminary Identification of Enteric Pathogens

1. *Motility test medium.* If the TSI agar slants have an acid butt and an alkaline slant (H_2S may or may not be present), remove some growth from the TSI slant using your inoculating needle, and make a stab inoculation of the motility test medium. Do this for each TSI with possible pathogen reactions. Incubate for 48 hours at 37°C. Motility can be seen macroscopically as a diffuse zone of growth spreading from the line of the stab inoculation. Some motile organisms diffuse from only one or two points of the stab, which appears as nodular outgrowths along the stab line. Others display a fanlike growth pattern near the top of the stab line. Use a comparative control to aid you in determining if motile organisms are present or not. If you have any doubt about the motility of the organism, make a wet mount from your TSI slant, and look for actively motile cells. Motility sulfide medium is also an excellent medium to demonstrate motility as well as hydrogen sulfide production. You may try this if you wish.

2. *Urease test tablet.* Read Module 37 on urea hydrolysis. Note that you used a broth culture inoculum. The incubation was shorter because the enzyme was already present in the broth environment.

Procedure using growth from solid media. Add 1 ml of distilled water to a test tube. Place one urease tablet in the water. With your inoculating loop, obtain a *large amount* of growth from your TSI slant, and aseptically transfer it to the water-urease test tablet mixture.

Incubate at 37°C for 2 to 6 hours. The appearance of a cerise color is a positive test for urea hydrolysis. If you get such a result, you have ruled out enteric pathogens. Note the increased incubation time. Why? (See Table 42-1.)

Figure 42-3 is a schematic summary of the inoculations described in all three of the previous activities.

After you have collected all your data, check your results with Table 42-1. Make a table similar to Table 42-1, and fill in your results from the particular mixed culture with which you started. From your data, conclude the genus and species name of the enteric pathogen, along with the names of other organisms in the original broth *if* you were able to determine them. Both the duplicate table and organism names are to be submitted to your file.

Remember that a stool specimen has many more genera and species of the genera than you have worked with in this module. This module is designed only to acquaint you with *some* media used to begin the identification of enteric pathogens. In a later module, you will be identifying unknown bacteria with less guidance. That is, you will not know the names of the bacteria in the original broth mixture. This module should help prepare you for how to proceed in the identification of unknowns, Module 53.

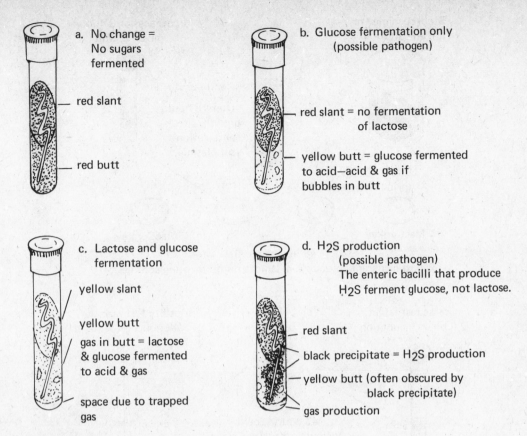

FIGURE 42-2
Summary of possible biochemical information to be obtained from TSI agar slants.

Related Experiences

1. In order to learn to identify both genera of enteric pathogens repeat all the activities in this module, using the other mixed broth culture.

2. Select the appropriate biochemical tests, and prepare the necessary media not included in this module to identify to the genus and species those organisms you did not identify in the mixed broth culture you used throughout this module, for example, the lactose fermenters on the MacConkey agar streak plate.

3. Collect your own stool specimen, and repeat the procedure in this module. Include added media for the study of significant biochemical activities necessary to identify to the genus and species two fecal organisms in your stool specimen. You may choose to identify a lactose fermenter if two lactose nonfermenters are not present.

4. If serological typing sera are available, do slide agglutination tests to confirm the indentification of the enteric pathogen that you identified biochemically in this module. Follow the instructions accompanying the typing sera to complete this related experience.

It is suggested that Related Experience 1 be your first choice since you will learn much about the behavior of a different enteric pathogen. You are encouraged to perform Related Experience 2 also.

When you feel ready, take the post test. Be sure that you score 100% on the test. Since this module is more complex, review it until you do score 100%. You will then be better prepared to identify the unknowns, Module 53, if required to do them.

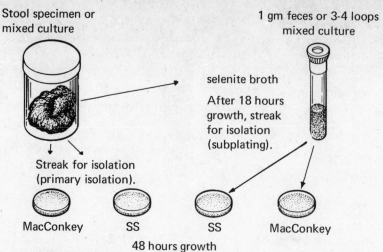

Stool specimen or mixed culture

1 gm feces or 3-4 loops mixed culture

selenite broth

After 18 hours growth, streak for isolation (subplating).

Streak for isolation (primary isolation).

MacConkey   SS   SS   MacConkey

48 hours growth
Pick transparent colonies (i.e. lactose nonfermenting colonies).

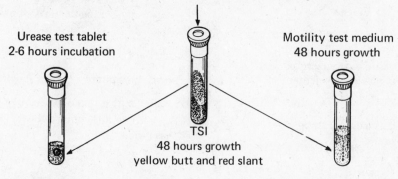

Urease test tablet
2-6 hours incubation

Motility test medium
48 hours growth

TSI
48 hours growth
yellow butt and red slant

FIGURE 42-3
Schematic summary of inoculations.

POST TEST

The post test is a self-evaluation. It is not used for a grade. It is designed only to let you decide if you have successfully completed this module.

Part I: True or False

_____ 1. Intestinal pathogens in the family *Enterobacteriaceae* are Gram negative bacilli.

_____ 2. All intestinal pathogens are motile.

_____ 3. EMB or MacConkey agar are more useful than SS agar or enrichment broth for the isolation of *Shigella*.

_____ 4. A streak plate after incubation should have isolated colonies.

_____ 5. An inoculating loop is used to pick a pure colony.

_____ 6. *Shigella* sp. are more often isolated from direct plating of the feces than from subplating.

_____ 7. Selenite broth is commonly called a differential medium.

_____ 8. If a pure colony does not ferment lactose on selective-differential media and produces H_2S on a TSI agar slant, it is one of the salmonellae.

_____ 9. Subplating as used in this module is a streak plate done from enrichment broth.

_____ 10. Selenite broth enhances the growth of salmonellae.

_____ 11. Selenite broth when subcultured to SS medium, in the case of carriers, yields more uncolored colonies than does the direct plating of the patient's stool specimen.

_____ 12. A positive urease test is manifested as a black precipitate.

_____ 13. A TSI slant inoculated with an enteric pathogen will have an acid slant and an alkaline butt after incubation.

_____ 14. Motility test medium inoculated with _Shigella_ will usually show a positive motility.

_____ 15. When urea is hydrolyzed by a bacterium, enteric pathogens have been ruled out.

Part II: Completion

1. List four diseases caused by enteric pathogens in the family _Enterobacteriaceae_.

   a. _____

   b. _____

   c. _____

   d. _____

2. Which of the previous four diseases is caused by _Shigella?_

   _____

3. Name the condition caused by _more_ species of _Salmonella._

   _____

4. List four genera of Gram negative bacilli that are part of the normal flora of human intestinal tracts.

   a. _____

   b. _____

   c. _____

   d. _____

5. List two of these genera that can be confused with enteric pathogens on selective-differential media.

   a. _____

   b. _____

   Why?

   c. _____

6. Describe the specimen of choice for bacteriological studies in enteric disease.

   _____

7. List three significant media used in the preliminary identification of enteric pathogens after their isolation on selective-differential media.

   a. _____

   b. _____

   c. _____

8. List three biochemical characteristics discernible from inoculated TSI agar slants.

   a. _____

   b. _____

   c. _____

9. Name two lactose nonfermenting bacteria that produce H$_2$S.

a. _____

b. _____

10. Among the organisms that you worked with, which will give an alkaline reaction throughout a TSI agar slant?

_____

11. Define direct plating. _____

_____

12. A stool specimen on arrival at the diagnostic laboratory is inoculated on three media. Which of the three is most helpful in detecting *Salmonella* carriers?

_____

13. Why is the incubation time for a urease positive test longer when inoculum from a solid media is used?

_____

_____

MODULE 43

Pathogenic
Staphylococcus aureus

PREREQUISITE SKILL

Successful completion of Module 8, "Aseptic Transfer of Microbes," Module 14, "Streaking for Isolation," Module 21, "Preparing a Bacterial Smear," and Module 23, "Gram Stain."

MATERIALS

slant cultures of
  Staphylococcus aureus
  Staphylococcus epidermidis
mannitol salt agar plates (2)*
plasma, rabbit (fresh)— commercially
  available
microscope slides (2)
blood agar plates (2)
3% hydrogen peroxide

For related experiences:
  sterile swab (1 to 2) tubed in 2 to 3 ml
    0.85% saline*
  mannitol salt agar plate (1 to 2)*
  blood agar plate (1 to 2)
  plasma, rabbit (fresh)—commercially
    available

*To be prepared by the student if the instructor so indicates.

OVERALL OBJECTIVE

Differentiate *Staphylococcus aureus* from *Staphylococcus epidermidis*.

Specific Objectives

Compare *Staphylococcus aureus* and *Staphylococcus epidermidis* in Specific Objectives 1 to 5.

1. Describe their Gram reactions and the characteristic groupings.
2. Describe the hemolytic differences, if any, on a blood agar plate.
3. Describe the colony morphology differences, if any, especially chromogenesis and size of colony.

4. Describe the coagulase reactions of both organisms.
5. Describe the mannitol fermentation reactions of both organisms.
6. List the significant ingredients and the results obtainable from the blood agar plate, coagulase test, and mannitol salt agar.
7. List five pathological conditions caused by *Staphylococcus aureus*.
8. State the primary pathological condition caused by *Staphylococcus epidermidis*.
9. List the areas of the human body to which *Staphylococcus* is indigenous.
10. Describe the handling of specimens collected on dry swabs.
11. Describe the principle of the coagulase test.

DISCUSSION

There are 3 genera in the family *Micrococcaceae*. This module is limited to the study of one of these genera, the genus *Staphylococcus*. This genus is more interesting than the other genera because it includes the pathogen commonly called "staph."

Bacteria in the genus *Staphylococcus* are Gram positive spherical cells that occur singly, occasionally in pairs, but most frequently in irregular clumps. These irregular grapelike clusters of cells are best observed by growing staphylococci in thioglycollate broth and then making a stained smear carefully, disturbing the cell arrangement as little as possible. The appearance of a Gram stained smear is usually sufficient to distinguish staphylococci from the streptococci because of this characteristic cell grouping (grapelike clusters). If in doubt, these two genera can be separated on the basis of the presence of the enzyme, catalase. Catalase will break down hydrogen peroxide to form water and oxygen. If you mix a loopful of staphylococci on a slide with 3% hydrogen peroxide, bubbles of oxygen will be visible to the naked eye. Streptococci do not form catalase, so no bubbles will be released. Both species of staphylococci are often normal or transient flora of your skin, mouth, or nasal cavity. Although these bacteria can be present in you and on you without causing any physical discomfort, they can also be responsible for several pathological conditions.

Staphylococcus aureus is more pathogenic than another member of this genus, *Staphylococcus epidermidis*. Disease processes associated with *Staphylococcus aureus* are numerous. The portal of entry is variable, since they gain access to the body via the skin, the respiratory tract, or the genito-urinary tract. Some of the infections of humans in which *Staphylococcus aureus* is the etiological agent are boils, carbuncles, impetigo, meningitis, osteomyelitis, urinary infections, and food poisoning. Staph food poisoning is caused by an exotoxin secreted into the contaminated food as the organisms actively multiply in the food before it is ingested. The staph exotoxin is commonly called an enterotoxin since the toxin remains in the intestinal tract, with the intestinal mucosa as the area of insult. The violent symptoms begin about 2 to 4 hours after the contaminated food is eaten. The onset is abrupt with nausea, vomiting, unbelievable intestinal cramps, and diarrhea. The symptoms usually subside within a day. Food is often contaminated by food handlers with pathogenic *Staphylococcus aureus* lesions on their hands, arms, faces, or via the respiratory secretions of staph carriers.

Staphylococcus epidermidis is not as invasive as *Staphylococcus aureus;* therefore, it primarily causes small abscesses. Occasionally, however, *Staphylococcus epidermidis* can also be responsible for some of the other, previously listed infections.

Fortunately, staphylococci are usually susceptible to most antibiotics. You found this to be true if you performed Module 20 on antibiotic sensitivity testing. Drug-resistant, mutant staphylococci do arise, however. When these drug-resistant strains occur, the patient is often plagued with chronic staph infections or may even become a healthy carrier of pathogenic staph. Pathogenic staphylococci carriers should not be allowed to work in hospital nurseries or be food handlers since they unknowingly cause the spread of staphylococci infections.

Since the infections caused by staph are so varied, the sample collected for culture may be a throat or lesion swab, a urine specimen, or even contaminated food. If the specimen is collected on a sterile swab, it should be cultured immediately or placed in a transport medium, since bacteria do not live very long on a dry swab. However, even in transport medium, delivery to the diagnostic laboratory must be rapid since normal flora can outgrow the pathogen.

You can differentiate the more pathogenic *Staphylococcus aureus* from *Staphylococcus epidermidis* and other Gram positive cocci by knowing a few specific characteristics of its enzymatic make-up. Some of these enzymatic characteristics of *Staphylococcus aureus* are:

1. Always produces the enzyme coagulase.
2. Almost always ferments mannitol.
3. Usually produces a golden pigment—hence the species name, *aureus.*
4. May produce the enzyme hemolysin.

Utilizing these species characteristics, you can positively identify pathogenic *Staphylococcus aureus.* The following activities are designed to show how this enzymatic individuality of *Staphylococcus aureus* can be used to incriminate it as the causative agent of suppurative lesions or food poisoning.

Precaution: Remember that these organisms have the potential to be pathogenic. Be especially cognizant of your aseptic technique. If your aseptic technique is correct, you cannot infect yourself.

ACTIVITIES

Activity 1: Gram Stains

Make a Gram stain of the slant cultures of *Staphylococcus aureus* and *Staphylococcus epidermidis.* On a separate sheet of paper, make a copy of Table 43-1, which will be used to collect all the data asked for in this module. When completed, submit the tabulated data to your file. Now make a careful microscopic examination of your Gram stains, and write your findings in the appropriate columns of the table.

Activity 2: Blood Agar Plate for the Study of Hemolysis and Chromogenesis

In this activity, you will be observing isolated staphylococci colonies on a blood agar plate. This medium contains a base similar to nutrient agar, to which 5% defibrinated sheep red blood cells have been added. Some bacteria have the genetic code that allows them to synthesize the enzyme hemolysin. Hemolysin is an exoenzyme that digests red blood cells. So if a colony of bacterial cells is producing hemolysin and secreting it into the medium, there will be a round, clear zone surrounding the colony because all the red blood cells (erythrocytes) in the medium have been lysed, and the hemoglobin (the red pigment in erythrocytes) has been digested. This clear area around the colony is called the zone of hemolysis. Some genera of bacteria are hemolytic, while others are not. Some species within a genus are hemolytic, while other species are not. Some strains of the species of bacteria are hemolytic, while others are not. For instance, *Staphylococcus aureus* cells can be hemolytic, but often they are not. It is for this reason that the hemolytic capability of *Staphylococcus aureus* is not used as an identification characteristic of pathogenic staph.

Now streak a blood agar plate for isolated colonies using the *Staph. aureus* slant prepared for you. Using another blood agar plate, do the same with *Staph. epidermidis.* Incubate these streak plates for 48 hours at 37°C. After the incubation period, inspect the individual colonies of each organism for hemolysis, chromogenesis, and other colony morphology differences, if any. Record your observations in the appropriate columns of your table.

Activity 3: Mannitol Salt Agar Plate, a Selective-Differential Medium

Mannitol salt agar is a selective medium for the isolation of staphylococci. The significant ingredients of this medium are the 7.5% sodium chloride (NaCl), mannitol, and the

TABLE 43-1 Significant Tests Leading to the Identification of Pathogenic *Staphylococcus aureus*

| Organism | Gram Stain | | Blood agar plate | | Mannitol salt agar | Coagulase test |
	Gram reaction	Cell arrangement	Hemolysis	Chromogenesis and other differences	Fermentation	
Staph. aureus						Positive
Staph. epidermidis						Negative

phenol red pH indicator. The selective component is the high salt concentration, since the growth of most other bacteria is inhibited by this abnormally high amount of salt, although staph is not. Phenol red is a pH indicator that gives a red appearance to the uninoculated medium. In the presence of acid, the phenol red turns yellow. Mannitol is a substrate for the enzyme systems of pathogenic *Staphylococcus aureus*, which metabolizes mannitol, converting it to acid waste products. The acid end products of mannitol fermentation diffuse into the medium, giving the medium a yellow appearance around the colonies due to the effect of the pH change on phenol red. *Staph. epidermidis* colonies are often smaller, and the color of the medium remains unchanged since mannitol is not fermented. This makes mannitol salt agar a differential medium also. The pathogenicity of mannitol positive staphylococci must be confirmed by the coagulase test that you will do in Activity 5.

Aseptically streak a mannitol salt agar plate for isolation, now, using the slant culture of *Staphylococcus aureus*. Using another mannitol salt agar plate, do the same with *Staphylococcus epidermidis*. You may use a heavy original inoculum on both plates because the high salt concentration has a slightly inhibitory effect even on the salt tolerant staph. Incubate the plates at 37°C for 48 hours. When you collect your data, look for mannitol fermentation as well as pigmentation of the colony. Record your data in the appropriate columns of your table.

Activity 4: Catalase Test

Place a drop of 3% hydrogen peroxide on a clean microscope slide. Aseptically add a visible amount of bacterial growth with your loop. Mix slightly and look for bubbles of oxygen. Repeat with *S. epidermidis*.

Activity 5: Slide Coagulase Test

Coagulase is an exoenzyme that causes the fibrin of blood plasma to clot. Pathogenic *Staphylococcus aureus* produces coagulase, while nonpathogenic strains are coagulase negative. Coagulase production and mannitol fermentation have a higher degree of correlation with pathogenicity than the production of hemolysin or chromogenesis.

Perform the coagulase test on *Staph. aureus* and *Staph. epidermidis* as follows, noting first these precautions.

Precautions: Take care not to contaminate yourself or your table top. When you have finished, discard the slides into a beaker of disinfectant.

Put a drop of water on a microscope slide. Next, using your inoculating loop, transfer *Staph. aureus* cells from the slant culture to the drop of water. Mix the growth into the water to make a heavy, cloudy suspension of cells. Using a medicine dropper, add one drop of plasma to the suspension, and mix with your loop for 10 to 20 seconds. If the test is coagulase positive, visible clumps will appear within 1 to 2 minutes. It may be necessary to hold the slide mixture up to a microscope lamp to see the granular appearance of the clumping fibrin. Repeat the coagulase test using *Staph. epidermidis*. Discard the slides in a beaker of disinfectant, and record your results in the appropriate columns of your table.

Inspect your data table, and note the similarities and differences between pathogenic *Staphylococcus* and a nonpathogenic *Staphylococcus*. It should be a simple task for you to differentiate between these two organisms in the future.

There is also a tube test to show coagulase production. The ingredients are the same as for the slide test, but it takes 12 hours before results can be obtained. Therefore, it is not practical to use during your laboratory period.

Related Experiences

1. Using a saline-moistened, sterile swab, wipe your face in the area of a pimple, if present. Use the skin-contaminated swab for the original inoculum, and streak for isolation on a blood agar and a mannitol salt agar plate. After incubation, Gram stain staph resembling colonies. If the Gram stain is indicative of staph, perform the coagulase test. Correlate your data, and conclude whether pathogenic staph is present on your skin. The value of dextrose utilization is pointed out in the accompanying dichotomous key, Figure 43-1.

2. Do the same tests as in Related Experience 1 on your nasopharyngeal swab if you are interested in knowing if you are a staph carrier. Insert a thin flexible wire swab through your previously cleared nasal passage until it meets definite resistance and then gently rotate the swab. Do not be surprised if you are a

DICHOTOMOUS KEY TO THE FAMILY MICROCOCCACEAE

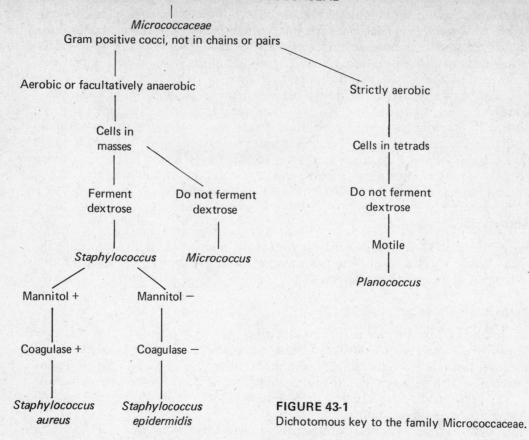

FIGURE 43-1
Dichotomous key to the family Micrococcaceae.

carrier since the carrier state is not uncommon. If you are a carrier, you may wish to do an antibiotic sensitivity test to see if the *Staphylococcus* indigenous to your nose is a drug-resistant mutant.

POST TEST

The post test is a self-evaluation. It is not used for a grade. It is designed only to let you decide if you have successfully completed this module.

Part 1: True or False

_____ 1. Bacteria in the genus *Staphylococcus* are always Gram positive.

_____ 2. Bacteria in the genus *Staphylococcus* are sometimes rod shaped.

_____ 3. Bacteria in the genus *Staphylococcus* usually group together in grapelike clusters.

_____ 4. Pathogenic *Staphylococcus aureus* is always hemolytic.

_____ 5. Hemolysin is an endoenzyme.

_____ 6. Hemolysin is produced in erythrocytes.

_____ 7. *S. aureus* often produces a gold pigment that is not water soluble.

_____ 8. Since the *S. aureus* pigment is not water soluble, it diffuses into the medium.

_____ 9. Coagulase is an enzyme produced by *S. epidermidis.*

_____ 10. Coagulase clots the fibrin in blood plasma.

_____ 11. Mannitol salt agar is considered to be a differential medium on the basis that the salt inhibits the growth of many other bacteria but not staphylococci.

_____ 12. A positive coagulase test and the fermentation of mannitol in the presence of high salt concentration are correlated more with the pathogenicity of staphylococci than are chromogenesis or hemolysis.

_____ 13. A mannitol positive colony on mannitol salt agar is surrounded by a red area.

_____ 14. Acid end products of mannitol fermentation are produced by *S. epidermidis.*

_____ 15. Specimens collected for culturing on a dry swab should be placed in sterile saline for preservation until transported to the diagnostic laboratory.

_____ 16. The round, clear zone surrounding a colony of *S. aureus* on a blood agar plate is called the zone of inhibition.

Part II

List the most significant ingredient or ingredients of the following, and describe the purpose of each significant ingredient.

		Significant ingredient	Purpose
A.	Blood agar plate	_____	_____
B.	Mannitol salt agar	_____	_____
		_____	_____
		_____	_____
C.	Coagulase test	_____	_____
		_____	_____

Part III: Completion

A. List three areas of the body to which staphylococci are indigenous.

1. _____

2. _____

3. _____

B. List five pathological conditions in humans caused by *Staphylococcus aureus.*

1. _____

2. _____

3. _____

4. _____

5. _____

C. Name the primary pathological condition caused by *Staphylococcus epidermidis.*

_____

MODULE 44

Pyogenic Streptococci

PREREQUISITE SKILL

Successful completion of Module 8, "Aseptic Transfer of Microbes," Module 14, "Streaking for Isolation," and Module 23, "Gram Stain."

MATERIALS

young trypticase soy broth cultures (18 hours or less) of
  Streptococcus pyogenes
  Streptococcus faecalis
  Streptococcus faecium (durans)
Gram stain reagents*
microscope slides (3)

blood agar plates (3)
Strep A disk (bacitracin) (3)
inoculating equipment
3% hydrogen peroxide

*To be prepared by the student if the instructor so indicates.

OVERALL OBJECTIVE

Differentiate between the pyogenic streptococci and the nonpyogenic streptococci.

Specific Objectives

1. Describe the three hemolytic differences of the genus *Streptococcus*.
2. Observe and describe the hemolytic differences of the streptococci used in this module from their appearance on blood agar plates.
3. Describe streptococcal susceptibility differences to the antibiotic bacitracin.
4. Describe the Gram reaction and cell arrangement of the streptococci used in this module.
5. List six diseases caused by beta hemolytic Group A streptococci.
6. List two pathological diseases that can be sequelae to beta hemolytic Group A streptococcal diseases.

7. Name the most serious pathological condition caused by non-Group A streptococci.

8. Give the genus and species name of a beta hemolytic Group A streptococcus and an alpha hemolytic non-Group A streptococcus.

9. Define the terms *sequelae, pyogenic, hemolysis, zone of hemolysis, nonsuppurative, enterococcus,* and *indifferent* (relating to hemolysis).

10. List and describe the tests or procedures routinely used to identify a disease-producing organism such as *Streptococcus pyogenes.*

DISCUSSION

The bacteria in the genus *Streptococcus* are Gram positive cocci normally occurring in short or long chains. However, pairs of cells also occur, and the cells may elongate and become ovoid. This is especially true on solid media. However, on transfer from a solid medium into a liquid medium, chains are found. The appearance of a Gram stained smear is usually sufficient to distinguish the streptococci from the staphylococci because of this tendency of the streptococci to form chains. If in doubt, these two genera can be separated on the basis of the presence of the enzyme catalase. Catalase will break down hydrogen peroxide to form water and oxygen. If you mix a loopful of staphylococci on a slide with 3% hydrogen peroxide, bubbles of oxygen will be visible to the naked eye. Streptococci do not form catalase, so no bubbles will form. Growth on agar surfaces is usually scant since streptococci range from microaerophilic to anaerobic. Isolated colonies are small, usually less than 1 mm in diameter; hence they are referred to as pinpoint colonies.

Many methods are used to group the streptococci. No one classification is absolute, and classification schemes overlap, which results in some confusion. Lancefield used serological tests to separate the streptococci into several different antigenic groups. She named the groups alphabetically, A through O. The beta hemolytic, pyogenic streptococci are usually long chained and are in Group A, according to Lancefield typing. Nutritional requirements and physiological characteristics have also been used in an attempt to classify streptococci. All these classification schemes are important to taxonomists, but they do not need to be of any real concern to you. In medical laboratories, only three simple tests are used to categorize pyogenic beta hemolytic Group A strep. They are the Gram stain, type of hemolysis, and the bacitracin sensitivity test.

No attempt to speciate the streptococci is necessary since about 90% of acute streptococcal infections in humans are caused by beta hemolytic streptococci of Lancefield Group A. Some of the infections caused by this pyogenic group are septic sore throat, erysipelas, scarlet fever, puerperal fever, bronchial pneumonia, meningitis, and wound infections. More serious, nonsuppurative complications, or sequelae, that may follow Group A streptococcal infections are rheumatic fever and glomerulonephritis. The other 10% of streptococcal infections in humans are caused by alpha hemolytic Group D streptococci (for example, *S. faecalis*) and alpha hemolytic (no Lancefield group) streptococci of the viridans group, specifically *S. mitis. Streptococcus faecalis* is a Group D enterococcus and grows in higher salt concentrations than *Streptococcus mitis.* Consult *Bergey's Manual* on this. Both *Streptococcus mitis* and *Streptococcus faecalis* are etiological agents of subacute bacterial endocarditis. In addition, *Streptococcus faecalis,* being an enterococcus, is often the cause of urinary infections. Fortunately most streptococci are susceptible to penicillin, and early treatment of an upper respiratory infection caused by streptococci prevents the development of sequelae.

In this module you will be using *Streptococcus pyogenes* as a representative of beta hemolytic Group A streptococci. You will compare it to the beta hemolytic non-Group A *Streptococcus faecium.* Demonstration of a few characteristics of streptococci (chain formation, hemolysis, pinpoint colonies, and sensitivity to bacitracin) and the fact that all are sensitive to penicillin make speciation unnecessary in the field of medicine.

ACTIVITIES

Activity 1: Gram Stain and Cell Arrangement

Make a smear and a Gram stain of the broth cultures of *Streptococcus pyogenes,* *Streptococcus faecium,* and *Streptococcus faecalis.* Be careful not to decolorize too much since streptococci lose their Gram positivity readily, especially in old cultures. Make drawings and give written descriptions of each species, and submit them to your file.

Activity 2: Type of Hemolysis

Streak each of the species of streptococci named in Activity 1 for isolation on a separate blood agar plate. Perform Activity 3 before incubating these plates at 37°C for 48 hours. Make observations and drawings of colony size and shape, as well as the type of hemolysis surrounding the colony. See Figure 44-1 for types of hemolysis. Observing the medium surrounding a colony with the low-power objective of your microscope is helpful in differentiating between alpha hemolysis and beta hemolysis. If no intact red cells are observed, then it is beta hemolysis. If a few red blood cells remain, it is partial or alpha hemolysis. Submit your findings to your file.

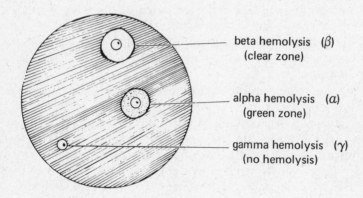

FIGURE 44-1
Types of hemolysis.

Activity 3: Bacitracin Sensitivity Test

Bacitracin is an antibiotic that inhibits the growth of beta hemolytic Group A streptococci but does not inhibit beta hemolytic non-Group A streptococci. So a bacitracin sensitivity test is done on all beta hemolytic streptococci in the diagnostic laboratory since most streptococcal diseases are caused by Group A streps.

Use a flamed forceps to aseptically place a Strep A disk (bacitracin) on the surface of the blood plates that you just streaked for isolation. Place the disk on the sector you used for the original inoculum since it is here that the growth will be confluent. Now incubate as instructed in Activity 2. Make drawings and give a written description of the zones of inhibition, if any, and submit this to your file.

Correlate your data from all three activities in table form for future reference and submit it to your file. Figure 44-2 will be helpful to you in your study of Streptococcaceae.

Activity 4: Catalase Test

Place two to three loopfulls of broth culture of *S. pyogenes* on a clean microscope slide. Add a drop of 3% hydrogen peroxide and mix slightly. Watch for bubbles of oxygen. Repeat with the other two stock cultures.

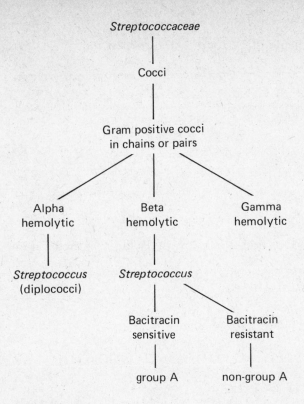

FIGURE 44-2
Dichotomous key to the Streptococci.

Related Experience

Demonstrate salt tolerance of various streptococci as noted in *Bergey's Manual* by adding the desired amount of NaCl to simple glucose-free heart infusion broth.

POST TEST

The post test is a self-evaluation. It is not used for a grade. It is designed only to let you decide if you have successfully completed this module.

Part 1: True or False

_____ 1. *Streptococcus faecium* is a beta hemolytic streptococcus.

_____ 2. *Streptococcus faecalis* is a beta hemolytic streptococcus.

_____ 3. *Streptococcus faecium* is a beta hemolytic Group A streptococcus.

_____ 4. Alpha hemolysis means that an exoenzyme from a colony has only partially destroyed the red blood cells in the medium surrounding the colony.

_____ 5. Alpha hemolysis is manifested as a clear zone around the colony.

_____ 6. In Gram stains made from old cultures, the streptococci appear Gram negative.

_____ 7. Streptococci are usually arranged in chains.

_____ 8. All beta hemolytic streptococci are in Lancefield's Group A.

_____ 9. Bacitracin inhibits the growth of beta hemolytic Group A streptococci.

_____ 10. A sequelae to a *Streptococcus pyogenes* infection is often endocarditis.

_____ 11. When bacteria are said to be pyogenic, it means that they are not pus producers.

_____ 12. The zone of hemolysis caused by indifferent streptococci is large and clear.

Part II

List six diseases caused by beta hemolytic Group A streptococci.

1. _____
2. _____
3. _____
4. _____
5. _____
6. _____

Part III

List two noninfectious complications that can follow pyogenic streptococcal infections.

1. _____
2. _____

MODULE 45

Pneumococci and Alpha Streptococci

PREREQUISITE SKILL

Successful completion of Module 8, "Aseptic Transfer of Microbes," Module 14, "Streaking for Isolation," and Module 24, "Capsule Stain."

MATERIALS

Todd-Hewitt broth cultures of
  Streptococcus pneumoniae (24 to 28
    hours)
  Streptococcus faecalis
blood agar plates (2)
P disks (Optochin)
desiccated oxgall (beef bile) or bile salts
Gin's stain reagents (capsule stain)
Gram stain reagents*

For related experience:
  Todd-Hewitt broth tubes (2)*
  Bromthymol blue indicator, dropper
    bottle†
  sodium hydroxide 0.5 N, dropper
    bottle*†
  Desoxycholate reagent, dropper bottle†
  1 ml pipets (2)
  clean, empty test tubes (2)

*To be prepared by the student if the instructor so indicates.
†Formulae are at the end of this module.

OVERALL OBJECTIVE

Differentiate pneumococci from alpha streptococci.

Specific Objectives

Use *Streptococcus pneumoniae* and *Streptococcus faecalis* in Specific Objectives 1 to 7.

1. Describe differences or similarities of Gram reaction and cell groupings.
2. Describe differences shown by the capsule stain and discuss the pathological significance of the capsule.
3. Describe differences or similarities of the type of hemolysis.

4. Describe differences in susceptibility to optochin.
5. Describe differences in bacterial cell lysis by a surface acting agent.
6. Summarize all the differences and similarities of pneumococci and alpha streptococci.
7. Summarize the procedures used to distinguish between pneumococci and alpha streptococci.
8. Name the disease caused by a pneumococcus that has a high mortality rate.
9. List five other pathological conditions caused by a pneumococcus.
10. Give the genus and species name of a pneumococcus and an alpha streptococcus.
11. Discuss a pathogenic condition caused by alpha streptococci.

DISCUSSION

Streptococcus pneumoniae are Gram positive, encapsulated diplococci, and they grow as small alpha hemolytic colonies on blood agar. The diplococci are said to be lancet-shaped, which means that in each pair the flattened surfaces oppose each other and the distal ends are somewhat pointed. See Figure 45-1a, for example. This lancet-shaped pair arrangement is seen more readily in a direct smear of sputum from a patient who has lobar pneumonia. Each pair is surrounded by a capsule. The organism may form short chains, in which case the entire chain is surrounded by a capsule with indentations between the pairs, Figure 45-1b. Since the organism may form chains and the capsule does not show in the Gram stain, and since colonies on blood agar are pinpoint and are surrounded by a zone of alpha hemolysis, it is difficult to distinguish them from other alpha hemolytic streptococci. The tests you will do in this module (capsule stain, optochin sensitivity, and bile solubility) are used to distinguish between pneumococci and other alpha streptococci.

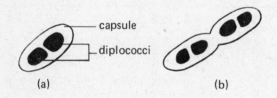

FIGURE 45-1
Pneumococci surrounded by a capsule.

The pneumococci can be part of the normal flora of the human throat. Yet *Streptococcus pneumoniae* is the most frequent cause of lobar pneumonia in humans and continues to be a major cause of mortality in older age groups. The pneumococcus is often a secondary invader (opportunist) after the lining of the respiratory tree has been predisposed by an influenza virus. The pneumococcus may also cause pleurisy, empyema, mastoiditis, and meningitis. Its pathogenicity is directly related to the antiphagocytic property of its polysaccharide capsule. The alpha hemolytic streptococci are indigenous to the throat *(Streptococcus mitis)* or the intestinal tract *(Streptococcus faecalis). Streptococcus faecalis,* a Group D enterococcus, grows in higher salt concentrations than *Streptococcus mitis.* Consult *Bergey's Manual* on this. (Read Module 44 on pyogenic streptococci.) The alpha streps are often the cause of subacute bacterial endocarditis (SBE). Therefore, SBE is an infection of endogenous (originating from within) origin. These indigenous alpha streps are usually introduced into the bloodstream by trauma and localize at a site of previous injury on the endocardium, usually the valve surfaces. The heart valves have been predisposed by a previous weakening of the valvular endocardium. For example, more than 75% of patients with SBE have a history of rheumatic fever with consequent cardiac damage.

SBE differs from a pneumococcal infection in that it is not communicable. Therefore, it is necessary to distinguish between these two organisms that have the same colonial appearance on blood agar. The following activities will make this distinction possible.

Precaution: In the following activities, you will be working with a potential pathogen. Absolute asepsis is necessary.

ACTIVITIES

Activity 1: Gram Reaction and Cell Arrangement

Gram stain both the pneumococcus and streptococcus broth cultures. Make drawings and descriptions of each, and submit them to your file. Also complete Table 45-1, and submit a duplicate of it to your file.

Activity 2: Demonstration of Capsule by Gin's method

As mentioned in the Discussion section, the capsules of *Streptococcus pneumoniae* are associated with its pathogenicity. Pneumococci may lose their capsules when artificially cultivated. When this happens, the organism becomes avirulent, and the colonies appear rough. Virulent colonies appear smooth because each pair of organisms is surrounded by a large polysaccharide capsule.

Now do a capsule stain of both the streptococcus and the pneumococcus from the broth cultures. Refer to Module 24 on capsule stains for this procedure, if necessary. Make drawings, give a written description of both organisms, and submit this to your file. Fill in Table 45-1 in the appropriate columns.

Activity 3: Type of Hemolysis and Optochin Sensitivity

Using the broth cultures of the two different organisms, streak each for isolation on separate blood agar plates. Aseptically place a P disk (optochin) on the surface of the sector that you used for the original inoculum. Incubate for 48 hours at 37°C. Remember that if your colonies are smooth, they are encapsulated and virulent. Record your data in Table 45-1.

Optochin is ethylhydrocupreine hydrochloride. This chemical inhibits the growth of *Streptococcus pneumoniae* but has no effect on other alpha streptococci.

Activity 4: Bile Solubility

Pneumococci not only tend to lyse spontaneously, but are also very sensitive to surface acting agents. Bile is a surface acting agent. Hence desiccated beef bile (oxgall) or bile salts added to an isolated *Streptococcus pneumoniae* colony will cause the colony to disappear before your eyes.

Open your petri dish of *Streptococcus pneumoniae,* place it on the stage of a dissecting microscope or under a good hand lens, and focus on one or two well-isolated colonies. Using a toothpick, add a very small amount of desiccated beef bile (oxgall) to the isolated colony that you have selected on your 48 hour blood agar plate of *Streptococcus pneumoniae* from Activity 3. Look at the colony through the microscope or hand lens until the powdered bile has dissolved. Do the same to a colony of *Streptococcus faecalis* on your other blood agar plate. Compare the results, and record them in Table 45-1. Consult Figure 45-2 as you correlate your data from these activities.

Another, more dramatic test for bile solubility is described in the related experience.

Related Experiences

1.   The disappearance of growth (turbidity) in the presence of bile or bile salts is visible macroscopically in the following bile solubility tube test. For this reason, you may find it useful to perform this related experience:

TABLE 45-1 Summary of the Differences and Similarities of *S, pneumoniae* and *S. faecalis*

| Organism | Gram stain | | Capsule preparation | | Blood agar plate | Optochin | | Bile soluble | |
	Gram reaction	Cell shape and arrangement	Present	Absent	Type of hemolysis	Sensitive	Resistant	Yes	No
Streptococcus pneumoniae (Diplococcus pneumoniae)									
Streptococcus faecalis									

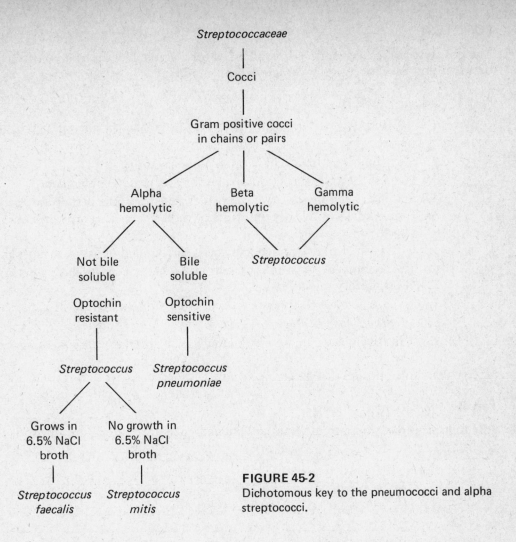

FIGURE 45-2
Dichotomous key to the pneumococci and alpha streptococci.

1. Inoculate one tube of Todd-Hewitt broth with *Streptococcus pneumoniae* and another tube with *Streptococcus faecalis.* Incubate at 37°C for 8 to 24 hours.

2. Aseptically pipet 1 ml of each broth culture into separate, clean, empty test tubes.

3. Add one or two drops of bromthymol blue indicator to each tube containing 1 ml of broth culture. The pH indicator will usually turn yellow because of the acid by-products of bacterial metabolism present in the culture.

4. Add 0.5 N sodium hydroxide drop by drop until the broth just retains a slight blue-green color. This neutralizes the acid by-products of metabolism so that no precipitate will form when you add the desoxycholate reagent.

5. Add two or three drops of desoxycholate reagent.

6. If the bacteria are pneumococci, the broth should clear within 15 minutes. The broth will remain turbid if the bacteria are other streptococci.

Trypticase soy broth is also a suitable medium for this test.

2. Demonstrate salt tolerance of various streptococci as noted in *Bergey's Manual* by adding the desired amount of NaCl to simple glucose-free heart infusion broth. A Gram stain may be necessary here for proof of growth.

POST TEST

The post test is a self-evaluation. It is not used for a grade. It is designed only to let you decide if you have successfully completed this module.

Part 1: True or False

_____ 1. *Streptococcus pneumoniae* are Gram positive, encapsulated, beta hemolytic cocci.

_____ 2. Most alpha streptococci are optochin resistant.

_____ 3. Pneumococci is the common name for *Klebsiella pneumoniae*.

_____ 4. *Streptococcus pneumoniae* can be described as alpha hemolytic, encapsulated, Gram positive cocci in pairs or short chains that are optochin sensitive.

_____ 5. Each individual pneumococcus cell is surrounded by its own capsule.

_____ 6. The pneumococcus is only present in the human throat during periods of respiratory infection.

_____ 7. Pathogenicity of the pneumococcus is directly related to the capsule.

_____ 8. *Streptococcus pneumoniae* are most often opportunists.

_____ 9. A relatively large amount of bile salts is required to cause pneumococci to lyse.

_____ 10. Bile salts are surface acting agents.

Part II

List four procedures used to differentiate pneumococci from other streptococci.

1. _____

2. _____

3. _____

4. _____

Part III

A. List six diseases caused by pneumococci.

1. _____ 4. _____

2. _____ 5. _____

3. _____ 6. _____

B. Which of these six diseases causes the most fatalities?

_____

C. Give the genus and species name of the bacterium used in this module that can cause SBE.

_____

FORMULAE FOR REAGENTS

1. **0.5N SODIUM HYDROXIDE**

sodium hydroxide (NaOH), C.P.	2.0 gm
distilled water	100.0 ml

Dissolve the NaOH in distilled water.

2. **0.04% BROMTHYMOL BLUE pH INDICATOR (6.0 to 7.6, yellow to blue)**

Bacto-bromthymol blue	0.1 gm
0.01 N sodium hydroxide (NaOH)	16.0 ml
add distilled water to make a total of	250.0 ml

Dissolve the bromthymol blue in the sodium hydroxide solution. Dilute to 250 ml with distilled water.

3. **DESOXYCHOLATE REAGENT**

sodium desoxycholate	10.0 gm
1:50,000 dilution merthiolate	100.0 ml

Dissolve the sodium desoxycholate in the dilute merthiolate solution. Store in the refrigerator when not in use.

MODULE 46

Dental Caries Susceptibility

PREREQUISITE SKILL

Successful completion of Module 8, "Aseptic Transfer of Microbes," and Module 9, "Aseptic Use of a Serological Pipet."

MATERIALS

paraffin cubes, approximately 1 cm x 1 cm

Snyder test agar short; 5 ml/tube (commercially available)*

1 ml pipet, sterile*

50 ml beaker, sterile*

glass stirring rod, sterile*

*To be prepared by the student if the instructor so indicates.

OVERALL OBJECTIVE

Relate the acid production of mouth bacteria to dental caries susceptibility.

Specific Objectives

1. List four microbes implicated in dental caries.
2. Relate the Snyder test to dental decay.
3. Describe the tooth decaying process.
4. List the significant ingredients of Snyder test agar.
5. List two common tests used to detect caries susceptibility.
6. Define the terms *homofermentative, heterofermentative, organic acid,* and *indigenous.*

DISCUSSION

It is generally agreed that dental decay is initiated by lactic acid or other organic acids. This acid production is a result of bacterial fermentation of carbohydrates on the tooth surface. The enamel of the tooth in the immediate vicinity of the bacterial fermentation process is slowly decalcified and softened by the continued presence of the organic acids, causing decay to begin.

Lactobacillus acidophilus has been considered to be the predominant organism responsible for dental caries. However, streptococci and diplococci of the mouth are also acid producers. More recently *Actinomyces odontolyticus* has also been isolated from diseased teeth. All these microbes can be indigenous flora of your mouth. We will study the presence of *Lactobacillus* since experimental evidence indicates that it is probably the most significant organism in tooth decay.

The bacteria comprising the genus *Lactobacillus* are nonmotile, non-spore-forming, Gram positive rods. As the genus name suggests, these bacteria all produce some lactic acid by fermentation of simple sugars. They can be either homofermentative or heterofermentative. The homofermentative species produce lactic acid as the only by-product of fermentation, while the heterofermentative species produce several end products such as lactic acid, acetic acid, alcohol, and CO_2. Some species can produce an acid environment of pH 2. All the lactobacilli range from microaerophilic to anaerobic. Therefore you would expect poor surface growth *in vitro.* This property is considered in the activity you will be doing in this module.

There are two common methods of determining the presence of lactobacilli in saliva: a *Lactobacillus* count and a Snyder test. The validity of a *Lactobacillus* count is questionable because susceptibility to tooth decay varies among individuals. That is, the teeth of certain people do not decay in the presence of the same acid that produces decay in others.

The Snyder test is a simple method of determining caries susceptibility. It is a measure of the rate of acid production from the metabolism of glucose by lactobacilli. This method employs Snyder test agar. The significant ingredients of this medium are glucose and bromcresol green. Important also is the fact that the pH of the uninoculated medium is adjusted to 4.8.

Lactobacilli are unusual mouth organisms in that they grow at a low pH of 4.7 to 5.0. As they grow in Snyder test agar, they utilize the glucose in the medium, converting it to organic acids, and thereby lower the pH even more. The same process occurs in your mouth to cause tooth decay. When the pH in certain pockets on or around your teeth becomes 4.4 or lower, the carious process begins.

Bromcresol green is a pH indicator and turns yellow if the pH of the Synder test agar becomes lower due to acid production by microbes. The susceptibility of an individual is determined by the time it takes for the medium to turn yellow. If the medium turns yellow in 24 to 48 hours, the individual is said to be susceptible to dental caries.

ACTIVITY

Activity 1: The Snyder Test

Allow a small piece of paraffin to soften under your tongue, and then chew it for three minutes. *Do not swallow your saliva.* Instead, collect all the saliva in the 50 ml beaker over the three-minute period.

Now vigorously stir the saliva sample with the stirring rod for 30 seconds to disperse the organisms evenly. Obtain your tube of Snyder test agar from the holding water bath, and aseptically pipet 0.2 ml of saliva into it. Disperse the saliva in the cooling agar by rotating the tube vigorously between your hands. (Refer to the figures in Module 11 on loop inoculated pour plates regarding this mixing procedure.) Incubate this inoculated tube at 37°C. Examine the tube every 24 hours to see if the pH indicator has changed to yellow. Study Table 46-1 to determine the degree of your dental caries susceptibility. This standardized table is used universally in conjunction with the Snyder test. Compare the amount of dental work you have had done or should have done with the results of your Snyder test, and submit this to your file.

Then take the post test.

TABLE 46-1 Standardized Table to Determine Dental Caries Susceptibility

Caries susceptibility	Medium turns yellow in		
	24 hours	48 hours	72 hours
Marked	Positive		
Moderate	Negative	Positive	
Slight	Negative	Negative	Positive
Negative	Negative	Negative	Negative

POST TEST

Part I

Define the following terms.

1. Indigenous _____

_____

2. Homofermentative _____

_____

3. Heterofermentative _____

_____

Part II: True or False

_____ 1. Bromcresol purple is the pH indicator in Snyder test agar.

_____ 2. When the pH indicator in Synder test agar turns yellow, it is negative.

_____ 3. The *Lactobacillus* count is more reliable than the Synder test.

_____ 4. Most bacteria grow at pH 4.8.

_____ 5. Carbohydrates are fermented to organic acids to produce tooth decay.

_____ 6. Acetic acid is an organic acid.

_____ 7. Hydrochloric acid is an organic acid.

_____ 8. In all individuals tooth decay begins at pH 4.4.

_____ 9. Interpretation of the Snyder test results depends solely on the amount of acid produced.

_____ 10. Acid-producing diplococci are not indigenous to your mouth.

Part III

List four microbes thought to be the cause of tooth decay.

1. _____

2. _____

3. _____

4. _____

Part III: 1. *Lactobacillus acidophilus*, 2. streptococci, 3. diplococci, 4. *Actinomyces odontolyticus*.

Part II: 1-F, 2-F, 3-F, 4-F, 5-T, 6-T, 7-F, 8-F, 9-F, 10-F.

3. more than one end product results from the anaerobic metabolism of a monosaccharide.

2. a single end product results from the anaerobic metabolism of a monosaccharide (such as glucose).

Part I: 1. normal, native, resident.

KEY

MODULE 47

Urine Culture

PREREQUISITE SKILL

Successful completion of Module 8, "Aseptic Transfer of Microbes," Module 14, "Streaking for Isolation," Module 20, "Effects of Antibiotics," Module 23, "Gram Stain," Module 40, "The IMViC Tests," Module 42, "Intestinal Pathogens in the Family *Enterobacteriaceae*," Module 43, "Pathogenic *Staphylococcus aureus*,"* and Module 44, "Pyogenic Streptococci."*

MATERIALS

Simulation of urinary tract infection: three midstream urine samples (labeled a, b, and c) each seeded with a differen bacterium commonly found in urinary tract infections

blood agar plate

MacConkey agar plate†

All student selected media that are necessary to complete Activity 2 and the related experiences†

*All or part of these modules may or may not be necessary as a prerequisite skill.
†To be prepared by the student if the instructor so indicates.

OVERALL OBJECTIVE

Identify the causative organism of a urinary tract infection and determine the antibiotics to which it is sensitive.

Specific Objectives

1. Describe two methods of collecting urine for bacteriological study.
2. Describe handling of the urine between the time it is collected from the patient and cultured in the laboratory.
3. List the names of nine bacteria that can be responsible for urinary tract infections, and give the Gram reaction and shape of each, as well as the medium used for their isolation.
4. Give the genus and species name of the bacterium most commonly found as the causative agent of urinary tract infections.

5. Name the two etiological agents of urinary tract infections that are the most resistant to treatment.
6. Define the terms *cystitis, pyelitis, glomerulonephritis,* and *midstream.*
7. List and discuss the media and tests you would select if the urine that you cultured grew only on blood agar.
8. List and discuss the media and tests you would select if the urine that you cultured grew on both blood agar and MacConkey agar.
9. Describe the procedure involved in determining the sensitivity of bacteria to antibiotics.

DISCUSSION

Normally urine in the bladder, ureter, and kidneys is sterile. However, the external genitalia of both sexes harbor several types of microorganisms such as staphylococci, streptococci, diphtheroids, yeasts, and intestinal organisms. As the urine passes out of the urethra, it may become contaminated with genitalia flora, especially in females. Therefore, the method of collecting urine for culture is most important. To avoid all possibility of the urine becoming contaminated from the microbes on the genitalia the urine can be collected in a sterile container by means of a sterile catheter. However, no matter how much care is taken there is always the danger of introducing bacteria into the urinary tract with the catheter. For this reason, techniques have been developed that avoid the use of a catherter. A common method of collecting urine without the use of a catheter is to "catch a midstream specimen." This is done by thoroughly cleansing the genitalia just prior to urination . The first portion of the urine passed functions to wash most contaminating organisms out of the urethra. The middle portion or midstream is collected in a sterile container and is satisfactory for culture.

Urine should be cultured shortly after collection. If this is not possible, it can be stored in a refrigerator for a short time. Failure to culture the urine promptly results in an excessive increase in microbial numbers because even a few contaminants from the lower urinary tract will reproduce at room temperature, and the results of culturing will be misleading.

Routinely, when the physician knows a patient's symptoms and the genus causing an infection, he begins therapy. Later he receives results of a standardized sensitivity test (Kirby-Bauer) giving him definitive information as to the drugs of choice. With this added information at hand, the physician may then give instructions for a change in antibiotic therapy.

Most of the bacteria responsible for urinary tract infections can be placed into two groups and are cultured accordingly.

1. Gram positive cocci—medium: blood agar
   a. Staphylococci
   b. Enterococci *(Streptococcus faecalis* and *Streptococcus faecium)*
   c. *Streptococcus pyogenes*
2. Gram negative bacilli—medium: MacConkey or EMB
   a. *Escherichia coli*
   b. *Proteus* sp.
   c. *Pseudomonas* sp.
   d. *Enterobacter-Klebsiella* group

One exception to this is *Alcaligenes faecalis,* which is a Gram negative coccus to coccobacillus and is cultured on blood agar.

Escherichia coli is most commonly found as the etiological agent of urinary tract infections. *Proteus* and *Pseudomonas* are the most difficult to treat because of their resistance to antibiotic therapy. *Mycobacterium tuberculosis* causes a rare kidney infection. If TB of the kidneys is suspected, an acid-fast strain of urine sediment is done and confirmed by culture.

When the urinary infection is limited to the urethra and bladder, the condition is called cystitis. If the ureters and the pelvis of the kidney are involved, the condition is called pyelitis. Glomerulonephritis, an inflammation and destruction of the renal corpuscles, may result from a more total involvement of the kidney by bacteria. In this condition, the glomeruli and the filtering surfaces of the Bowman's capsules are destroyed. Consequently the patient's urine will be loaded with red blood cells, and a wet mount of the urine sediment is helpful.

Since Gram positive cocci and Gram negative colon bacilli are found in urinary tract infections, it is necessary to streak for isolation on a blood agar plate and on a selective-differential medium such as MacConkey or EMB for the Gram negative bacilli. Fortunately, most urinary tract infections are caused by a single genus and species of bacteria, which makes bacteriological studies and identification much easier. However, other organisms may be present from improper cleansing and specimen collection. Therefore, identification of the causative organism depends upon the type of colony that *predominates*. The necessary biochemical tests are determined by the type of medium on which the organism grew and the appearance of the colonies.

ACTIVITIES

Activity 1: Streaking Seeded Urine Specimen for Isolation

Choose one of the seeded urine samples. Make a note of the letter on the specimen bottle since your instructor has a key and knows which organism you should isolate from the particular bottle you selected. Using your best aseptic technique, streak the seeded urine specimen for isolated colonies on a blood agar plate and on a MacConkey agar plate. Obtaining isolated pure colonies will be most important to you in doing the following activities. Incubate the isolation plates at 37°C for 48 hours.

You will not be able to make your media selection for Activity 2 until your next lab session.

Activity 2: Identification of the Causative Organism

Examine your streak plates from Activity 1. If the causative organism is a Gram positive coccus, you will have significant growth on the blood agar plate only. If the causative organism is a Gram negative rod, there will be growth on both the blood agar plate and the MacConkey agar plate. Be prepared to explain these differing growth characteristics.

If you obtained growth on the blood agar only, do a Gram stain and the appropriate tests to determine the genus and species name of the organism. Review Module 43, "Pathogenic *Staphylococcus aureus*," and Module 44, "Pyogenic Streptococci," for selection of the appropriate media and tests necessary for the identification of the predominating pure colony type that you isolated.

Reminder: Be sure to make all the inoculations of the significant media that you selected from a single, pure colony.

If there is growth on both the blood agar plate and the MacConkey agar plate, do a Gram stain and the appropriate tests to determine the genus and species name of the causative organism. Review Module 40, "The IMViC Tests," Module 42, "Intestinal Pathogens in the Family *Enterobacteriaceae*," and Module 35, "Carbohydrate Fermentation." You may also wish to review other modules listed in the Table of Contents under Part VI: "Selected Physiological Reactions of Bacteria" to help you select the necessary identifying tests.

It is to your advantage to pick out the tests that you would need for the identification of both Gram positive cocci and Gram negative bacilli. This module is designed to help prepare you for the identification of the unknowns, Module 53, if you do this investigative review.

When you have looked at your Gram stain and completed the significant tests of *your* selection, correlate your data in tabular form. Submit to your file this data table and the genus and species name of the organism causing the simulated urinary tract infection. Check your results with your instructor. Your instructor has a key and knows the organism that you should have identified. He will let you know if you have been successful.

You are probably very aware of the fact that this module gives you little procedural guidance. This was done deliberately to stimulate you to use scientific reasoning. At this point in the course, you should be able to choose the appropriate media and tests, correctly interpret your results, and be able to conclude the genus and species names of causative organisms. Therefore, your choice of identifying media and tests is dependent upon knowledge you have accumulated in previous modules. To become a microbiologist you must develop techniques much like those of a detective. That is, you must be able to uncover clues and put the clues together to identify the organisms.

Take the post test now if you feel ready.

Do the related experiences if you have time. You will learn much more microbiology.

Related Experiences

1. Repeat the activities in this module, and identify the causative organism in each of the other two seeded urine specimens. This is excellent preparation for identifying unknowns in Module 53.
2. After careful cleansing, collect your own midstream urine specimen, and repeat the parts of this module that are significant to you.

POST TEST

The post test is a self-evaluation. It is not used for a grade. It is designed only to let you decide if you have successfully completed this module.

Part I: True or False

_____ 1. Urine for bacteriological studies should always be collected in a sterile container via a sterile catheter.

_____ 2. Urine held at room temperature for a few hours before culturing can give misleading results.

_____ 3. When it is not possible to culture urine immediately, the specimen can be refrigerated for a short time.

_____ 4. The most effective therapeutic drug for a urinary tract infection is determined by a sensitivity test done on the predominating organism.

_____ 5. *Proteus* and *Enterobacter* are the urine bacteria most resistant to treatment.

_____ 6. *Escherichia coli* is found most often in urinary tract infections.

_____ 7. Pyelitis is often a bacterial infection of the urethra and bladder.

_____ 8. If organisms from a urine culture grow on blood agar and MacConkey agar, they are Gram positive cocci.

_____ 9. Drugs of choice are determined by animal inoculation.

_____ 10. If urine organisms grow on your selective-differential medium, you can by inspection know if they ferment lactose or not.

Part II: Completion

A. List the names and Gram reactions of seven bacteria that can be responsible for urinary tract infections.

Name   Gram reaction

1. _____   _____

2. _____   _____

3. _____   _____

4. _____   _____

5. _____   _____

6. _____   _____

7. _____   _____

B. If the coloneis from urine growing on blood agar but not on MacConkey agar were large, yellow, beta hemolytic colonies, what two significant tests or media would you choose for definitive identification?

1. _____

2. _____

C. If the urine bacteria produced uncolored colonies on MacConkey agar, which two genera would you suspect?

1. _____

2. _____

D. If the colonies on MacConkey agar were red, which two genera would you suspect?

1. _____

2. _____

Why?

3. _____

E. If the colonies on MacConkey agar were uncolored, what rapid test should you choose to perform first?

1. _____

Why?

2. _____

F. What conditions of growth on MacConkey agar would necessitate that you perform all or part of the IMViC tests?

1. _____

Why?

2. _____

G. Name the two most significant characteristics upon which you base your choice of media for Activity 2.

1. _____

2. _____

Why?

3. _____

MODULE 48

Throat Culture

PREREQUISITE SKILL

Mastery of Module 8, "Aseptic Transfer of Microbes," and Module 14, "Streaking for Isolation."

Completion or careful reading of Module 35, "Carbohydrate Fermentation," Module 40, "The IMViC Tests," Module 42, "Intestinal Pathogens in the Family *Enterobacteriaceae,*" Module 43, "Pathogenic *Staphylococcus aureus,*" Module 44, "Pyogenic Streptococci," and Module 45, "Pneumococci and Alpha Streptococci."

MATERIALS

sterile swab*
tongue depressor
blood agar plates (2)
MacConkey agar plate*
candle jar
trypticase soy agar plates (2 to 3)*†
Gram stain reagents†

All media that you select as significant for the identification of the throat organisms*††

*To be prepared by the student if the instructor so indicates.
†To be used 48 hours after primary inoculations.
††To be used 96 hours after primary inoculations.

OVERALL OBJECTIVE

Obtain a throat culture specimen, streak for isolated colonies on appropriate selective and differential media, and identify two potentially pathogenic organisms from the specimen, if present.

Specific Objectives

1. Describe the accepted method for obtaining a throat culture specimen.
2. List six organisms that are part of the normal flora of the human throat.
3. List four organisms that are pathogens of the respiratory tract which can also be found as part of the normal throat flora in the absence of pathogenic conditions.

4. Discuss the significance of the predominance of a specific organism in a throat culture.

5. Describe the proper treatment of a throat culture specimen until it can be inoculated.

6. Explain the reason for incubating 1 BAP in a 10% CO_2 environment.

7. List four characteristics that can be observed from your original BAP as guides to the final identification of the organisms.

8. Describe the inoculation of your original isolation plates from a throat swab.

DISCUSSION

Although the mucous membranes of the mouth and throat are often sterile at birth, alpha hemolytic streptococci of the *viridans* group *(Streptococcus mitis* and *S. salivarius)* establish themselves within 4 to 12 hours after birth and remain the most prominent members of the normal flora for life. The viridans streptococci are soon joined by various bacteria, such as micrococci, *Staphylococcus epidermidis, Staphylococcus aureus,* enterococci, *Neisseria* sp., diphtheroids *(Corynebacterium* sp.), lactobacilli, *Escherichia coli, Enterobacter aerogenes, Klebsiella pneumoniae,* and *Proteus* sp., some of which are potential pathogens.

Other organisms that may occur as normal or transient flora of the healthy human throat include *Streptococcus pneumoniae, Streptococcus pyogenes, Neisseria meningitidis,* and *Hemophilus influenzae.* These organisms are also potential pathogens, as are *Staphylococcus aureus* and *Klebsiella pneumoniae. Bordetella pertussis* and *Corynebacterium diphtheriae* are also pathogens that can be isolated from the throat. Classical meningitis, which begins in the throat, is caused by *Neisseria meningitidis. Hemophilus influenzae* is the primary cause of fatal meningitis in infants. Two former scourges of childhood, whooping cough and diphtheria, are the result of infection by *Bordetella pertussis* and *Corynebacterium diphtheriae,* respectively. All the above mentioned pathogens can be isolated from the throat.

Pseudomonas aeruginosa is the most common contaminant of inhalation therapy equipment, and many serious lung infections have been seeded by the use of this equipment. A lung pathology caused by *Pseudomonas* is very resistant to antibiotic treatment. This same organism *(P. aeruginosa)* is a commonly called "the bacillus of blue pus." It is also a frequent and fatal opportunist in the exposed tissues of burn patients. Therefore, *Pseudomonas,* like many bacteria, can cause infections at various sites of the body.

Much can be determined by the approximate number of colonies of a specific organism cultured from a throat specimen. For example, a few colonies of beta hemolytic streptococci indicate that they are probably transients in the throat without pathogenic significance. On the other hand, in "strep throat" the predominant colony type would be beta hemolytic Group A streptococci. A large number of colonies of *Streptococcus pneumoniae, Klebsiella pneumoniae, Pseudomonas aeruginosa,* or *Proteus* sp. may not contribute to pathology in the throat but rather indicate sinusitis or pathology in the lungs or bronchioles. An inordinately large number of *Staphylococcus aureus* in the throat can be a reflection of a staph infection anywhere in the body. In the evaluation of throat cultures, any unusual predominance of a particular organism, even one of the normal flora, may have significance and should be reported to the doctor.

Specimens for throat cultures are collected with one or more sterile swabs that are inserted through the mouth to the tonsillar area without touching the tongue or any other oral surface. A tongue depressor is generally used to hold the tongue down and to assure easy access to the throat. Any visible lesions or crypts in the throat or tonsillar area should be explored with the swab vigorously enough to penetrate the surface and to obtain material from the lesion or crypt. You should constantly rotate the swab as you obtain the specimen. If no lesions are apparent, rotate the swab vigorously over the tonsillar surface and other surfaces, giving special attention to any inflamed areas.

Once you have taken a "good" specimen as described, it is vital that you do not allow the swabs to dry out before the organisms can be cultured. To prevent drying, the swab should be placed in a tube with 1 ml of sterile trypticase soy broth or Stuart's transport medium immediately after collection and taken to the lab as soon as possible.

When the swabs arrive at the clinical laboratory, they are used to streak the *original inoculum sector* on plates of appropriate selective and differential media, then the original inoculum is streaked for isolation with the inoculating loop. If two swabs are available, a direct smear is made on a clean, dry microscope slide. The smear is then Gram stained and may be helpful as a guide to which media would be most useful. Remember, however, that a direct smear at best is only indicative and is not reliable as a basis for any conclusions about causative organisms.

Normally, swabs from a simple sore throat are rotated over a small area (sector 0) of the surface of two blood agar plates and a MacConkey agar plate. The inoculating loop is then used to streak this original inoculum for isolated colonies. One BAP is incubated in 10% CO_2 environment at 37°C, and the other plates (one BAP and one MacConkey) are incubated aerobically at 37°C. The CO_2 environment is useful for the cultivation of *Neisseria meningitidis* and *Hemophilus influenzae*. Both grow better on chocolate agar (hemolyzed blood agar). *H. influenzae* is so fastidious that it requires supplement A* to be added to the chocolate agar. Some strains of streptococci, under increased CO_2 tension (microaerophilic conditions), develop larger, more mucoid colonies than the pinpoint colonies they produce aerobically.

After 24 to 48 hours of incubation, all predominating hemolytic colony types are picked for Gram staining and inoculation onto appropriate differential media for identification. For example, if you isolated large white to gold colonies that are beta hemolytic on blood agar, you would pick and subculture one of these colonies onto mannitol salt agar and do a coagulase test to confirm the identification as *Staphylococcus aureus*. Often these pure colonies are used to inoculate plates for sensitivity tests at the same time the media for identification are inoculated.

The presence of small, pinpoint hemolytic colonies wold be suggestive of streptococci or diplococci. If the colonies are beta hemolytic, one should be picked and streaked on a fresh BAP and a Strep A disk added to distinguish *Streptococcus pyogenes* from beta hemolytic non-group A streps. Alpha hemolytic colonies would be treated with optochin and/or bile salts to determine if they are *Streptococcus pneumoniae*.

If diphtheria is suspected. the throat culture specimens are treated differently. One swab is used to make two direct smears, one of which is Gram stained and the other stained for metachromatic granules with Albert's stain or Loeffler's alkaline methylene blue. The second swab is used to inoculate a Loeffler's serum slant, a tellurite agar plate, and a BAP. Clinically a chocolate agar plate grown in 10% CO_2 is always included for throat cultures primarily for the isolation of *H. influenzae* and more luxuriant growth of *N. meningitidis*.

In any case, colony size, color, shape, effect on blood, and Gram stain reaction are all guides to tentative identification and suggest the means of final identification. The isolation plates must be examined carefully because colonies of some of the organisms may be very small.

If the growth on the MacConkey agar plate is quite dense or appears significant for any reason, it will be necessary to utilize the IMViC tests, the urease test, and other tests appropriate to the identification of Gram negative rods.

You will probably be identifying staphylococci and streptococci colonies unless the growth on your MacConkey agar plate is significant. This means that you will be working mostly from your blood agar plates.

Difco Manual, 9th ed. (Detroit: Difco Laboratories, 1953).

ACTIVITIES

Activity 1: Primary Inoculations from Throat Culture Specimens

Have another microbiology student or your instructor obtain a throat culture specimen from you as described. Remember to rotate the swabs vigorously over the tonsillar area, and be careful not to touch the tongue or other oral surfaces. Now take the swab with your throat specimen, and streak and rotate it over the original inoculum sector of two BAP and one MacConkey agar plate. Discard the swab in a container of disinfectant. Flame your loop, and streak out the original inoculum for isolation. Once again your goal here is representative growth with isolated colonies for study. You are limited to these media for practical purposes.

Label the plates carefully. Invert and incubate one BAP and your MacConkey plate aerobically at 37°C for 48 hours. Place your other BAP in the candle jar, and incubate it at 37°C for 48 hours after the candle burns out. That is, place the candle jar and the plates it contains in the 37°C incubator.

Activity 2: Selection and Reisolation of Throat Flora for Identification

After incubation, examine your blood agar plates carefully for hemolytic colonies, both alpha and beta. You will be looking primarily for staphylococci and streptococci, so it may be useful to review Module 43, "Pathogenic *Staphylococcus aureus*," and Module 44, "Pyogenic Streptococci," for the tests needed to identify staph and strep.

Not all pathogenic staph are hemolytic but most are chromogenic. So if you find large chromogenic colonies with or without hemolysis, review Module 43 on staphylococci. If there are many pinpoint colonies producing alpha or beta hemolysis, review Module 44, "Pyogenic Streptococci," and Module 45, "Pneumococci and Alpha Streptococci," and make your selection of tests necessary for definitive identification of the organism.

For organisms from your MacConkey agar plate, you will need to review Module 40, "The IMViC Tests," Module 35, "Carbohydrate Fermentation," and Module 42, "Intestinal Pathogens in the Family *Enterobacteriaceae*," for test selections leading to the identification of Gram negative bacilli.

With your inoculating needle, now pick the colonies that you selected to identify, and streak for isolation on separate trypticase soy agar plates. Work carefully, and remember to touch only the center of the colony. Incubate your plates at the optimum oxygen requirement at 37°C for 48 hours. This step is necessary to be absolutely sure that you are working with pure colonies when you inoculate your significant media for final identification. TSA instead of blood agar is used here since it is enriched enough to support the growth of fastidious organisms yet is inexpensive and easy to prepare.

If enough growth remains of the colonies that you just picked, you should make smears and Gram stain them. If there is not enough growth remaining, make the Gram stains from the TSA plate that you streaked for pure colonies (after incubation). In either case, examine the stained slides carefully, and make a note of their Gram reaction, shape, grouping, and other significant characteristics. The Gram stain will be most helpful in your choice of identifying tests and media.

Select and prepare your significant media for identification as previously described.

Activity 3: Inoculation of Significant Media for Identification of Selected Throat Flora

Examine your trypticase soy agar plates, and note any pigment production or other significant colonial characteristics. Correlate these observations with the Gram reaction, if possible.

Working with isolated, pure colonies from your TSA plates, inoculate the significant media that you selected and prepared for the identification of your throat organisms. Label your subcultures carefully, and incubate them at 37°C for 48 hours or the appropriate time for the specific tests that you are doing.

After incubation, gather your data, and record them in tabular form. Draw a conclusion from these data as to the genus and species of your two significant throat organisms, if present. Submit these names, along with your table of significant data, to your file. Once again, it would be to your advantage to complete this module to the best of your ability in preparation for unknowns.

Take the post test now.

Related Experience

If you isolate normal flora organisms only, it is strongly recommended that you identify them.

POST TEST

The post test is a self-evaluation. It is not used for a grade. It is designed only to let you decide if you have successfully completed this module.

Part I: True or False

_____ 1. The presence of *Streptococcus pyogenes* in a throat culture always indicates "strep throat."

_____ 2. A good throat culture specimen can be obtained by gently rotating the swab over the tonsillar area.

_____ 3. The viridans streptococci are the most prominent organisms of normal throat flora throughout life.

_____ 4. A predominance of a specific organism in a throat culture usually indicates that it is responsible for a pathogenic condition in the throat.

_____ 5. If a throat swab cannot be cultured immediately, it should be placed in a tube of transport medium to prevent drying.

_____ 6. When the original culture is done from a throat swab, one BAP should be incubated in a 10% CO_2 environment for the possible detection of *Neisseria meningitidis*.

_____ 7. The throat swab is used to streak the entire surface of your original isolation plates.

_____ 8. Colony size, color, hemolysis type, and Gram reaction are your most useful indicators of means necessary for the final identification of an organism from a throat culture.

_____ 9. The predominance of *Staphylococcus aureus* in a throat culture may be indicative of pathology elsewhere in the body, instead of in the throat.

_____ 10. Some streptococci produce larger, more mucoid colonies on BAP when incubated in the candle jar.

Part II

A. List six organisms that are part of the normal flora of the human throat.

1. _____   4. _____

2. _____   5. _____

3. _____   6. _____

B. List seven organisms that can be transients or pathogens in the human throat.

1. _____

2. _____

3. _____

4. _____

5. _____

6. _____

7. _____

PART EIGHT

Selected Serological Tests

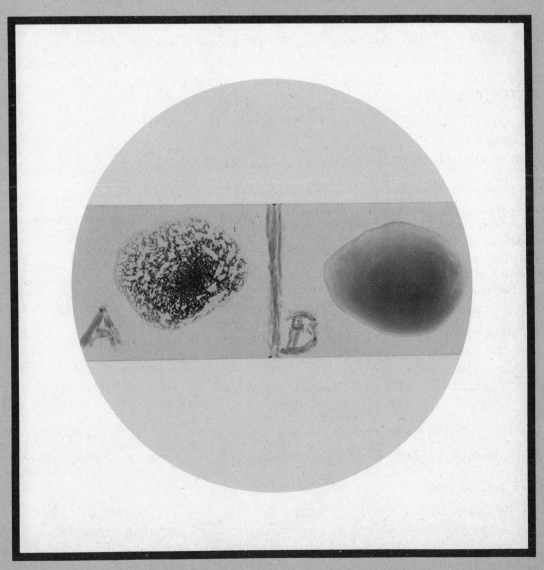

An agglutination reaction showing type A blood.

MODULE 49

Bacterial Agglutination Tests

PREREQUISITE SKILL

Successful completion of Module 8, "Aseptic Transfer of Microbes."

MATERIALS

Salmonella typhimurium H antigen
(flagellar antigen)*

Salmonella typhimurium H antiserum
(flagellar antibodies)* (reconstitute
with sterile 0.85% saline)

Salmonella typhimurium O antiserum
(somatic antibodies)* (reconstitute
with sterile 0.85% saline)

slant cultures of
Salmonella typhimurium
Escherichia coli

0.85% saline

microscope slides (2)

wax glass-marking pencil

inoculating equipment

toothpicks

container of disinfectant

For related experience:
glass plate with rings 14 mm in
diameter

Salmonella typhimurium H antiserum
(flagellar antibodies)* (reconstitute
with sterile 0.85% saline)

Salmonella typhimurium H antigen
(flagellar antigen)*

0.2 ml pipets graduated in 0.01 ml

toothpicks

*Commercially prepared and purchased.

OVERALL OBJECTIVE

Learn the bacteriological use and principles of agglutinating antigen-antibody reactions.

Specific Objectives

1. Define the term *agglutination reaction.*
2. Define the terms *serology, immunology, agglutinogen, agglutinin, typing serum, serum antibody quantitation, Widal test,* and *particulate antigen.*
3. List three examples of particulate antigens.

4. Describe the supporting role of the agglutination test in identifying the causative organism in bacterial disease.

5. Describe how commercially available typing sera are prepared.

6. Describe how agglutinating antigens are used to determine the bacterial disease of a patient.

7. Describe the slide test used to determine the antibody titer of a patient's serum.

8. Describe your experience in this module with antigen-antibody specificity.

9. Give the interpretation of an agglutination persisting at the 1:160 dilution of a patient's antibody titer in the Widal test for a year or more.

10. Define the O and H antigens of salmonellae.

DISCUSSION

Serology is a specialty in the field of immunology. Both are vast subjects, and one could spend a lifetime on the study of each. In a general sense, immunology refers to the immune response (protective mechanisms) of the body. Serology is an *in vitro* study of antigen-antibody reactions. Several types of serological tests are used to demonstrate the presence of antibodies or to identify an antigen. The general names of a few of these tests are agglutination, precipitation (flocculation), and complement fixation tests.

By definition, an agglutination test requires that the antigen be particulate, that is, of particle size. Bacterial cells, red blood cells, and latex particles coated with antigen are examples of particulate antigens. When a particulate antigen is mixed with its specific antibodies *in vitro*, large, three-dimensional lattice aggregates of cells and antibodies occur, and we call this agglutination. Agglutination aggregates are visible macroscopically.

Agglutination tests are frequently used in the identification of the causative organism in bacterial diseases. Rapid slide agglutinations are often employed to confirm the biochemical identification of a bacterium. For example, you can purchase typing sera that contain the antibodies (agglutinins) for specific bacteria. These typing sera are commercially prepared by injecting laboratory animals with a pure culture of attenuated bacteria. The bacterial injection is an agglutinogen (particulate antigen), and the animal responds immunologically by producing specific antibodies to the bacterium. The serum of the lab animal containing the antibodies is collected, processed, and sold as typing serum. In this module, you will be doing a slide agglutination test using *Salmonella typhimurium* typing serum that was prepared in this way.

It becomes apparent now that if a drop of *Salmonella typhimurium* antiserum is mixed with an unknown organism and agglutination occurs, the identification of the organism has been confirmed, since only *S. typhimurium* cells will be agglutinated by the *S. typhimurium* antibodies. Typing sera are available for many genera and species of bacteria and are used routinely as an aid in the identification of Gram negative enteric pathogens. In this use, the known antibodies are employed to identify the antigen, which is the bacterial cells isolated from the patient's specimen.

Conversely, *in vitro* agglutination reactions are also used to identify the antibodies in a patient's serum. Commercially prepared antigens for many bacteria are also available, or you can make your own antigens by heat treating and phenolizing a suspension of bacteria. Agglutination occurs if the patient's serum contains the antibodies specific for the antigen, and the bacterium causing the patient's disease has been identified. There are only rare exceptions to this antigen-antibody specificity. One is the Weil-Felix agglutination test for identification of Richettsial species.

Quantitation of the amount of antibodies present in a patient's serum can be done by a rapid slide test or a tube test. In either case, decreasing amounts of a patient's serum are mixed with a constant amount of antigen. This is called an antibody titer and is explained more completely in the related experience. In this use of the agglutination test, known antigens are employed to determine the amount of antibodies in the patient's serum. Determination of the amount of antibodies at various time intervals aids in following the course of the disease.

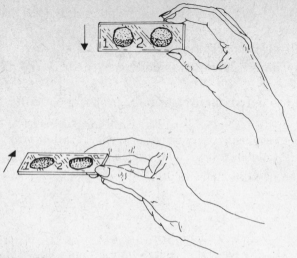

FIGURE 49-1
Proper mixing of a bacterial slide agglutination test. Hold the slide by the edges only. Rock the mixture back and forth. Do not allow the mixture to flow outside the wax circles.

ACTIVITIES

Activity 1: Slide Agglutination Test with Known Antigen and Antibody

With your wax pencil, draw two circles about the size of a nickel on a microscope slide. Number the circles 1 and 2. To Circle 1 add one drop of 0.85% saline and one drop of *S. typhimurium* H antigen. Mix well with a toothpick, taking care not to splash any of the fluid out of the circle. Discard the toothpick in a container of disinfectant. To Circle 2 add one drop of *S. typhimurium* H antigen and one drop of *S. typhimurium* H antiserum. Mix with a toothpick, and discard the toothpick in the same container of disinfectant. Now pick up the slide, and hold it by the edges, as shown in Figure 49-1. Take care not to get the mixture on your fingers. Rock the slide back and forth gently, and look for agglutination. Circle 1 is a negative control and resembles a negative agglutination test since the antigen particles are dispersed.

Make a drawing of the two circles. Accompany the drawing with a written description of the agglutination results, and relate the reaction in both circles to disease. Discard the slide in the container of disinfectant. Submit your drawings and descriptions to your file.

Activity 2: Slide Agglutination Showing Specificity of the Antigen-Antibody Reaction

Draw and number two circles on a microscope slide, as you did in Activity 1. Use your inoculating loop to aseptically obtain *E. coli* cells from the slant culture. Mix these cells into a drop of O antiserum in Circle 1. Use enough cells to make a cloudy (heavy) suspension. Be sure to flame your loop after using it for mixing.

Make another heavy suspension in Circle 2 by mixing *S. typhimurium* cells into a drop of O antiserum. Flame your loop after mixing. Pick up the slide, rotating it back and forth with care as you did in Activity 1. Refer to Figure 49-1 again.

Now make a drawing of your slide agglutination results. Since O agglutination aggregates are smaller, you may have to examine the slide under the low-power objective of your microscope. Include a written description of the results of the antigen-antibody relationship in each circle, and submit both to your file. Perform and/or read the related experience before taking the post test.

Related Experience: Simulation of a Patient's Antibody Titer Using the Rapid Slide Test Method

Consider the commercially prepared *Salmonella typhimurium* H antiserum to be serum of a patient who has a *S. typhimurium* infection.

Use a glass slide (2 x 3 inches) with ceramic rings that are 14 mm in diameter, or use a glass plate and draw five circles the size of a nickel with your wax pencil, as you did in Activities 1 and 2. Number the rings 1 through 5. Once you begin this titer, prepare to work rapidly since the small amounts you will be placing in each ring will dry out quickly. Read through this exercise first so that you *can* work rapidly.

Using a 0.2 ml pipet graduated in 0.01 ml, add the amounts of *S. typhimurium* H antiserum (representing the patient's serum) listed in Table 49-1. To each of the five rings, immediately add 0.03 ml of commercially prepared H antigen. The antigen dropper will deliver this amount. Stir the mixture in each ring with a different tooth-pick, and discard them in disinfectant. Rock the glass plate back and forth three or four times, and note the ring in which agglutination terminates. Record the dilution of this ring, and discard the slide in the container of disinfectant.

The highest dilution at which agglutination can be seen macroscopically is considered the antibody titer of the patient's serum. Antibody titers are of significance in certain diseases. They are clinically significant especially in the Widal test, which is specific for *Salmonella typhi* and the paratyphoids. For example, if a patient has had typhoid fever and his antibody titer remains at 1:160 or higher for a year or more after recovery, the patient is probably a chronic carrier. However, rapidly increasing titer indicates an active case of the disease and as the disease subsides, the titer returns to normal.

In this experience you have used an organism, *S. typhimurium,* that causes only a temporary gastroenteritis so antibody titers are of no concern to the physician. Nevertheless, the exercise that you have just performed demonstrates the principle of serum antibody titers nicely, and it can be equated with the Widal test. Also it is an excellent macroscopic demonstration of bacterial agglutination since flagellar antigens (H antigens), especially, produce large, macroscopic, agglutinating aggregates.

Write a description of your results and their significance, and submit this to your file. Then take the post test.

Suggestion: If agglutination occurs in all five rings, dilute the H antiserum 1:10 and 1:100 if necessary and repeat exactly as above with the diluted serum.

TABLE 49-1 Amounts of H antiserum to be used.

Ring number	Amount of H antiserum	Pipet readings	Dilution of patient's serum
1	0.08 ml	0.08 ml	1:20
2	0.04 ml	0.12 ml	1:40
3	0.02 ml	0.14 ml	1:80
4	0.01 ml	0.15 ml	1:160
5	0.005 ml	0.155 ml	1:320

POST TEST

The post test is a self-evaluation. It is not used for a grade. It is designed only to let you decide if you have successfully completed this module.

Part I: True or False

_____ 1. Immunology is the *in vitro* study of antigen-antibody reactions.

_____ 2. An agglutinogen is an agglutinating antibody.

_____ 3. The Widal test is an agglutination test for *Salmonella typhi* and the paratyphoids.

_____ 4. Antibodies are found in the fluid portion of blood.

_____ 5. Routinely, bacterial diseases are conclusively identified by using the agglutination test without using other bacteriological routines.

_____ 6. Commercially prepared typing sera contain the antigens harvested from the blood of lab animals.

_____ 7. Bacterial cells can be used to determine the bacterial disease of a patient.

_____ 8. Quantitation of antibody levels in a patient's serum is always done by the rapid slide test.

_____ 9. The antibodies in a patient's serum that arise due to a *Salmonella typhimurium* infection will agglutinate *Escherichia coli* cells when mixed together.

_____ 10. H antigens are agglutinogens located in the flagella of some motile bacteria.

_____ 11. A patient is suspected of being a chronic carrier if his antibody titer remains at 1:160 or higher for a year or more after he has recovered from the symptoms of typhoid fever.

_____ 12. An agglutinin is a particulate antigen.

Part II: Completion

A. List three examples of particulate antigens.

1. _____

2. _____

3. _____

B. List the ingredients of a bacterial slide agglutination test.

1. _____

2. _____

KEY
Part I: 1-F, 2-F, 3-T, 4-T, 5-F, 6-F, 7-T, 8-F, 9-F, 10-T, 11-T, 12-F.
Part II: A. 1. bacteria, 2. erythrocytes, 3. antigen-coated latex particles.
B. 1. bacterial cells (antigen), 2. patient's serum (antibodies).

314   Module 49

MODULE 50

Blood Groups and Rh Factor

PREREQUISITE SKILL

None.

MATERIALS

microscope slides (2)
blood typing sera:
  anti-A
  anti-B
Rh typing serum (anti-D)

70% alcohol (isopropyl)
blood lancet
cotton sponges
toothpicks
wax glass-marking pencil

OVERALL OBJECTIVE

Learn the agglutination principles involved in blood transfusion reactions and the *in vitro* typing of blood.

Specific Objectives

1. Explain why blood typing and Rh factor determination are agglutination reactions.
2. Describe why different blood types cause transfusion reactions.
3. Name the source of blood typing serum for both anti-A and anti-B.
4. Describe the laboratory procedure performed to determine a person's blood type.
5. Name the four most common agglutinogens associated wtih blood groups.
6. Name the investigators involved in the unfolding of the enigma of transfusion reactions.
7. Give the derivation of the symbol Rh and the percentage of humans who are Rh positive.
8. Describe how Rh typing serum is obtained.
9. Describe why transfusion reactions occur when an Rh negative person is transfused with Rh positive blood.
10. Define the term *erythroblastosis fetalis* and describe in detail the cause of the disease.

DISCUSSION

Although determination of human blood types does not directly pertain to microbiology, it is another good example of the agglutination test. Here again, the particulate antigen is intact cells, that is, human red blood cells.

In 1900, Karl Landsteiner showed that the red blood cells of humans were not all antigenically alike. In fact, he determined that within the human population there were four antigenic types. Landsteiner named the four blood types A, B, AB, and O. Besides showing that the erythrocytes of humans were antigenically different, he also found that the plasma in which the erythrocytes float contained different antibodies. These antibodies had no effect on the red blood cells in a single individual. However, if an individual had Type A blood cells, for example, the antibodies in his serum would not be compatible with the red blood cells of Type B blood, and an antigen-antibody reaction would take place in his circulatory system if he were transfused with Type B blood. This explained the enigma of blood transfusion reactions that were so common in Landsteiner's time.

In summary, Landsteiner found that the four antigenic types of red blood cells contained the following antibodies in their plasma:

Antigenic type of red blood cell (agglutinogens)	Antibodies in plasma or serum (agglutinins)
A	B (anti-B typing serum)
B	A (anti-A typing serum)
AB	none
O	A and B

This shows that only two typing sera are needed to determine all four blood types, and that the typing sera are named for the erythrocytic type they will agglutinate.

Type B blood would not be compatible with Type A blood since antigen-antibody reactions would take place *in vivo*. This same phenomenon is used to determine blood types *in vitro*. The slide agglutination test done to determine a person's blood type can be accomplished by using serum from a Type A individual that contains B antibodies. A macroscopically visible agglutination takes place if these anti-B antibodies are mixed with Type B blood cells, demonstrating that this person is Type B.

Conversely, the serum of a Type B person when mixed with Type A cells will show a distinct antigen-antibody agglutination, showing that the person is Type A. The previous list of antigenic types and antibodies, plus Figure 50-1, show the results of all four types of blood group agglutinations.

Many other erythrocytic antigens have been found, such as Duffy, Kell, Jansky, MNSs, and others. In this module, your activities will be limited to the major

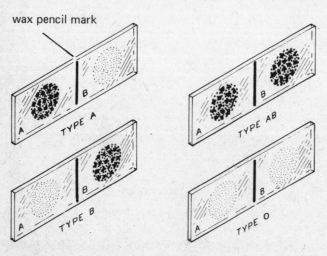

FIGURE 50-1
Four types of blood group agglutinations.

antigens of the four blood groups and the Rh factor.

In 1940, Landsteiner and Wiener were comparing blood groups in rhesus monkeys to blood groups in humans. They found that if they injected the blood cells from rhesus monkeys into a rabbit, the rabbit produced antibodies (agglutinins) which agglutinated the erythrocytes of 85% of the humans tested, as well as the rhesus monkeys' cells. The antigen (agglutinogen) was called Rh_O for the rhesus monkey at first; later its designation was changed to D.

In these early studies, the sensitized rabbit's serum was shown to contain the Rh agglutinins. The serum was harvested and used as anti-D typing serum. Individuals who possessed the Rh agglutinogens were designated as Rh positive. Neither Rh+ nor Rh- individuals (15% of the human population) have Rh agglutinins in their blood plasma. However, the erythrocyte Rh antigen, if transfused into an Rh negative individual, sensitizes the person. This means that the individual produces antibodies to the Rh factor (D) which accumulate in his serum. Hence, subsequent Rh+ blood transfusions may cause intravascular reactions because the individual's serum now contains agglutinins for the Rh factor. Today, rabbits are still used for the commercial production of Rh typing sera.

This discovery led to an explanation of erythroblastosis fetalis (hemolytic jaundice in newborns). The fetus developing from an Rh positive father and an Rh negative mother genetically can have Rh positive erythrocytes. These Rh positive fetal erythrocytes may pass through the placenta into the maternal circulation, thus stimulating the formation of Rh antibodies by the mother. These Rh factor antibodies can recross the placenta into the fetus. An antigen-antibody reaction then takes place, producing anemia, jaundice, and sometimes death in the newborn.

Several other Rh factors are known which are designated as C, E, c, e, and others. We will deal only with Rh factor D since it is clinically most important by virtue of its more potent antigenicity.

ACTIVITIES

Activity 1: Blood Typing

Divide a microscope slide in half by drawing a line down the middle of it transversely, and label one side A and the other side B, as shown in Figure 50-1. Wipe the tip of the finger located next to your little finger with a cotton sponge soaked in 70% alcohol. Allow the alcohol to evaporate. Aseptically unwrap the blood lancet, and pierce the disinfected finger. Allow a drop or two of blood (depending on the size of the drop) to fall on each side of the microscope slide. Place one drop of anti-A typing serum on the side of the slide marked A and a drop of anti-B typing serum on the half marked B. Mix each side with a different toothpick, spreading the mixture out to the size of a nickel, and rock the slide and mixtures back and forth for two minutes. Use a warm, slide-rocking box if available. Look for agglutination, and record your blood type. If agglutination takes place on Side A of the slide, you are Type A. If agglutination takes place on Side B, then you are Type B. If agglutination takes place on both sides, you are AB; if no agglutination takes place, you are Type O. Again refer to Figure 50-1.

Activity 2: Rh Slide Agglutination Test

Place one drop of anti-D typing serum on a clean glass slide. Add *two large drops* of whole blood obtained from your finger as you did in Activity 1. Mix thoroughly with a toothpick, spreading the mixture over most of the slide. Rock the slide gently back and forth, and examine it for agglutination. Use a warm viewing box if available, and rock back and forth over a period not exceeding two minutes. If macroscopic agglutination occurs, you are Rh positive. If the red blood cells do not agglutinate, you are Rh negative. Record your results, and submit them to your file.

Precaution: Incorrect blood type and Rh factor determinations often result from an inadequate ratio of blood to typing serum. Always use a large drop of blood.

Related Experience

For more practice, take a few drops of blood from a classmate, and determine his or her blood type and Rh factor by repeating the activities in this module. Do this without being preinformed as to the person's blood type and Rh factor.

     Now take the post test.

POST TEST

The post test is a self-evaluation. It is not used for a grade. It is designed only to let you decide if you have successfully completed this module.

Part I: True or False

_____ 1. Human red blood cells are antigenic only in other individuals of a different blood type.

_____ 2. A person with Type A agglutinogens has Type A antibodies in his serum.

_____ 3. The typing serum for Type A can be obtained from the serum of a Type B person.

_____ 4. The typing sera used to determine all four blood types are called anti-A and anti-B.

_____ 5. The Rh factor is not a particulate antigen.

_____ 6. Landsteiner and Wiener discovered the four blood groups.

_____ 7. The Rh factor is an agglutinogen.

_____ 8. The symbol Rh is derived from the name of a monkey type.

_____ 9. Rh typing serum is composed of commercially harvested antibodies.

_____ 10. Of the human population, 15% is Rh negative.

_____ 11. Rh transfusion reactions occur only in Rh negative individuals.

_____ 12. Rh negative individuals, if transfused with Rh positive blood, becomes sensitized by forming Rh agglutinins.

Part II: Completion

1. Give the scientific name for hemolytic jaundice in newborns.

_____

2. Hemolytic jaundice in newborns arises only from an Rh _____ father and an Rh _____ mother.

3. Place the following list in order of the steps that must take place for hemolytic jaundice in newborns to occur.

    a.   Maternal circulation forms Rh antibodies.

    b.   An intravascular, antigen-antibody reaction causes anemia, jaundice, and death.

    c.   Rh positive erythrocytes from the fetus pass through the placenta into the mother's blood stream.

    d.   The fetus inherits Rh positive erythrocytes from the father, while the mother carrying the fetus is Rh negative.

    e.   Antibodies recross the placenta into the fetus.

MODULE 51

Precipitin Tests

PREREQUISITE SKILL

Successful completion of Module 9, "Aseptic Use of a Serological Pipet," and Module 49, "Bacterial Agglutination Tests."

MATERIALS

bovine globulin antiserum (rabbit)*

bovine serum, normal (dilute 1:50 with 0.85% NaCl solution)*

rabbit serum, normal (dilute 1:50 with 0.85% NaCl solution)*

Nobel agar plate, poured ¼ inch thick (1)†

cork borer, #3

0.2 ml or 0.5 ml pipets (3)

Kahn tubes (8 mm x 75 mm) (2)

Kahn tube rack

For related experience:
  C-protein antiserum (Difco)*
  C-protein standard (Difco)*
  normal human serum

*Commercially available.
†To be prepared by student if instructor so indicates.

OVERALL OBJECTIVE

Perform two types of precipitin tests and understand the principles involved in each.

Specific Objectives

1. Explain how a precipitin test differs from an agglutination test.
2. Describe the procedure of the ring test.
3. List three factors that could interfere with the formation of a precipitate in the ring test.
4. Describe a titration for the precipitin content of a serum.
5. Discuss the formation of precipitin antibodies.
6. Describe the precipitin test in an agar gel.
7. List three uses for the ring test.
8. Define the terms *precipitin reaction, antiserum,* and *antibody specificity.*

DISCUSSION

Precipitin tests differ from agglutination tests principally in the form of the antigen. The precipitin test *antigen* is *soluble* rather than particulate.

The precipitin reaction is often called the ring or interfacial test, from the method by which it was originally performed. In a small Kahn tube, the soluble antigen is gently layered over a corresponding antiserum (antibodies) to form a sharp interface between the two solutions. As the two substances diffuse into each other, a precipitate is formed at the interface. You should remember that a *specific antibody* reacts with the antigen to produce the visible precipitate. For example, in the original experiment with precipitin reactions in 1897, a bacteria-free filtrate of a bacterial culture was mixed with its corresponding antiserum, and a flocculent precipitate resulted. When a different antiserum was mixed with the same filtrate, there was no precipitation.

Precipitin reactions are readily inhibited by the use of excess antigen. For this reason, to perform a titration of the precipitin content of a serum, a constant amount of serum is usually mixed with serial dilutions of the antigen, as opposed to the procedure for most other serologic reactions in which serial dilutions of the serum are mixed with a constant amount of antigen. The precipitin concentration of the serum is expressed as the greatest dilution of antigen precipitated.

The pH must be near neutrality, and salt (an electrolyte) must be present for precipitin reactions to occur. The precipitation rate is faster at higher temperatures.

Precipitin antibodies are often produced during the course of an infection, usually in response to soluble microbial substances released as a consequence of a degenerative process. The presence of these antibodies is demonstrated by the precipitin ring test or by a precipitin test incorporating an agar gel base.

The tests using agar gel are done quite extensively for qualitative and semi-quantitative studies. A small, flat-bottomed plate is poured with Noble agar, which is nonnutritive, and small holes are cut in the solidified agar. Into these holes, measured amounts of antigen and antiserum are placed. As the reagents diffuse through the gel, opaque precipitation bands are formed where optimum levels of antigen and antibodies have accumulated. See Figure 51-1 for an example.

The general or ring test for precipitin reactions is, however, still functional. It is used for Lancefield grouping of streptococci, identification of blood and flesh (forensic medicine), and the diagnosis of trichinosis.

Blood can be identified by layering an extract of a blood sample (antigen) over antisera of various animal species and observing for precipitation. The antisera are produced by repeatedly inoculating laboratory animals (usually a rabbit) with blood proteins or fractions of blood protein (especially globulin) from another animal (for example, a horse, cow, or human) to stimulate antibody production. The antisera are then harvested from the rabbits and, when mixed with blood (antigen) from the particular animal species corresponding to the antiserum, will cause a precipitate to form. For example, the serum of a rabbit that has been repeatedly injected with bovine serum globulin contains antibodies against bovine serum globulin and is called bovine globulin antiserum (rabbit). When this antiserum is collected from the rabbit and overlaid with bovine serum, normal, a precipitate forms at the interface. If the rabbit serum (bovine globulin antiserum) is overlaid with horse serum, no precipitate results.

The precipitin test in agar gel is used to diagnose and to follow the course of San Joaquin Valley Fever infections in patients.

ACTIVITIES

Activity 1: Precipitin Test in Agar Gel

Obtain a plate of Noble agar. With a #3 cork borer, cut three small holes in the agar at equal distances 4 mm apart, as shown in Figure 51-1. Lift the agar disks out of the plate with your inoculating loop, taking care not to cut the agar. On the bottom of the plate, label the holes 1, 2, and 3. Place a drop of diluted bovine serum, normal,

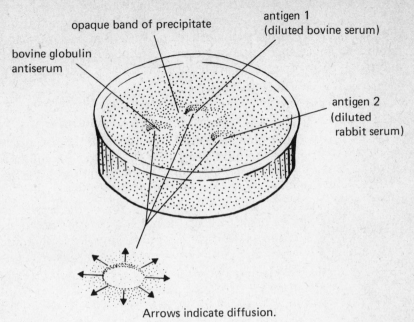

FIGURE 51-1
Precipitin test in agar gel.

into Hole 1. Put a drop of diluted rabbit serum, normal, into Hole 2 and a drop of bovine globulin antiserum (rabbit) into Hole 3. Refer to Figure 51-1 again for a schematic presentation of this procedure. Seal the plate with masking tape to retain moisture and incubate it right side up at 37°C for 24 hours or longer. Examine the plate carefully for the presence of one or two opaque white bands of precipitate between the holes.

Record your results, and submit them to your file.

Activity 2: The Ring Test

Gather together two Kahn tubes and a rack to support them. Label the tubes 1 and 2. With a 0.2 or 0.5 ml pipet, introduce 0.2 ml of bovine globulin antiserum into each tube. Dilute the antigens (bovine serum, normal, and rabbit serum, normal) 1:50 with 0.85% NaCl solution. With a fresh pipet, carefully overlay the antiserum with 0.2 ml of the diluted bovine serum, normal, in Tube 1. Using another clean pipet, carefully overlay the antiserum with 0.1 ml of diluted rabbit serum, normal, in Tube 2. It is necessary to hold the tube containing the antiserum almost horizontally and let the antigen run down the tilted side of the tube close to the antiserum to get a good overlay.

Incubate the tubes at 37°C for 30 minutes, and then observe them frequently for a ring of precipitate at the interface of the liquids. Examine the tubes with the light from a microscope lamp or a desk lamp shining through them. Observe the interface from a slight angle so that the light does not shine directly into your face.

Record your results, and submit them to your file. Then take the post test.

Related Experience

C-reactive protein (CRP) is a beta globulin not found in normal human sera. It was first reported in the blood of patients during the acute phase of pneumococcal pneumonia. CRP forms a precipitate with the somatic G-polysaccharide of *Streptococcus pneumoniae.* It was later found that CRP is formed in the sera of patients with a variety of other conditions and so is nonspecific for any particular disease. The C-reactive protein test is still useful, however, as a very sensitive, nonspecific test for the presence of inflammation or tissue injury of infectious or noninfectious origin. Changes in the amount of CRP quickly indicate the efficacy of therapy.

C-reactive protein tests are useful in the diagnosis of many viral and bacterial infections, acute rheumatic fever, rheumatoid arthritis, and most other collagen diseases, acute myocardial infarction, and various metastatic malignancies. Its most extensive clinical application, however, is in the evaluation of progress of the disease and therapeutic measures in acute rheumatic fever. A decrease of CRP level indicates adequate therapeutic response.

If the reagents are available, do a CRP test as follows.

1. Dip a capillary tube (0.8 mm x 100 mm) into a vial of CRP antiserum (available commercially), and allow the antiserum to rise about 1/3 the length of the tube. Place your finger over the top of the tube (like a pipet) before you remove it from the antiserum, and carefully wipe off the excess clinging to the outside of the capillary tube.

2. Still holding your finger over the top of the capillary tube, dip it into a tube of normal serum (or patient's serum), and draw up an equal volume. *Precaution:* Take care to avoid air bubbles, which will prevent proper mixing. Plasma is not satisfactory for this test.

3. Place your finger over the top of the capillary tube again, and carefully invert the tube several times to mix the two sera. Wipe off excess serum and finger marks with a tissue, and place the tube upright in plasticine or modeling clay, leaving an air space between the bottom of the serum and the clay.

4. Incubate the tube at 37°C for 2 hours and in the refrigerator overnight.

5. Repeat this procedure using CRP antiserum and a positive control serum.

Examine your tubes against a black background for the presence of a precipitate that may appear in as short a time as 15 minutes. A final reading should not be made, however, until overnight incubation is complete.

Record your results, and submit them to your file.

A modification of the CRP test uses polystyrene latex particles coated with CRP antiserum and is done as a slide test. By coating the latex particles with soluble antibodies, the active units become particle size, and an agglutination occurs instead of precipitation.

POST TEST

The post test is a self-evaluation. It is not used for a grade. It is designed only to let you decide if you have successfully completed this module.

Part I: True or False

_____ 1. A precipitin ring test is done in a ceramic ring on a 2 inch x 3 inch glass slide.

_____ 2. The formation of a precipitate is inhibited by an excess of antigen.

_____ 3. A titration for precipitin content of serum is done with serial dilutions of the serum and a known concentration of antigen.

_____ 4. Serum is incorporated into a liquid agar medium and allowed to solidify in order to perform a precipitin test in an agar gel medium.

_____ 5. The primary difference between a precipitin reaction and an agglutination reaction is one of antigen size.

_____ 6. The precipitin test antigen is particulate.

_____ 7. The ring test is used for Lancefield grouping of streptococci.

_____ 8. Precipitin antibodies are formed in response to exposure to a soluble microbial antigen during the course of an infection.

_____ 9. The antigen is carefully layered over antiserum for the ring test.

_____ 10. Precipitate will not form if an electrolyte is not present in the solution.

Part II

Define the following terms.

1. precipitin reaction _____

_____

2. antiserum _____

_____

3. antibody specificity _____

_____

FORMULA
NOBLE AGAR

Noble agar	1 gm
0.85% saline solution	100 ml

Heat to boiling. Cool to pouring temperature and dispense directly into clean petri dishes so that the medium is approximately ¼ inch thick. (This is a comparatively thick plate.) Noble agar plates should be prepared on the day of use.

MODULE 52

VDRL, a Serological Test for Syphilis

PREREQUISITE SKILL

Successful completion of Module 9, "Aseptic Use of a Serological Pipet," Module 49, "Bacterial Agglutination Tests," and Module 51, "Precipitin Tests."

MATERIALS

Rotating machine adjusted to 150 rpm, circumscribing a circle ¾ inch in diameter on a horizontal plane (if available). The slides can be rotated by hand with some success.

Flat slides with paraffin or ceramic rings approximately 14 mm in diameter. Slides must be clean, and ceramic rings must be high enough to prevent spillage at prescribed rotation.

18-gauge hypodermic needle without bevel for the delivery of antigen. A beveled needle can be filed off until it will deliver 60 drops ±2 drops of antigen suspension per milliliter when checked with a 1 ml or 2 ml syringe held vertically. These needles can be purchased from biological supply houses, or you may wish to calibrate your own by filing the bevel and counting the drops in 1 ml of antigen.

30 ml flat-bottomed, round, glass-stoppered, narrow-mouthed bottle.

1 ml or 2 ml Luer syringe.

VDRL antigen and buffered saline, both commercially available as a set.

0.2 ml serological pipets graduated in 0.01 ml (2)

syphilitic serum 4+ = positive control*

human serum normal = negative control*

*Commercially available.

OVERALL OBJECTIVE

Demonstrate the use and describe the principle of the VDRL qualitative slide test for syphilis.

Specific Objectives

1. Explain why a VDRL is a precipitation type antigen-antibody test.
2. Name four screening tests for syphilis other than the VDRL.
3. Explain in detail what is meant by a nonspecific test for syphilis.
4. Give the full meaning of the abbreviations VDRL, RPR, STS, BFP, and TPI.
5. Give the genus and species name, shape, and motility of the syphilis organism.
6. List four diseases that give a VDRL false positive test.
7. Name a specific test for syphilis and explain why is is not used in the average clinical laboratory.
8. Describe how the quantitative VDRL test differs from the qualitative VDRL test.
9. Describe how a blood sample is normally treated from the time it is taken from the patient until the VDRL test is begun.
10. List the steps of the VDRL test procedure.
11. Describe the appearance of a negative VDRL and a positive VDRL at 100x magnification.

DISCUSSION

Several flocculation (precipitation type) tests are used to diagnose syphilis: VDRL, Kahn, Kline, Mazzini, and others. These tests differ primarily in the amounts of the different components of the antigen, which are cardiolipin, cholesterol, and lecithin. The VDRL flocculation test is most commonly used today since it is the approved test of the California State Department of Public Health. The letters VDRL are a designation for the laboratory in which the test was developed, the Venereal Disease Research Laboratory. Most recently, the California Public Health Lab has also approved the RPR (Rapid Plasma Reagin) card test as another accepted serological test for syphilis. For years the VDRL was the only approved test for prenatal and premarital syphilis examination. In time, perhaps the RPR may completely replace the VDRL since the commercially prepared antigen is ready for use, and the tests are read macroscopically.

The previously mentioned serologic tests for syphilis (STS) are not specific. You have already noticed in the preceding paragraph that the antigen is not the syphilis organism (*Treponema pallidum*) or an extract of it, as it usually is in bacterial antigen-antibody reactions. Therefore, most serologic tests for syphilis are based on a coincidental relationship between lipid extracts of tissue, that is, the cardiolipin-cholesterol-lecithin antigen and a nonspecific antibody called reagin that develops in the serum of persons with syphilis infections. The coincidental relationship is that the nonspecific, lipid tissue extract antigen precipitates the tissue antibodies, reagin, and these tissue antibodies happen to be produced in increasing amounts in the blood stream of humans when the syphilis organism is present. The nonspecificity of the STS is magnified by the fact that people without syphilis infection can have a positive flocculation test. So normal persons can have reagin in their serum. These people are called biologic false-positives (BFP). In addition to this, a temporary increase in reagin is found in people with certain diseases or recent vaccinations. A few of these diseases are other nonvenereal treponemal infections such as yaws, as well as malaria, leprosy, infectious mononucleosis, and others. Here again false-positive tests for syphilis do occur.

In California and in most other states, all positive VDRL sera are sent to the State Public Health laboratory for definitive diagnosis. The state lab does a specific test for syphilis such as the TPI (*Treponema pallidum* immobilization) test on the suspected serum. This test uses the living *T. pallidum* spirochete as the antigen and therefore is specific for the syphilis antibodies in the patient's serum. The syphilis organism does not grow on artificial media and must be propagated in rabbit testes. The propagation of virulent syphilis spirochetes and their preparation for the TPI test are too comlex for the average clinical laboratory. Thus, it is through the state laboratory that false-positive tests can be ruled out by a specific antigen-antibody reaction such as the TPI or other specific tests, e.g., the fluorescent antibody (FTA) test.

Therefore, you will do a VDRL test, which is one of the many nonspecific, rapid-screening tests for syphilis. We call this a qualitative test. If a patient has a positive qualitative VDRL (screening test), a quantitative VDRL is done routinely to determine the antibody titer. Once again, quantitation is done by dilutions of the patient's serum with a constant amount of antigen. (Refer to the related experience in Module 49, "Bacterial Agglutination Tests.") The quantitative VDRL allows a physician to follow the course of the disease and the response of the patient to therapy.

The activity you will be doing in this module is the qualitative VDRL slide test.

ACTIVITY

Activity 1: Qualitative VDRL Test

Your instructor may prepare the antigen for you. If not, the directions for antigen preparation appear at the end of this module.

Since you probably do not know of a patient with syphilis, you will not need to draw a blood sample, allow it to clot, centrifuge, and remove the clear serum as you normally would. Instead, treat your 4+ positive and normal control sera as if you had removed them from the clotted blood sample of a syphilitic patient and a nonsyphilitic patient, respectively. Heat both sera in a 56°C water bath for 30 minutes before testing. Sera to be tested more than 4 hours after the original 30 minute heating period should be reheated at 56°C for 10 minutes.

Note that the antigen is referred to as particles. This can be misleading, so do not let this be confused with particulate antigens! The cardiolipin-cholesterol-lecithin antigen gives this *appearance,* but these short, rod-shaped antigen "particles" are actually an emulsion of cardiolipin (beef heart extract), cholesterol, and lecithin. This VDRL antigen emulsion is indeed a *soluble* antigen since the components are soluble in some liquid solvents. A positive VDRL gives the appearance of floccules; hence the term flocculation test is often applied to it.

Since it takes 30 minutes to heat the sera before testing, this allows you enough time to study the procedure of the qualitative test, given below, and to gather together the equipment necessary to perform it. Put the sera in the water bath now, and begin the 30 minute heating period.

Perform the test procedure as follows:

1. Pipet 0.05 ml of each heated serum into a separate ceramic ring. Use a different 0.2 ml pipet for each serum.
2. Add one drop (1/60 ml) of antigen emulsion into each serum with the calibrated 18 gauge needle.
3. Immediately rotate the slides containing the two mixtures for 4 minutes on the rotating machine. If rotation is by hand, a circle 2 inches in diameter should be circumscribed 120 times per minute for 4 minutes.
4. Read the tests with the low-power objective (10x) of your microscope immediately after rotation. If you are using the large, ceramic-ringed slides, you will have to remove the mechanical stage from your microscope.

The antigen *appears* as short, rod-shaped "particles" at 100x magnification if the test is negative. If there is reagin in the serum, the antigen rod "particles" will aggregate, and the test is a presumptive positive for syphilis. Draw the microscope field of both normal and syphilitic serum. Describe and record your findings as in Table 52-1, and submit them to your file.

Remember that, in a clinical laboratory, all sera producing reactive results or weakly reactive results in the qualitative VDRL slide test must be quantitatively retested to an end point titer. The dilutions of positive sera can be made in tubes and

TABLE 52-1 Summary of Reactions to VDRL Test

Reading	Record and report
No clumping or very slight roughness	Nonreactive (N)
Small clumps	Weakly reactive (WR)
Medium and large clumps	Reactive (R)

then transferred to the ceramic circled slide, or the dilutions can be made directly on the ceramic circled slide. Remember also that the sera you use in this activity are used routinely in a clinical laboratory as positive and negative controls along with an antigen-saline control. Take the post test when you are ready.

POST TEST

The post test is a self-evaluation. It is not used for a grade. It is designed only to let you decide if you have successfully completed this module.

Part I: True or False

_____ 1. The antigen emulsion used in the VDRL test is easy to prepare.
_____ 2. The VDRL is a precipitation type test because the antigen is particulate.
_____ 3. Reagin is a nonspecific antibody.
_____ 4. Reagin appears in the bloodstream only if a person has a syphilis infection.
_____ 5. The syphilis organism is a motile, Gram negative bacillus.
_____ 6. The TPI test is a specific test for syphilis.
_____ 7. The antigen "particles" of a VDRL emulsion are rod shaped.
_____ 8. A positive VDRL shows clumping of antigen "particles."
_____ 9. The syphilis organism requires enriched media when grown artificially.
_____ 10. The antigen is always dispensed with a 0.2 ml pipet graduated in 0.01 ml.

Part II: Completion

1. Give the genus and species name of the syphilis organism.

_____

2. List four diseases that give false positive VDRL tests.

   a. _____

   b. _____

   c. _____

   d. _____

3. List four screening tests for syphilis other than the VDRL.

   a. _____

   b. _____

   c. _____

   d. _____

   Which test was recently approved as an accepted STS by the California Public Health laboratory?

   e. _____

5. Give the meaning of the following abbreviations.

   a. VDRL _____

   b. RPR _____

   c. STS _____

   d. TPI _____

   e. BFP _____

6. What is the antigen for the TPI test?

   a. _____

   Where is it propagated?

   b. _____

7. What is the shape of the syphilis organism?

FORMULAE PREPARATIONS

1. VDRL ANTIGEN EMULSION

The temperature of reagents at the time of preparation should be in the range of 23° to 29°C.

Procedure:

1. Pipet 0.4 ml of buffered saline solution to the bottom of a clean, 30 ml round, glass-stoppered or screw-cap bottle.

2. Add 0.5 ml of antigen (from the lower half of a 1.0 ml pipet graduated to the tip) directly into the 0.4 ml saline solution while continuously but gently rotating the bottle on a flat surface.
   Note: The antigen must be added drop by drop, but rapidly, so that the 0.5 ml of antigen is delivered in approximately 6 seconds. Also, the pipet should remain in the upper third of the bottle to avoid splashing saline on it while rotating the bottle. The outer edge of the bottle should circumscribe a 2 inch diameter circle approximately three times per second.

3. Blow the last drop of antigen from the pipet without touching the pipet to the saline solution.

4. Continue rotation of the bottle for 10 more seconds.

5. Add 4.1 ml of the buffered saline solution from a 5 ml pipet.

6. Place the top on the bottle, and shake it from bottom to top, back and forth, approximately 30 times in 10 seconds.

7. The antigen emulsion is now ready to use and may be used for one day.

8. If the antigen is allowed to stand, mix it gently by swirling before using.

Antigen emulsion may be stabilized by adding benzoic acid and can be used for a period of four weeks if stored in a refrigerator at 6° to 10°C. Allow the antigen to stand at room temperature 30 minutes each day before using. Swirl gently to mix and use.

2. BENZOIC ACID, 1.0% ALCOHOL SOLUTION

Dissolve 1.0 gm of reagent grade benzoic acid in 100 ml of absolute ethyl alcohol. Store this in the refrigerator.

To stabilize one volume of VDRL antigen emulsion add 0.05 ml of this 1.0% benzoic acid immediately after the antigen emulsion is prepared. Shake the benzoic acid antigen emulsion from bottom to top gently for 10 seconds.

Remove the day's aliquot from this stabilized antigen, and return the stock bottle to the refrigerator.

PART NINE

Independent Investigation

"It is a capital mistake to theorize before you
have *all* the facts. It biases your judgment."

MODULE 53

Identifying Unknown Bacteria from a Mixed Culture

PREREQUISITE SKILL

Mastery of the techniques and concepts presented in this course.

MATERIALS

---

unknown mixed broth culture:
  of two different genera of bacteria
Gram stain reagents*
trypticase soy agar (TSA) slants (4)*
  (tryptic soy agar)
trypticase soy agar (TSA) plates (3)*
azide blood agar base plates (2)*
MacConkey agar plates (2)*

any other significant media, stains, and
  reagents *necessary* to identify your
  unknown organisms†

*To be prepared by the student if the instructor
so indicates.
†To be prepared by the student.

OVERALL OBJECTIVE

Isolate and identify to the genus and species the two organisms in a mixed culture.

Specific Objectives

1. Demonstrate ability to streak for isolation from a mixed broth culture by obtaining different colony types.
2. Demonstrate aseptic technique by maintaining stock cultures without contaminating them.
3. Demonstrate Gram stain technique by obtaining the correct Gram reactions.
4. Demonstrate the correct use of all-purpose, selective and selective-differential media, and determine why trypticase soy agar, azide agar, and MacConkey agar were suggested as representatives of these medium types.
5. Demonstrate an understanding of the individuality of bacteria by selecting only

the most significant media and reagents to use in studying the characteristic biochemical activities of two unknown bacteria.

6. Collect data in a logical sequence and correlate these data to give the correct genus and species names of two unknown bacteria.

DISCUSSION

For this module you will be given a mixed broth culture of two different genera of bacteria. The successful completion of the activities, which are carefully designed to initiate your independent investigation, should allow you to identify both unknown bacteria in the mixed culture to their genus and species name. Following the initial activities, it will be necessary for you to employ all the techniques and to demonstrate an understanding of the enzyme systems of the various bacteria you have used during this laboratory course. (Your instructor may choose to add a few other easily identifiable types of bacteria.)

It is the intent of this module to give you enough guidance to avoid frustration, yet permit you to do enough independent reasoning to prove the worth of the self-instructional approach to learning. Albert Eistein once said, "The search for truth is more valuable than its possession." Discovering information for yourself is more interesting than simply having the answers given to you. You will select the appropriate procedures and, by the process of elimination, the correct answer will emerge.

Since this module is a teaching device, it will take approximately two weeks to identify both your unknowns. This differs from the clinical laboratory approach in that treatment may be instituted by a physician on the basis of a patient's symptoms and/or microscopic examination of the specimen before definitive laboratory identification of the causative bacterium is complete. It should also be pointed out that the clinical laboratories use rapid methods to do the same tests for which you are using the more lengthy, classical approach. In either case, eventual identification of the causative agent is important—clinically for epidemiological purposes and confirmation of the presumptive treatment, and classically for a more thorough presentation of the principles involved.

You have been given a mixed culture for three reasons:

1. To make this investigation resemble certain actual infections in which more than one organism is involved as a causative agent.
2. To demonstrate your ability to successfully use the basic microbiological techniques you have learned during this course.
3. To demonstrate your understanding of the structural and enzymatic individuality of bacteria, and how these differences can be used to identify bacteria.

Since this is a lengthy module, and many media will be inoculated, certain precautions should be brought to your attention before you begin the activities.

Precautions:

1. Always make your inoculations from your working stock culture slant and never from the streak plate.
2. Keep your reserve stock culture slant as a backup stopgap in case your working stock slant becomes contaminated.
3. Gram stain a young working slant culture to get a true Gram reaction and bacterial shape. It may be necessary to inoculate a nutrient broth or thioglycollate broth culture to determine cell arrangement (see Module 15 "Cultural Characteristics"). *Do not* Gram stain from differential or selective media because the inhibitory ingredients of these media may give a false Gram reaction and distort the cell shape.
4. Do the primary inoculations (in Activity 1) in duplicate to determine the optimum growth temperature for both unknown bacteria. Continue to use the optimum growth temperature of each organism for all subsequent inoculations.

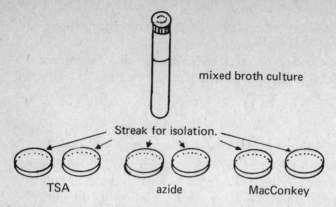

mixed broth culture

Streak for isolation.

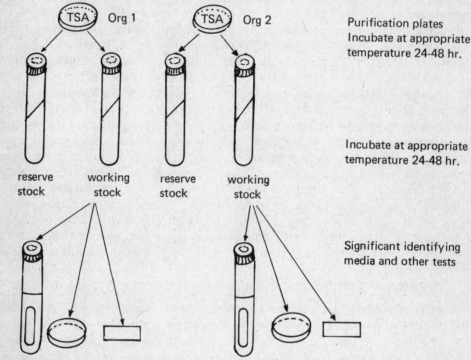

TSA   azide   MacConkey

Primary isolation
Incubate one plate of each
medium at 25°C and one at
37°C for 24-28 hr.

Select two different appearing colonies, or colonies with different
growth characters, and streak for purification on TSA.

TSA Org 1   TSA Org 2

Purification plates
Incubate at appropriate
temperature 24-48 hr.

reserve   working   reserve   working
stock   stock   stock   stock

Incubate at appropriate
temperature 24-48 hr.

Significant identifying
media and other tests

FIGURE 53-1
Suggested inoculation scheme.

5. Purify the primary streak plates, if necessary (in Activity 2), so that you will be absolutely sure of making your stock cultures from a pure colony. Purifying a streak plate means to pick and restreak for isolation a single colony from the original plate so that the new isolated colonies are certain to be free of the other bacterium present in the original, mixed broth culture.

6. The suggested times of incubation on the flow sheet, as shown in Figure 53-1, are 24 to 48 hours. Usually a 24 hour incubation is sufficient, since the organisms will be in the logarithmic growth phase and physiologically very active. Therefore, if possible, do each successive set of inoculations after 24 hours since this will decrease the length of time necessary to make the final identification and complete this module. There are some exceptions to the 24 hour incubation period. A few examples are the slow-growing streptococci and the final reading of a medium such as litmus milk.

7. There are bacterial variants. Therefore, do not rule out an organism on the basis of one test that does not conform to the classical reaction. The test that does not conform may also be due to faulty reagents or media. Check your reagents and media. Make them up fresh if necessary.

8. Finally, the value of careful collection and recording of the results for each inoculated medium and/or test *cannot be overemphasized.*

9. It is advisable to check your results from the MacConkey agar plates by inoculating another medium designed to demonstrate lactose fermentation. *Enterobacter-Klebsiella* group is mucoid but not pink on MacConkey agar even though lactose is fermented.

10. You must use deductive reasoning to identify your unknown organisms correctly. That is, collect all your data before drawing a conclusion. As Holmes said to Watson, "When you rule out the impossible, whatever remains, however improbable, is the truth." Facts first and then conclusions.

Reread these precautions as you proceed through the following activities to avoid repeating inoculations and other tests. You are now ready to play the role of a detective by using Sherlock Holmes techniques to carefully put clues together to identify and incriminate the two unknown culprits in your mixed broth culture. Good luck!

ACTIVITIES

Activity 1: Primary Isolation from Mixed Broth Culture

Record the code number of the unknown mixed broth culture assigned to you. Your instructor will keep a record of this also. From the mixed broth culture, streak for isolation in duplicate on an all-purpose medium, a selective medium, and a selective-differential medium. The suggested media are trypticase soy agar, azide agar, and MacConkey agar, respectively. Incubate one set of plates at 37°C and the other at room temperature. Be able to discuss why these three media are the media of choice for primary isolation. Gram stain your mixed broth culture. Record the Gram reaction and cell grouping of both organisms.

Reminder: It is important that you understand why you are using these suggested media. Review the purpose of each. Occasionally it is possible that your organisms will grow only on TSA. If so, you should find two colony types that you will proceed to identify from your TSA plates.

Activity 2: Restreaking to Ensure Pure Colonies

You may feel that this activity is not necessary. If you choose to ensure obtaining pure colonies, carefully inspect all six streak plates from Activity 1. Select two different-appearing colonies or two organisms growing on different media, and restreak each for isolation on a TSA plate to purify each colony. Use numbers or descriptive symbols for differential labeling of the two unknown bacteria. Careful labeling of the two bacteria once separated is most important to your accurate recording of results.

If enough of the colonies that you selected for purification remain, Gram stain the remainder to compare with the results you obtained from the mixed broth culture in Activity 1. Incubate the TSA purification plates at the optimum temperature determined for each organism in Activity 1.

Activity 3: Inoculation and Maintenance of Stock Cultures

Inoculate two TSA slants for both organisms using a single colony of each organism from the purified TSA plates, if you found it necessary to purify. One TSA slant of each organism will be your reserve stock culture; the other will be your working stock culture. Incubate all four slants at their respective, appropriate temperatures for 24 to 48 hours. Keep the reserve stock culture refrigerated, and remember that it is to be opened only if your working stock culture becomes contaminated. Refrigerate your working stock also between inoculations. Gram stain the working stock of each organism to check for purity and to compare with previous stains.

TABLE 53-1 Selected Characteristics of Some Bacteria Used as Unknowns*

Organism	Gram stain	Cell shape and grouping	Pigment	Glucose	Lactose	Sucrose	Mannitol	Gelatin hydrolysis	Indole
Staphylococcus epidermidis	+	Cl	–	A	V	A	–	+	
Staphylococcus aureus	+	Cl	+	A	A	A	A	+	
Streptococcus pyogenes	+	Ch	–	A	A	A	–		
Streptococcus faecalis	+	Ch	–	A	A	V	A		
Streptococcus lactis	+	Ch	–	A	A	V	V		
Escherichia coli	–	r	–	AG	AG	V	AG	–	+
Enterobacter cloacae	–	r	–	AG	AG	A	AG	+	–
Klebsiella pneumoniae	–	r	–	AG	AG	A	A	–	–
Proteus vulgaris	–	r	–	AG	–	AG	–	+	
Pseudomonas aeruginosa	–	r	+	–	–	–	–	+	
Salmonella typhimurium	–	r	–	AG	–	–	AG	–	
Shigella flexneri	–	r	–	A	–	V	V	–	
Serratia marcescens	–	C	+	A	–	A		+	
Bacillus subtilis	+	rCh	–	A	V	A	A	+	
Corynebacterium xerosis	+	rp	–	A		A	A	–	
Shigella dysenteriae	–	r	–	A	–	–	–	–	–
Streptococcus faecium (durans)	+	Ch	–	A	A	V	V		
Bacillus polymyxa	V	rCh	–	AG	AG	AG	AG	+	–
Micrococcus luteus	+	Cl	+	–	–	V	V	–	–
Micrococcus roseus	V	Cl	+	V	V	V	V	V	
Neisseria sicca	–	DC	–	A		A			
Alcaligenes faecalis	–	C	–	–	–	–	–	–	–

Code for cell shape and grouping:

    Cl = cocci in clusters
    Ch = cocci in chains
    r = rods
    rCh = rods in chains
    rp - rods in palisades
    C = cocci or coccobacilli, single or in pairs
    DC = diplococci, flattened sides adjacent

Code for carbohydrate fermentation:

    - = no change
    A = acid
    AG = acid and gas
    V = variable

Methyl red	Voges-Proskauer	Citrate	Hydrogen sulfide	Nitrate reduction	Urea hydrolysis	Hemolysis	Motility	Capsules	Salt tolerance	Litmus milk reactions
							−	−	+	A
						β	−	−	+	Ac
						β	−	−		A
						α	−		+	AcR
							−		−	AcR
+	−	−	−	+			V	−		AcGR
−	+	+	−	+			+	−		AcGR
−	+	+	−	+			−	+		A
			+	+	+		+			AlkP
				+	+		+			AlkPR
			+	+			+			Alk
			−	+			−			Alk
			−	+			+			Ac
					+		V			AlkP
					−		−			NC
			−	+			−			Alk
						α-β	V		+	AcR
+					+		+			AcG
					−		−			A
					+		V			NC-Alk
					−			V		
V	−						+	V		Alk

Code for litmus milk reactions:
   Table 53-1 includes combinations of these reactions.

   NC = no change
   Alk = alkaline
   AlkP = alkaline and peptonization
   Ac = acid and curd
   AcG = acid, curd, and gas
   A = acid, *no* curd
   R = reduction

These are abbreviations of the descriptive terms used in Module 38, "Litmus Milk Reactions." Remember, organisms growing in litmus milk often cause a combination of these reactions to occur.

Empty spaces denote that the test is not significant in the identification of the bacterium.

*Consult R.S. Breed, et al., *Bergey's Manual of Determinative Bacteriology,* 7th ed. (Baltimore: Williams & Wilkins, 1957), for other identifying tests if necessary. Check with your instructor for name changes for a few organisms. Example: *Streptococcus pneumoniae* is *Diplococcus pneumoniae* in *Bergey's* 7th ed.

Activity 4: Inoculation of Significant Media from Working Stock Cultures

From the Gram reactions, media on which the organisms grew (primary isolation), and characteristic colony appearance, if any, begin your systematic inoculations of significant media. Perform other indicative tests to demonstrate the biochemical activities of your unknowns. Perform *only* those inoculations and tests *necessary* to identify your unknowns.

You are now "on your own," and your selection of significant media and tests will be considered by your instructor in the final analysis of your *degree* of success in identification of your unknowns. For the maximum degree of success, unnecessary testing should be avoided. Table 53-1 lists the selected characteristics of some bacteria used as unknowns. Do not use all the media included in the table, however; use only those media and tests that logically apply to the identification of your organism.

After you have completed Activity 1, review the modules that you have used during this course to select significant media and tests. That is, after you know the Gram reaction and growth characteristics of both unknown bacteria, you will know which modules to review.

Remember that you are "on your own," which means that you are expected to use scientific reasoning without consulting your instructor for the logical selection of significant media or tests. You will now be demonstrating the extent of the knowledge you gained in the laboratory during this course. After collecting your results from all inoculated media and other tests, summarize all your data in tabular form for both unknown organisms. Submit the table or tables, the code number of the mixed broth culture, and the genus and species name of both organisms to your instructor for verification.

In this module the correct identification of both unknowns, as well as the media and/or tests used, will be considered as successful completion of the self-evaluating post test.

If you have identified both your unknown organisms correctly, then you have passed the post test and the unknown part of the laboratory portion of this course with 100%. *Congratulations.* You have *indeed* learned a lot about microbiology!

Suggested References

Brock, T.D., and K.M. Brock, 1973. *Basic Microbiology with Applications.* Englewood Cliffs, NJ: Prentice-Hall, Inc.

*Buchanan, R.E. et al., 1974. *Bergey's Manual of Determinative Bacteriology.* 8th ed. Baltimore: The Williams & Wilkins Company.

Burdon, K.L., and R.P. Williams, 1968. *Microbiology.* 6th ed. New York: The Macmillan Company.

Davidson, Israel, and J.B. Henry, 1969. *Clinical Diagnosis by Laboratory Methods.* 14th ed. Philadelphia: W.B. Saunders Company.

Difco Laboratories, 1953. *Difco Manual of Dehydrated Culture Media and Reagents for Microbiological and Clinical Laboratory Procedures,* 9th ed. Detroit: Difco Laboratories, Inc.

Frobisher, Martin, 1974. *Fundamentals of Microbiology.* 9th ed. Philadelphia: W.B. Saunders Company.

Frobisher, Martin, and Robert Fuerst, 1969. *Microbiology in Health and Disease.* 13th ed. Philadelphia: W.B. Saunders Company.

Pelczar, Jr., M.J., and R.D. Reid, 1972. *Microbiology.* 3rd ed. New York: McGraw-Hill, Inc.

Smith, Alice L., 1969. *Principles of Microbiology.* 6th ed. St. Louis: The C.V. Mosby Company.

Volk, Wesley A. and Margaret F. Wheeler, 1974. *Basic Microbiology.* 3rd ed. Philadelphia. J.B. Lippincott.

Wilson, M.E., and H.E. Mizer, 1969. *Microbiology in Nursing Practice.* New York: The Macmillan Company.

Wistreich, G.A., and M.A. Lechtman, 1973. *Microbiology and Human Disease.* Beverly Hills: Glencoe Press.

*Breed, Robert S., *Bergey's Manual of Determinative Bacteriology,* 7th ed. is more useful to a first course microbiology student.

Index